Functional Biomaterials and Digital Technologies in Dentistry: From Bench to Bedside

Functional Biomaterials and Digital Technologies in Dentistry: From Bench to Bedside

Editors

Ping Li
Guojiang Wan
Shulan Xu
An Li

MDPI

Basel • Beijing • Wuhan • Barcelona • Belgrade • Novi Sad • Cluj • Manchester

Editors
Ping Li
Guangzhou Medical University
Guangzhou
China

Guojiang Wan
Southwest Jiaotong University
Chengdu
China

Shulan Xu
Southern Medical University
Guangzhou
China

An Li
Southern Medical University
Guangzhou
China

Editorial Office
MDPI AG
Grosspeteranlage 5
4052 Basel, Switzerland

This is a reprint of articles from the Special Issue published online in the open access journal *Journal of Functional Biomaterials* (ISSN 2079-4983) (available at: https://www.mdpi.com/journal/jfb/special_issues/2L58Q9V605).

For citation purposes, cite each article independently as indicated on the article page online and as indicated below:

Lastname, A.A.; Lastname, B.B. Article Title. *Journal Name* **Year**, *Volume Number*, Page Range.

ISBN 978-3-7258-1765-8 (Hbk)
ISBN 978-3-7258-1766-5 (PDF)
doi.org/10.3390/books978-3-7258-1766-5

Cover image courtesy of Ping Li

Contents

Journal of *Functional Biomaterials*

MDPI

Editorial

Functional Biomaterials and Digital Technologies in Dentistry: From Bench to Bedside

Ping Li [1,*], Guojiang Wan [2], Shulan Xu [3] and An Li [4,*]

1 Department of Prosthodontics, School and Hospital of Stomatology, Guangzhou Medical University, Guangzhou 510182, China
2 Key Laboratory of Advanced Technologies of Materials (Ministry of Education), School of Materials Science and Engineering, Southwest Jiaotong University, Chengdu 611756, China; guojiang.wan@home.swjtu.edu.cn
3 Center of Oral Implantology, Stomatological Hospital, School of Stomatology, Southern Medical University, Guangzhou 510280, China; xushulan_672588@smu.edu.cn
4 Department of Periodontology, Stomatological Hospital, School of Stomatology, Southern Medical University, Guangzhou 510280, China
* Correspondence: pingli@gzhmu.edu.cn (P.L.); ali9714ew@smu.edu.cn (A.L.)

This Special Issue, "Functional Biomaterials and Digital Technologies in Dentistry: From Bench to Bedside", highlights the integration of advanced materials science and digital technologies in dental and maxillofacial applications. We sincerely appreciate the Journal of Functional Biomaterials and the committed researchers for making this Special Issue possible through their contributions.

This Special Issue consists of 11 articles that include 6 original research studies (contributions 1–6) and 5 reviews (contributions 7–11) focusing on a broad spectrum of functional biomaterials and digital technologies in the field of dental applications. The research presented encompasses various applications, ranging from metallic and polymer biomaterials to organic and composite biomimetic materials. Additionally, this Special Issue emphasizes advancements in digital technologies, particularly in computer-aided imaging (CAI), computer-aided design (CAD), and computer-aided manufacturing (CAM).

1. Why We Focus on the Topic of "Functional Biomaterials and Digital Technologies in Dentistry: From Bench to Bedside"

The thematic focus of "Functional Biomaterials and Digital Technologies in Dentistry: From Bench to Bedside" is deliberate and dictated by the revolutionary developments in the field. Dentistry is currently at the cusp of a transformative epoch, fueled by material science, engineering, and computer science. These disciplines are pivotal in sculpting the future of dental science. Functional biomaterials, recognized for their enhanced biocompatibility, robust mechanical properties, and tailored biofunctionality, are paving the way for groundbreaking developments in oral and maxillofacial treatments. Furthermore, digital technologies have catalyzed a revolutionary shift in various healthcare sectors, including dentistry. Sophisticated methodologies such as CAD/CAM, 3D printing, bioprinting, and dental virtual patients have emerged as powerful tools, substantially improving the accuracy of procedures, patient experiences, and efficiency of dental care.

Citation: Li, P.; Wan, G.; Xu, S.; Li, A. Functional Biomaterials and Digital Technologies in Dentistry: From Bench to Bedside. *J. Funct. Biomater.* **2024**, *15*, 107. https://doi.org/10.3390/jfb15040107

Received: 19 March 2024
Accepted: 11 April 2024
Published: 17 April 2024

2. Special Issue of "Functional Biomaterials and Digital Technologies in Dentistry: From Bench to Bedside"

2.1. Functional Biomaterials in Dentistry

Within the field of dentistry, a diverse range of biomaterials—including metals, polymers, ceramics, and composites—is commonly utilized in creating dental implants, dental resin, osteosynthesis implants, and scaffolds for both bone and soft tissue regeneration [1–3]. Ongoing developments and investigations of new materials are propelled by their intrinsic biofunctional properties, which hold the promise to enhance clinical outcomes significantly [4,5].

- Surface modification of dental implant abutment: In this Special Issue, Wen et al. (contribution 5) investigated the impact of UVC (100–280 nm) pre-treatment on human gingival fibroblasts (HGFs) and *Porphyromonas gingivalis* (*P. gingivalis*) interactions with Ti-based implant surfaces. The results indicated that UVC pre-treatment improves HGF adhesion and proliferation while diminishing *P. gingivalis* colonization on smooth Ti substrates relative to untreated controls.
- Optimization of dental resin: Within this Special Issue, Bourgi et al. (contribution 10) investigated the influence of varying temperatures of warm air on solvent evaporation from dental adhesives and its subsequent effect on the bond strength of resin-based materials to dental and nondental substrates. The systematic review of in vitro studies identified that the optimal temperature range for a warm air stream to promote solvent removal and enhance adhesion to dentin is between 50 and 60 °C.
- Biodegradable metals for osteosynthesis implant: Zinc (Zn)-based biodegradable materials are emerging for use in osteosynthesis implants, despite inconsistencies between in vitro and in vivo biocompatibility [6,7]. In this Special Issue, Liu et al. (contribution 9) examined the cytotoxic potential of Zn and its alloys, revealing no harmful effects under specific conditions. Of note, the systematic review highlighted significant inconsistencies in cytotoxicity testing approaches.
- Scaffolds for bone tissue regeneration: Bone tissue regeneration employs a synergistic combination of biocompatible scaffolds, specialized cells, and growth factors, meticulously engineered to support and accelerate new bone formation, meeting both mechanical and biological needs [8]. Research by Yang et al. (contribution 6) highlighted the osteoinductive potential of RADA16 nanofiber scaffold hydrogel-wrapped concentrated growth factors for treating alveolar bone loss, while Huang et al. (contribution 8) reviewed the application of stem cell-derived extracellular vesicles in periodontal osteogenesis therapy, focusing on the RANKL/RANK/OPG pathway's role.
- Scaffolds for soft tissue regeneration: Soft tissue regeneration aims to repair or replace damaged tissues such as muscle and skin through biocompatible scaffolds, growth factors, and cells, ensuring functional restoration and integration. In this Special Issue, Li et al. (contribution 2)'s research depicts an enhancement of the bioactivity and biocompatibility of mesoporous silica nanospheres with liquid crystal, demonstrating the potential of the hydrogel for effective soft tissue regeneration. Concurrently, Dong et al. (contribution 1) probe the cytotoxic effects of black phosphorus nanosheets on vascular endothelial cells for tissue regeneration, uncovering a mechanism driven by excessive reactive oxygen species production and mitochondrial dysfunction leading to cell apoptosis.

2.2. *Digital Technologies in Dentistry*

Digital dentistry revolutionizes patient care by enhancing diagnostics with CAI and CBCT, and improving restorations via CAD/CAM for quicker, more precise outcomes [9]. Integrated digital workflows boost treatment predictability and efficiency, while 3D printing, artificial intelligence (AI), and robotic surgery offer customized treatments, significantly enhancing patient outcomes and experiences.

- Computer-aided imaging

CAI technology in dentistry enhances diagnosis and treatment planning through advanced digital imaging, such as intraoral scans and CBCT [10,11]. This technology provides precise visual data for better patient outcomes. Saravi et al.'s study evaluated the use of computerized optical impression making (COIM) in partially edentulous jaws, highlighting potential deviations in prosthetics involving mucosal tissues. Yuan et al. reported on the successful integration of stomatognathic data from various devices for creating dynamic virtual patients, showcasing the advancements and considerations in CAI applications (contribution 11).

- Computer-aided design and manufacturing

CAD/CAM technology in dentistry represents a transformative approach to dental restoration, offering a streamlined process from design to production. In this issue, Wang et al. (contribution 4) investigated the adhesion characteristics of a 3D-printed polyetheretherketone (PEEK) material. Their findings suggested that surface texture alone does not exclusively dictate the adhesion qualities of additively produced PEEK. Additionally, Chen et al. (contribution 7) provided a comprehensive review of the evolution of obturator fabrication for oronasal fistulas following cleft palate repair, tracing its progression from manual craft to the implementation of advanced digital methodologies.

3. Summary

Through curating this Special Issue, we place a spotlight on the fusion of functional biomaterials and digital advancements as a decisive force propelling dental care forwards. Offering treatments that are more personalized, efficient, and effective, these technologies align with the broader transition towards precision medicine. The customization of treatments to meet individual needs, utilizing biomaterials with superior biocompatibility and mechanical properties, is expected to improve health outcomes significantly. Additionally, the incorporation of digital technologies in dentistry is set to transform the intricacies of diagnostics and treatment, making dental care more accessible and cost-effective. This Special Issue not only provides an overview of current research and development in the field but also shines a light on potential shifts in dental care. Looking forwards, the intersection of functional biomaterials and digital technology is poised to redefine the future of dental and maxillofacial treatments, ensuring advanced care for patients worldwide.

Author Contributions: Conceptualization, A.L. and P.L.; writing—original draft preparation, P.L.; writing—review and editing, A.L.; supervision, P.L. and G.W.; project administration, P.L. and S.X.; funding acquisition, A.L. and P.L. All authors have read and agreed to the published version of the manuscript.

Funding: This research was supported by a grant from the National Natural Science Foundation of China [grant number 82301134], Guangdong Basic and Applied Basic Research Foundation [grant numbers 2021A1515111140 and 2021B1515120059], Science and Technology Projects in Guangzhou [grant number 202102080148], Guangzhou Health Commission Clinical Major Technology Project [grant number 2023C-ZD07], and Clinical Research Initiation Plan of the Stomatological Hospital, Southern Medical University [grant number KQIIT2021001].

Data Availability Statement: The data presented in this paper are available upon request from the corresponding author.

Conflicts of Interest: The authors declare no conflicts of interest.

List of Contributions

1. Dong, H.; Wen, Y.; Lin, J.; Zhuang, X.; Xian, R.; Li, P.; Li, S. Cytotoxicity Induced by Black Phosphorus Nanosheets in Vascular Endothelial Cells via Oxidative Stress and Apoptosis Activation. *J. Funct. Biomater.* **2023**, *14*, 284.
2. Li, X.; Wan, L.; Zhu, T.; Li, R.; Zhang, M.; Lu, H. Biomimetic Liquid Crystal-Modified Mesoporous Silica−Based Composite Hydrogel for Soft Tissue Repair. *J. Funct. Biomater.* **2023**, *14*, 316.
3. Saravi, B.; Ilbertz, J.; Vach, K.; Kohal, R.J.; Patzelt, S.B. Accuracy of Computerized Optical Impression Making in Fabrication of Removable Dentures for Partially Edentulous Jaws: An In Vivo Feasibility Study. *J. Funct. Biomater.* **2023**, *14*, 458.
4. Wang, D.; Han, X.; Luo, F.; Thieringer, F.M.; Xu, Y.; Ou, G.; Spintzyk, S. Adhesive Property of 3D-Printed PEEK Abutments: Effects of Surface Treatment and Temporary Crown Material on Shear Bond Strength. *J. Funct. Biomater.* **2022**, *13*, 288.

5. Wen, Y.; Dong, H.; Lin, J.; Zhuang, X.; Xian, R.; Li, P.; Li, S. Response of Human Gingival Fibroblasts and Porphyromonas gingivalis to UVC-Activated Titanium Surfaces. *J. Funct. Biomater.* **2023**, *14*, 137.
6. Yang, R.; Chen, J.; Wang, D.; Xu, Y.; Ou, G. Self-Assembling Peptide RADA16 Nanofiber Scaffold Hydrogel-Wrapped Concentrated Growth Factors in Osteogenesis of MC3T3. *J. Funct. Biomater.* **2023**, *14*, 260.
7. Chen, J.; Yang, R.; Shi, B.; Xu, Y.; Huang, H. Obturator Manufacturing for Oronasal Fistula after Cleft Palate Repair: A Review from Handicraft to the Application of Digital Techniques. *J. Funct. Biomater.* **2022**, *13*, 251.
8. Huang, X.; Li, Y.; Liao, H.; Luo, X.; Zhao, Y.; Huang, Y.; Zhou, Z.; Xiang, Q. Research Advances on Stem Cell-Derived Extracellular Vesicles Promoting the Reconstruction of Alveolar Bone through RANKL/RANK/OPG Pathway. *J. Funct. Biomater.* **2023**, *14*, 193.
9. Liu, Q.; Li, A.; Liu, S.; Fu, Q.; Xu, Y.; Dai, J.; Li, P.; Xu, S. Cytotoxicity of Biodegradable Zinc and Its Alloys: A Systematic Review. *J. Funct. Biomater.* **2023**, *14*, 206.
10. Bourgi, R.; Hardan, L.; Cuevas-Suárez, C.E.; Scavello, F.; Mancino, D.; Kharouf, N.; Haikel, Y. The Use of Warm Air for Solvent Evaporation in Adhesive Dentistry: A Meta-Analysis of In Vitro Studies. *J. Funct. Biomater.* **2023**, *14*, 285.
11. Yuan, Y.; Liu, Q.; Yang, S.; He, W. Four-Dimensional Superimposition Techniques to Compose Dental Dynamic Virtual Patients: A Systematic Review. *J. Funct. Biomater.* **2023**, *14*, 33.

References

1. Kabir, H.; Munir, K.; Wen, C.; Li, Y. Recent research and progress of biodegradable zinc alloys and composites for biomedical applications: Biomechanical and biocorrosion perspectives. *Bioact. Mater.* **2021**, *6*, 836–879. [CrossRef] [PubMed]
2. Xu, Y.; Xu, Y.; Zhang, W.; Li, M.; Wendel, H.-P.; Geis-Gerstorfer, J.; Li, P.; Wan, G.; Xu, S.; Hu, T. Biodegradable Zn-Cu-Fe alloy as a promising material for craniomaxillofacial implants: An in vitro investigation into degradation behavior, cytotoxicity, and hemocompatibility. *Front. Chem.* **2022**, *10*, 860040. [CrossRef] [PubMed]
3. Dai, X.; Heng, B.C.; Bai, Y.; You, F.; Sun, X.; Li, Y.; Tang, Z.; Xu, M.; Zhang, X.; Deng, X. Restoration of electrical microenvironment enhances bone regeneration under diabetic conditions by modulating macrophage polarization. *Bioact. Mater.* **2021**, *6*, 2029–2038. [CrossRef] [PubMed]
4. Li, P.; Zhang, W.; Spintzyk, S.; Schweizer, E.; Krajewski, S.; Alexander, D.; Dai, J.; Xu, S.; Wan, G.; Rupp, F. Impact of sterilization treatments on biodegradability and cytocompatibility of zinc-based implant materials. *Mater. Sci. Eng. C.* **2021**, *130*, 112430. [CrossRef] [PubMed]
5. Badkoobeh, F.; Mostaan, H.; Rafiei, M.; Bakhsheshi-Rad, H.R.; Rama Krishna, S.; Chen, X. Additive manufacturing of biodegradable magnesium-based materials: Design strategies, properties, and biomedical applications. *J. Magnes. Alloys* **2023**, *11*, 801–839. [CrossRef]
6. Zhuo, X.; Wu, Y.; Ju, J.; Liu, H.; Jiang, J.; Hu, Z.; Bai, J.; Xue, F. Recent progress of novel biodegradable zinc alloys: From the perspective of strengthening and toughening. *J. Mater. Res. Technol.* **2022**, *17*, 244–269. [CrossRef]
7. Wang, N.; Ma, Y.; Shi, H.; Song, Y.; Guo, S.; Yang, S. Mg-, Zn-, and Fe-based alloys with antibacterial properties as orthopedic implant materials. *Front. Bioeng. Biotechnol.* **2022**, *10*, 888084. [CrossRef] [PubMed]
8. Liu, Y.; Du, T.; Qiao, A.; Mu, Y.; Yang, H. Zinc-based biodegradable materials for orthopaedic internal fixation. *J. Funct. Biomater.* **2022**, *13*, 164. [CrossRef] [PubMed]
9. Goracci, C.; Juloski, J.; D'Amico, C.; Balestra, D.; Volpe, A.; Juloski, J.; Vichi, A. Clinically relevant properties of 3D printable materials for intraoral use in orthodontics: A critical review of the literature. *Materials* **2023**, *16*, 2166. [CrossRef] [PubMed]
10. Hou, X.; Xu, X.; Zhao, M.; Kong, J.; Wang, M.; Lee, E.S.; Jia, Q.; Jiang, H.B. An overview of three-dimensional imaging devices in dentistry. *J. Esthet. Restor. Dent.* **2022**, *34*, 1179–1196. [CrossRef] [PubMed]
11. Jain, S.; Choudhary, K.; Nagi, R.; Shukla, S.; Kaur, N.; Grover, D. New evolution of cone-beam computed tomography in dentistry: Combining digital technologies. *Imaging Sci. Dent.* **2019**, *49*, 179. [CrossRef] [PubMed]

Journal of
Functional Biomaterials

MDPI

Review

Obturator Manufacturing for Oronasal Fistula after Cleft Palate Repair: A Review from Handicraft to the Application of Digital Techniques

Jiali Chen [1,†], Renjie Yang [2,†], Bing Shi [1], Yichen Xu [3,*] and Hanyao Huang [1,*]

[1] State Key Laboratory of Oral Diseases & National Clinical Research Center for Oral Diseases, Department of Oral and Maxillofacial Surgery, West China Hospital of Stomatology, Sichuan University, Chengdu 610041, China
[2] State Key Laboratory of Oral Diseases & National Clinical Research Center for Oral Diseases, Eastern Clinic, West China Hospital of Stomatology, Sichuan University, Chengdu 610051, China
[3] State Key Laboratory of Oral Diseases & National Clinical Research Center for Oral Diseases, Department of Oral Prosthodontics, West China Hospital of Stomatology, Sichuan University, Chengdu 610041, China
* Correspondence: yichen.xu@scu.edu.cn (Y.X.); huanghanyao_cn@scu.edu.cn (H.H.)
† These authors contributed equally to this work.

Abstract: An oronasal fistula (ONF) is an abnormal structure between the oral and nasal cavities, which is a common complication of cleft palate repair due to the failure of wound healing. When some patients with ONF are unsuitable for secondary surgical repair, the obturator treatment becomes a potential method. The objectives of the obturator treatment should be summarized as filling the ONF comfortably and cosmetically restoring the dentition with partial function. The anatomy of patients with cleft palate is complex, which may lead to a more complex structure of the ONF. Thus, the manufacturing process of the obturator for these patients is more difficult. For performing the design and fabrication process rapidly and precisely, digital techniques can help, but limitations still exist. In this review, literature searches were conducted through Medline via PubMed, Wiley Online Library, Science Direct, and Web of Science, and 122 articles were selected. The purpose of this review was to introduce the development of the obturator for treating patients with ONF after cleft palate repair, from the initial achievement of the obstruction of the ONF to later problems such as fixation, velopharyngeal insufficiency, and infection, as well as the application of digital technologies in obturator manufacturing.

Keywords: additive manufacturing; oronasal fistula; palatoplasty; 3D printing

Citation: Chen, J.; Yang, R.; Shi, B.; Xu, Y.; Huang, H. Obturator Manufacturing for Oronasal Fistula after Cleft Palate Repair: A Review from Handicraft to the Application of Digital Techniques. *J. Funct. Biomater.* **2022**, *13*, 251. https://doi.org/10.3390/jfb13040251

Academic Editors: Nileshkumar Dubey and Vijay Kumar Thakur

Received: 29 September 2022
Accepted: 15 November 2022
Published: 17 November 2022

1. Introduction

Cleft palate is a common congenital anomaly of the oral cavity, resulting from no fusion or partial fusion of the lateral palatal eminences and nasal septum during the period of embryonic development [1], and it occurs unilaterally or bilaterally accompanied by cleft lip [2], ranging in frequency from 0.125% to 0.167% [3]. Typical symptoms include a palatal defect, inability to suck, swallowing difficulties, abnormal articulation, and severe malocclusion [3,4]. Possible causes include inheritance [1,2,5], nutrition [6], drugs [7,8], and tobacco and alcohol [3]. Surgical treatment is the first choice for cleft palate [3]. However, oronasal fistula (ONF) can happen as a complication after cleft palate repair [9,10].

An ONF is an abnormal communication between the oral and nasal cavities, clinically manifesting as a defect ranging from the alveolar process to the uvula. In addition to arising as a complication of cleft palate repair [9,10], it can also occur due to trauma, tumors, infections, and many other factors [11–15]. The rate of postoperative ONF ranges from 2.4% to 55%, related to the cleft width, Veau type, and surgical method [16–20]. ONF can result in nasal leakage, speech disorders, impaired hearing, and food reflux, significantly impacting the patient's quality of life [21,22]. Surgical treatment is the first choice for palatal fistula repair, but for large ONF (L-ONF) it may be unavailable for some patients due to the factors including the operation, such as age and cost [21,23]. The recurrence rate after ONF repair is also high, ranging from 0% to 43% [16,19,20]. The limitations of surgical conditions and the high incidence of morbidity and recurrence of ONF make the obturator an attractive option.

The ONF obturator should fill the ONF comfortably, be stabilized in position, and not obstruct speech function. Meanwhile, the design and fabrication process of the ONF obturator should be rapid and precise. In patients with cleft palate, the structure of ONF can be complex, e.g., in those patients with bilateral complete cleft palate, when ONF occurs, the front lesion can include the lateral hard palatines and the alveolar process near the cleft, along with the isolated premaxillary space with deformed and displaced incisors, making the design and fabrication of the obturator difficult [24–26]. Thus, the manufacturing process of an obturator for ONF in patients with cleft palate should be more intricate.

The history of a technique's development can reveal the thinking for solving problems, as with any medical problem, and the old problem may recur when a new technique is developed. In this review, we aimed to demonstrate the development of an obturator for treating patients with ONF after cleft palate repair, as well as to discuss the associated problems along with the development and their solutions. Meanwhile, with the development of digital technology in dentistry, new strategies—such as intraoral scanning and additive manufacturing—have been applied to improve the fabrication of dental prostheses [27,28]; thus, we also introduce the application of these techniques in the field of obturator manufacturing for patients with ONF after cleft palate repair.

2. Materials and Methods

A literature search was conducted through Medline via PubMed, Wiley Online Library, Science Direct, and Web of Science. Only articles in the English language were considered. A total of 314 articles were found, but 187 were not related to our purpose, so they were excluded. Abstracts, short communications, and company literature were also excluded. Finally, 127 articles were found to be relevant to this review.

In the "Development of ONF obturator" section, because we intended to explore the history and the development of oronasal fistula, we selected the publications from 1953 to 2022 using the following keywords: oronasal fistula, cleft palate, oronasal fistula obturator, and fabrication of obturator. Based on evidence from the currently available literature, Ackerman was the first to report the oronasal obturator in 1953 [29]. We selected an additional 95 articles from that point onwards. With the development of technology, a new method of fabricating obturators emerged recently [21].

In the "Digital process of ONF obturator" section, 32 articles were selected to explore the problem of the acquisition of accurate 3D images, precision fabrication of ONF obturators, and materials of ONF obturators using the following keywords: CBCT, 3D printing, oronasal fistula obturator, and material. We selected articles that introduced the process of ONF obturator fabrication clearly and in detail to explain a digital ONF obturator and the steps needed to fabricate it.

3. Development of ONF Obturator

3.1. Achievement of ONF Obstruction with the Obturator

In 1953, Ackerman introduced a maxillofacial prosthesis, which laid the foundation for expertise in this field [29]; two years later, the use of maxillofacial prostheses after cancer surgery was presented [30]. In the subsequent decades, maxillofacial prostheses were applied to patients with maxillary defects [31], oral and maxillofacial cancer patients [32], and patients after maxillectomy [33].

In 1984, Jacobson and Rosenstein formally reported the use of an obturator in newborn patients with cleft palate, along with its manufacturing process [34]. Firstly, they obtained a model of the patient's maxillary situation using trays and alginate impression materials, after which a plaster working model was made. Based on this plaster working model, the obturator was designed and fabricated using hard and soft acrylics. Soft acrylic was applied on the nasal side to be more comfortable, while self-curing hard acrylic was used to cover the soft acrylic in the lateral maxillary segment and the medial region (Figure 1).

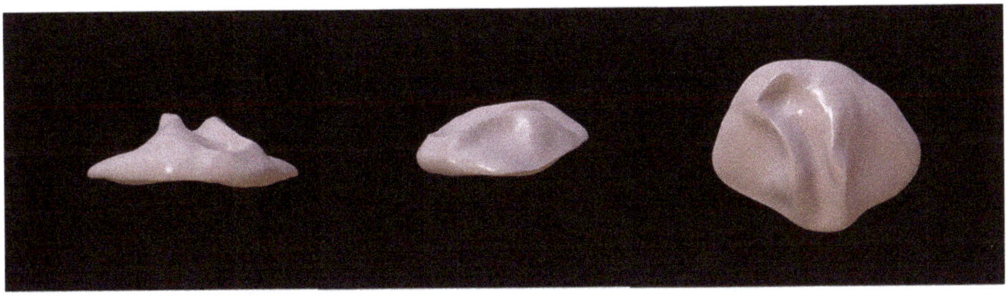

Figure 1. The original oronasal fistula obturator model [34]. The original oronasal fistula obturator was fabricated from hard and soft acrylics. Soft acrylic was applied on the nasal side, while self-curing hard acrylic was used to cover the soft acrylic.

3.2. *Improvement for More Stable Retention of the Obturator*

As the use of ONF obturators has increased, the placement of the obturator has been found to be essential in the outcome of the repair. In 1967, Pielou et al. used an obturator for prosthetic treatment in patients suffering from Pierre Robin syndrome [35]. To solve the problem of unstable retention, the front end of the obturator was extended to the outside of the oral cavity to form a wing-like shape, which was designed to attach to the tape. The distal end of the obturator was extended into the epiglottis, which relieved the symptom of retroflexion of the tongue back into the pharynx. However, there were other problems with this design, including the failure to achieve perfect retention when the patients had excessive oral mobility, as well as the possibility of causing other problems due to the persistent opening of the oral cavity.

Sullivan reported an adjustable "U"-shaped spring ONF obturator to address this problem in 1990, consisting of two components connected by a "U"-shaped spring that could be adjusted to place the obturator in the correct position (Figure 2) [36]. They also considered that the repair of ONF in infancy was a gradual process, with various treatment modalities being used as the fistula decreased until its closure [36]. During that time, the ONF obturator needed to be updated at different stages of treatment. This adjustable ONF obturator also eliminated the need for constant updating [36]. However, the long-term effect of the ONF obturator should be studied further. In patients with cleft palate, the anatomical structure is sometimes extremely complex, as no suitable fixation can be found; this merits more attention to improve obturator retention.

In recent years, two-piece, claspless, and implant-fixed ONF obturators have become hotspots for the shared goal of achieving better retention. There are three main types of two-piece ONF obturator, including ONF obturators with silicone bulbs (Figure 3), ONF obturators with embedded magnets (Figure 4), and ONF obturators with indenters (Figure 5) [37]. The two-piece obturator solves the problem of the obturator's insertion and removal, providing good comfort to the patients. The obturator without clasp fixation, as reported by Murakami et al. in 2020, consists of two parts: a palatal plate and a hollow obstruction, made of cold-curing resin using a compressed vacuum injection unit [38]. After the resin had cured, the palatal plate and the obstruction were attached with five magnetic attachments. To completely close the fistula, the rim of the obstruction was partially extended so that its posterior edge was in contact with the oral side of the soft palate at rest, and a silicone-type soft lining material was chosen for its durability. Studies have demonstrated that this innovative design increased the retention of the ONF obturator and was influential in solving food debris buildup and nasal reflux [38,39]. The implant-fixed ONF obturator designed by Buurman et al. also greatly improved retention due to the placement of the implant [40]. However, these new ONF obturators have been studied with small sample sizes and short follow-up times, and there is insufficient theoretical evidence and clinical data to prove their effectiveness—especially in patients with ONF after cleft palate repair.

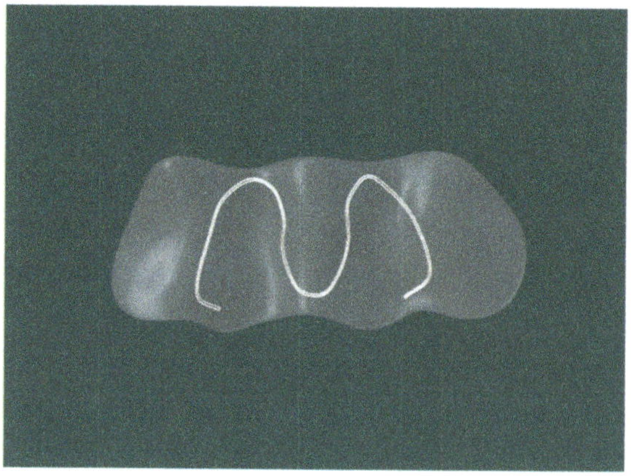

Figure 2. The "U"-shaped spring oronasal fistula obturator model [36]. The central part was made of silicone, and a spring was used to adjust the retention of the obstruction according to the size of the oronasal fistula.

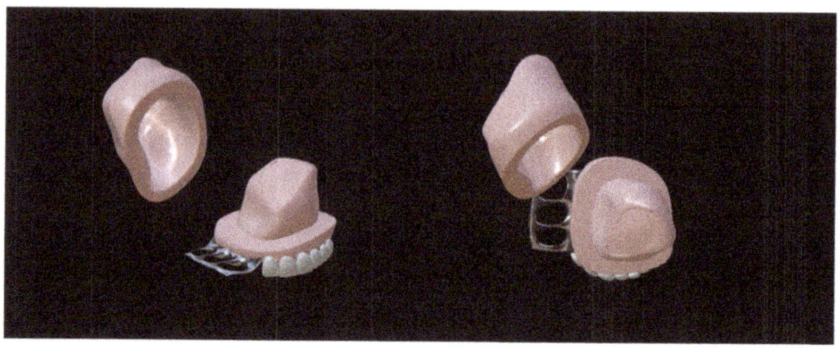

Figure 3. Two-piece oronasal fistula obturator with silicone bulb [41]. The silicone cap was placed over the maxillary defect, and the other part was inserted into the silicone cap.

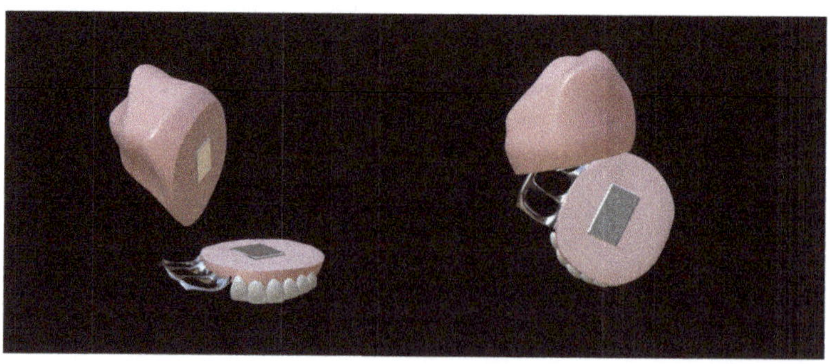

Figure 4. Two-piece oronasal fistula obturator model with embedded magnets [41]. The two parts of the obturator were joined together by magnets.

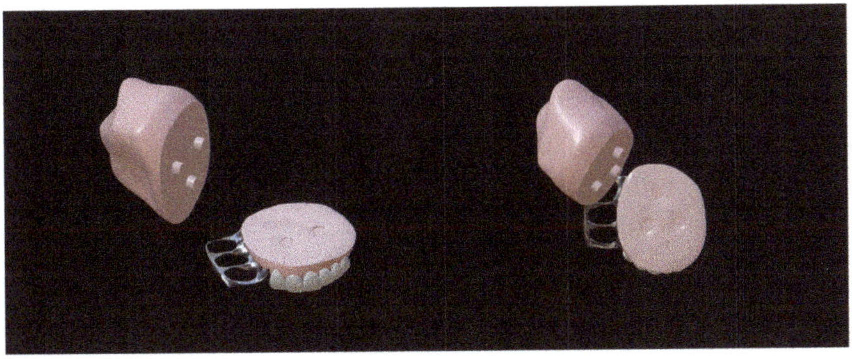

Figure 5. Two-piece oronasal fistula obturator model with indenters [41]. The two parts of the obturator were combined via a plug and a hole corresponding to one another.

3.3. Restoration of Speech Problems with the Obturator

The primary purpose of the ONF obturator is to obstruct abnormal communication between the oral and nasal cavities. However, in patients with cleft palate, velopharyngeal insufficiency (VPI) can also be the cause of speech problems and affect their quality of life [42–49]. Both ONF and VPI can contribute to the speech dysfunction of the patient [49–51], so both aspects should be considered during the manufacture of the obturator.

In patients without VPI—such as patients after tumor excision—when the obturator is tightly integrated with the mucosa (or implanted flap), speech and swallowing functions can be restored [52]. Initially, to improve the retention of the obturator by reducing its weight, obturators were designed to be hollow, which enabled them to engage the remaining tooth and tissue bearers and extend into the defect effectively (Figure 6). Different methods have been invented, including the technique of hollowing and rejoining directly [53], the salt-losing technique [54], and the dual-processing technique [55]. In the application of hollow obturators, an interesting finding was that the design of the hollow cavity aided speech resonance, increased speech intelligibility, and gradually improved speech function [53]. Since then, numerous studies have been focused on this aspect [56,57]. An article described the clinical and laboratory procedures of a hollow bulb obturator that was used in a hemimaxillectomy patient, demonstrating that it aided speech resonance [57].

Figure 6. Hollow obturator model [53]. The obturator was designed to be hollow to reduce its weight and provide good retention.

VPI can occur after the insufficient elevation of the soft palate to the pharyngeal wall, which remains as a port allowing airflow leakage [58–60]. After obstructing the ONF, the VPI problem remains to be solved. Blakeley was the first to use a speech bulb to improve velopharyngeal closure in a patient with cleft palate [61]. In subsequent years, studies focused on how the size and position of the speech bulb improved dysphonia [62–64]. However, the use of speech bulb obturators in the treatment of VPI became popular in the 20th century, partially due to techniques that permitted direct visualization of the velopharyngeal mechanism. In 1979, a palatal lift prosthesis was used to treat palatopharyngeal incompetence [65], and it was suggested that this was an effective method of improving articulation. However, this method was indicated for selected patients who have anatomically normal palates that are dysfunctional.

The speech bulb obturator was reported in 1993 [66], consisting of a custom-made dental appliance with an extension and advocated for use in individuals with severe pharyngeal articulation problems. Studies on speech bulb reduction later became popular once more. A case of speech function enhancement using a speech bulb was reported in detail by Bispo et al. [67]. The patient in this case report also had speech training before wearing a speech bulb obturator, but the recovery of speech function was poor. After the consultation, the authors created a removable obturator consisting of an acrylic front part with a fixed clip, a pharyngeal bulb part based on the shape of the palatopharyngeal gap, and a middle part connecting these two parts (Figure 7). During the treatment, they gradually reduced the size of the speech bulb and trained the patient, eventually improving the patient's speech function. This case suggested that a speech bulb obturator could significantly improve the patient's speech disorder. In addition, Elangovan [68], Fen-Huey Lin [69], Agrawal [70], and others have also reported good therapeutic outcomes with speech bulbs in ONF restoration.

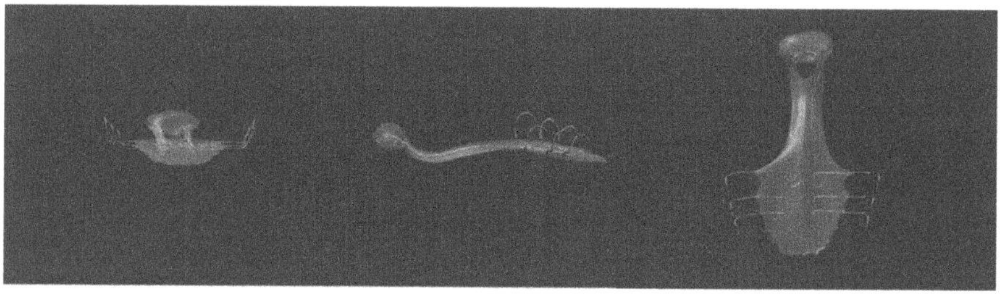

Figure 7. Speech bulb obturator model [67]. The obturator consisted of an acrylic front part with a fixed clip, a pharyngeal bulb, and a middle part connecting these two parts. The speech bulb improved speech resonance.

3.4. Resolution of Infections

Infections related to implanted devices are mainly caused by *Staphylococcus* spp., especially *S. epidermidis* and *S. aureus* [71–73]. The incidence of fungal infections related to implanted devices is lower but more severe, most commonly caused by pathogenic *Candida* species—especially *C. albicans* [74–76]. Furthermore, moving the obturator prostheses frequently increases the risk of infection because they may traumatize the oral mucosa [77]. Silicones and acrylic resins are the most commonly used materials at present [78,79].

In a study by Wieckiewicz et al. [80], *Candida* adhered well to silicone, and *Candida* on obturator prostheses made of silicone and on oral mucosa was found to be the leading cause of inflammation in patients after tumor resection. In 2009, Mattos et al. [81] reported that patients using acrylic ONF obturator frequently developed stomatitis, and found that the oral mucosa under the obturator was more susceptible to *Candida albicans* infections. Higher silicone porosity and a reduced degree of acrylic polymerization have been reported to contribute to the colonization of microorganisms, leading to the development of infections [82–84]. These reports suggest that the choice of material is also crucial for effective antimicrobial resistance and for preventing postoperative infections and complications.

The growth of fungi has been shown to destroy the lining surface, leading to irritation of the oral tissues. Possible causes include increased surface roughness and high levels of secreted enzymes and metabolic products produced by fungal cells [85,86]. Batches of methods were tried to avoid and reduce the adhesion of these microorganisms so as to reduce related infections, including the addition of antifungal agents or antiseptics in materials, the modification of surface physicochemical characteristics, and the use of different materials.

The method of decreasing biofilm formation by incorporating various antimicrobial materials into the obturator has been the focus of many studies [87–89]. In 2012, Jingwei He et al. incorporated quaternary ammonium salts into methyl methacrylate (MMA) to form a quaternary ammonium methacrylate that maintained its antibacterial activity without sacrificing its mechanical properties [90]. Other attempts were made to modify the surface characteristics of the materials, including electronegativity, wettability, and roughness, so as to reduce microorganism adhesion [91–93]. Nikawa introduced a thermocycling process to a fabricated maxillofacial prosthesis in 2001 and observed *Candida albicans* growing on it [91]. The results suggested that the materials exhibited antifungal effects because the surface of these materials was made hydrophobic by this method. In 2007, Khalaf modified silicone elastomer surfaces with different surface roughness and porosity and concluded that a smoother, less porous surface exhibited a lower adhesion of microorganisms [93]. It was reported that parylene coating reduced the adhesion and aggregation of *C. Albicans* on the surface of silicone and improved the wettability of the silicone [92], while titanium offered good biomechanical properties, low weight, and high corrosion resistance, and bacteria were not able to penetrate the surface [94]. The titanium surface was also polished so that the microorganisms could not adhere easily. In one study, nanostructured materials were reported to show a slight decrease in microorganism adhesion [95].

4. Digital Process of ONF Obturator

Digital ONF obturators have become a new treatment modality, shortening the production process and making the ONF obturators more precise, allowing for personalized treatment. The digital ONF obturator is a 3D-printed obturator based on a model obtained via intraoral scan, CBCT technology, or other imaging techniques.

4.1. Acquisition of Accurate 3D Images

Advances in radiological imaging technology have facilitated the creation of 3D imaging methods, with CT being the first technique to present stereoscopic hard and soft tissues of maxilla through the acquisition capability of multiple consecutive cross-sectional images [96]. Dental cone-beam CT (CBCT) was born and became a widely used imaging tool in dental diagnostic treatment due to the changing demand [97]. Kuijpers et al. [98], in a systematic review, stated that CT, CBCT, MRI, stereophotogrammetry, and laser surface scanning were the 3D techniques most commonly used for patients with cleft lip and palate—mainly for soft tissue analysis, bone graft evaluation, and craniofacial skeleton changes. CBCT has the advantages of high spatial resolution, low radiation dose, small size, and low cost compared with conventional medical CT. A digital ONF obturator takes advantage of its ability to be reconstructed to provide a 3D view of the patient's oral cavity. Many studies have focused on the precision of CBCT [99,100], which is the basis of the digital manufacturing [101]. Because its low-density resolution, poor soft-tissue imaging, and metal artifacts hamper the achievement of accurate intraoral images, MRI is required to obtain clear soft-tissue images to compensate for nasal-side CBCT images [102]. A patient-specific low-cost ONF obturator was explored by Bartellas et al. [27]. A CT scan of the patient's maxilla was performed, which was visualized and rendered using OsiriX Lite after creating a model of the maxilla and then using Meshmixer to design the ONF obturator.

4.2. Precision Fabrication of ONF Obturators

CBCT can obtain realistic and accurate 3D oral images to assess ONF before repair. Three-dimensional (3D) printing technology can accurately print a model of the patient's palate and a computer-designed ONF obturator. Choi et al. [103] used the 3D scanning procedure of CBCT to create 3D digital images of the patient's palate, which were then exported to a computer for 3D analysis in a standard language format of surface subdivision. Following this analysis, they created a model of the patient's palate using 3D printing and, by measuring it, they concluded that this technique could accurately simulate the patient's palate condition. A detailed step-by-step description of how to design an obturator using dental CAD software (exocad DentalCAD 3.0) and produce the obturator using 3D printing was given by Krämer Fernandez et al. [104], but they stated that this method was limited to small defects. CAD/CAM systems have also been developed to manufacture fixed and removable obturators [105,106]. In 2000, Bibb [107] reported using CAD/CAM technology to fabricate prostheses and stated that CAD could be applied to produce accurate physical models based on careful acquisition of 3D scan data.

4.3. Further Improvements in Materials of ONF Obturators

Materials for dental treatment are also constantly being updated thanks to the advent of 3D printing. In particular, titanium and its alloys are suitable for 3D printing technology. The titanium created by 3D printing has high yield strength, ultimate strength, excellent ductility, and low solubility, which could resolve safety issues caused by the dissolution of metal ions [108]. Studies on obturator materials have been stimulated by the increasing demand for high esthetic restoration [109–111]. Recently, Schonhoff et al. [111] reported the mechanical properties of thermoplastic polymer materials by 3D printing, and the results showed that the mechanical properties were affected due to 3D printing, while the 3D printing parameters employed for the additive manufacturing of thermoplastic polymer material specimens require further optimization.

We also found that the following materials can potentially be used: polymethylmethacrylate (PMMA) is used widely in obturators due to its low density, aesthetics, cost-effectiveness, and stability [112,113]. In addition to mechanical properties, the material's biocompatibility is also an important factor to consider, representing the most important biological property [114]. Studies have demonstrated that PMMA has lower toxicity [114,115]. Properly cured PMMA materials have good biocompatibility due to low amounts of monomers, such as in heat-cured and microwave-cured PMMA [116]. As mentioned above, titanium and titanium alloys are currently commonly used as 3D-printing materials in the production of ONF obturators. The titanium alloys applied in 3D printing have good biocompatibility [117]. Titanium and its alloys have good mechanical properties, corrosion resistance, and a high strength-to-weight ratio [118–120], so they are widely used, including in the manufacture of ONF obturators. Meanwhile, other biodegradable metal materials such as Zn-Cu-Fe alloy can also be promising [121].

Improving the antimicrobial properties of 3D-printed materials is also a current research topic that can help to reduce bacterial adhesion and inflammation after wearing ONF obturator. However, although few antimicrobial materials have been used for ONF obturators to date, we can get a glimpse of other antimicrobial dental materials for 3D printing. In the study by Chen et al. [122], the authors demonstrated that the composite resin added TiO_2 showed good antibacterial properties compared to pure PMMA resin. Herrmann and Ren et al. [123] reported a 3D-printed polymeric resin containing antimicrobial positively charged quaternary ammonium groups, and the results showed that the quaternization of the material greatly improved its antimicrobial resistance, while the mechanical properties were similar to those of other materials. Understanding the application of other antimicrobial dental materials will allow us to apply them to ONF obturator fabrication in future studies. Moreover, anti-inflammation strategies such as cationic scavenging should also be considered [124,125].

4.4. Fabrication of a Digital ONF Obturator

Recently, the process of producing a digital ONF obturator was described in detail [21]. The first step was to obtain an intraoral situation using a confocal laser scanner, and then the probe was inserted as deep as possible to get a more complete model. Then, an ONF model was designed using CAD software, during which they first filled the ONF and then made the filled surface smooth because it could only be developed on a closed surface. Once the model had been created, the ONF obturator was designed and then hollowed out. After that, the designed ONF obturator was printed in two parts—the hexagonal anti-rotation cap and the perforated bottom, and then welded and polished. To sum up, the process of producing a digital ONF obturator is as follows (Figure 8).

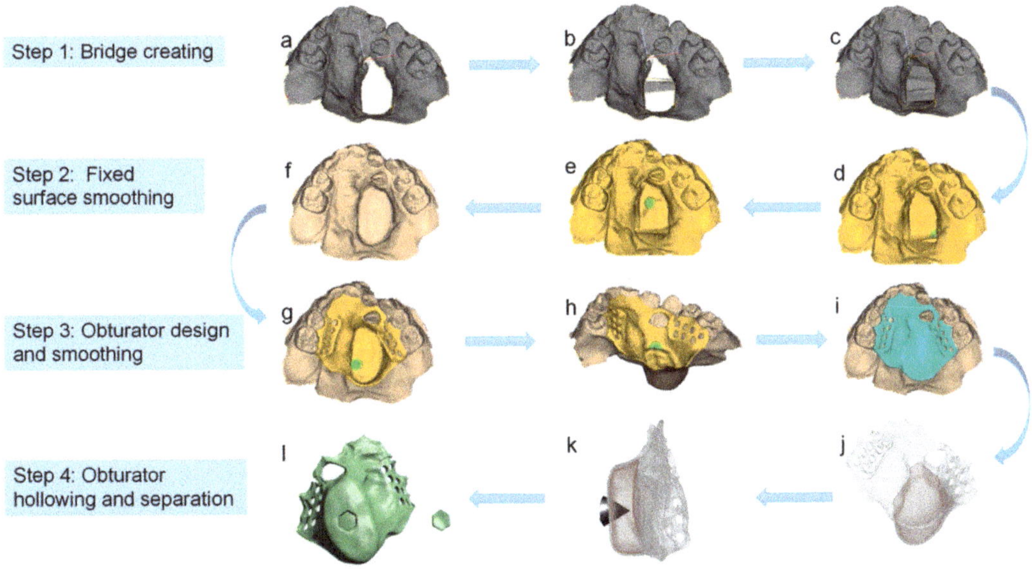

Figure 8. The process of manufacturing a digital ONF obturator [21] The whole process consists of four steps: bridge creation (**a–c**), fixed-surface smoothing (**d–f**), obturator design and smoothing (**g–i**), and hollow and separation (**j–l**).

5. Summary of the Important Designs during the Development

In 1953, Ackerman et al. reported a maxillofacial prosthesis for the first time, which laid the foundation for this field of study [29]. Then, acrylic was used in ONF obturators, and soft acrylic was used on the nasal side to increase comfort in 1984 [34]. Newly designed obturators have been used to improve the retention, including "U"-shaped [36] and two-piece ONF obturators [37]. In 2011, a hollow ONF obturator and speech ball obturator were used to aid in speech function [57,67]. Recently, digital ONF obturators have become a research hotspot; Yichen Xu et al. introduced the manufacture of ONF obturator clearly [21]. The important designs were shown in Table 1.

Table 1. The important designs during the development.

ONF Types	Time	Materials	Highlights	Authors
Maxillofacial prosthesis	1953	——	Laid the foundation for this field	Ackerman et al. [29]
Acrylic ONF obturator	1984	Acrylic	Used soft acrylic on the nasal side to improve comfort	Jacobson et al. [34]
"U"-shaped ONF obturator	1990	Silicone and metal	Improved ONF obturator retention with spring adjustability	Sullivan et al. [36]
Two-piece ONF obturator	2015	Acrylic, silicone, magnets	Used different forms of bonding to solve the problem of obturator insertion and removal, as well as to provide good comfort to the patients	Dholam et al. [37]
Hollow ONF obturator	2011	Acrylic	The hollow design aided the speech resonance and improved the retention	Bhasin et al. [57]
Speech ball obturator	2011	Silicone	The speech ball improved the patients' speech function	Bispo et al. [67]
Digital ONF obturator	2022	Acrylic and Ti-6Al-4V alloy	Clearly introduced the manufacturing process of a digital ONF obturator	Yichen Xu et al. [21]

6. Summary

In patients with cleft palate, the structure of ONF is much more complex than that in patients with other problems, making the design and fabrication of obturators difficult [24–26]. Currently, digital ONF obturators appear to be destined to become the trend, and for patients with cleft palate their fabrication is likely to be more challenging.

There are many advantages to digital technology in the treatment of ONF, such as avoiding the inevitable errors associated with the impressions, plaster revisions, and restoration of cusp misalignments in conventional manufacturing. In addition, the ONF obturator can be fabricated in a much shorter time due to the removal of tedious steps. In the future, digital techniques will be more widely used in this field, where virtual-reality design can interact with 3D printing. Doctors may directly perform the 3D design of the restoration in the virtual world, observe the 3D restoration products to better estimate the feasibility of the products, and reduce the wastage of time and resources.

In addition to good retention and antimicrobial properties, the perfect digital ONF obturator further achieves aesthetic restoration. Today, smart biomaterials and advanced stem cell culture technology, coupled with 3D printing, provide an excellent basis for patient-tailored treatments. The 3D printing technique has the advantage that it can produce various geometries to perfectly fit any tissue defect as well as mimicking complex inner tissue architecture and heterogeneity via the precise positioning of different materials or cell types. This technology is already being used for the manufacturing of periodontal and gingival tissues [126,127], and in the future it will likely be used for the manufacturing of palatal mucosa for better functional and aesthetic restoration.

However, since acquiring accurate 3D images and 3D printing are the basis of the manufacturing process, studies on obtaining accurate 3D images and 3D printing materials should be advanced to improve the mechanical properties and reduce the gap between 3D-printed models and human structures—especially for patients with ONF after cleft palate repair. Most 3D-printed materials lack the realism to adequately mimic soft human biological tissue and its great mechanical properties, and post-processing is often required to soften the printed structure. On the other hand, materials that can be applied to 3D printing are limited. Therefore, none of the currently available materials can fully mimic elastic biological tissue, which should also be investigated in the future.

Author Contributions: Conceptualization, H.H., Y.X., B.S. and J.C.; methodology, H.H. and J.C.; validation, H.H., Y.X., B.S. and J.C.; formal analysis, J.C. and R.Y.; investigation, J.C.; resources, H.H. and Y.X.; data curation, J.C. and R.Y.; writing—original draft preparation, J.C. and R.Y.; writing—review and editing, H.H.; visualization, H.H.; supervision, H.H. and B.S.; funding acquisition, H.H. and Y.X. All authors have read and agreed to the published version of the manuscript.

Funding: This research was supported by the Sichuan Science and Technology Program (No. 2021YFS0085 and 2022NSFSC0743); the Research and Development Program of West China Hospital of Stomatology, Sichuan University (No. RD-02-202107 and RD-02-202114), Sichuan Postdoctoral Science Foundation (TB2022005); and Research Funding from West China School/Hospital of Stomatology, Sichuan University (No. RCDWJS2021-13).

Data Availability Statement: No applicable.

Conflicts of Interest: It should be understood that none of the authors have any financial or scientific conflict of interest pertaining to the research described in this manuscript.

References

1. Paradowska-Stolarz, A. MSX1 gene in the etiology orofacial deformities. *Postep. Hig. Med. Dosw.* **2015**, *69*, 1499–1504.
2. Dixon, M.J.; Marazita, M.L.; Beaty, T.H.; Murray, J.C. Cleft lip and palate: Understanding genetic and environmental influences. *Nat. Rev. Genet.* **2011**, *12*, 167–178. [CrossRef] [PubMed]
3. Vyas, T.; Gupta, P.; Kumar, S.; Gupta, R.; Gupta, T.; Singh, H.P. Cleft of lip and palate: A review. *J. Fam. Med. Prim. Care* **2020**, *9*, 2621–2625. [CrossRef] [PubMed]
4. Roguzińska, S.; Pelc, A.; Mikulewicz, M. Orthodontic-care burden for patients with unilateral and bilateral cleft lip and palate. *Dent. Med. Probl.* **2020**, *57*, 411–416. [CrossRef] [PubMed]

5. Nasroen, S.L.; Maskoen, A.M.; Soedjana, H.; Hilmanto, D.; Gani, B.A. IRF6 rs2235371 as a risk factor for non-syndromic cleft palate only among the Deutero-Malay race in Indonesia and its effect on the IRF6 mRNA expression level. *Dent. Med. Probl.* **2022**, *59*, 59–65. [CrossRef]
6. Crider, K.S.; Bailey, L.B. Defying birth defects through diet? *Genome Med.* **2011**, *3*, 9. [CrossRef]
7. Bateman, B.T.; Hernandez-Diaz, S.; Straub, L.; Zhu, Y.; Gray, K.J.; Desai, R.J.; Mogun, H.; Gautam, N.; Huybrechts, K.F. Association of first trimester prescription opioid use with congenital malformations in the offspring: Population based cohort study. *BMJ* **2021**, *372*, n102. [CrossRef]
8. Wilcox, A.J.; Lie, R.T.; Solvoll, K.; Taylor, J.; McConnaughey, D.R.; Abyholm, F.; Vindenes, H.; Vollset, S.E.; Drevon, C.A. Folic acid supplements and risk of facial clefts: National population based case-control study. *BMJ* **2007**, *334*, 464. [CrossRef]
9. Alonso, V.; Abuin, A.S.; Duran, C.; Gomez, O.; Miguez, L.; Molina, M.E. Three-layered repair with a collagen membrane and a mucosal rotational flap reinforced with fibrine for palatal fistula closure in children. *Int. J. Pediatr. Otorhinolaryngol.* **2019**, *127*, 109679. [CrossRef]
10. Honnebier, M.B.O.M.; Johnson, D.S.; Parsa, A.A.; Dorian, A.; Parsa, F.D. Closure of Palatal Fistula with a Local Mucoperiosteal Flap Lined with Buccal Mucosal Graft. *Cleft Palate-Craniofacial J.* **2000**, *37*, 127–129. [CrossRef]
11. Abuabara, A.; Cortez, A.L.; Passeri, L.A.; de Moraes, M.; Moreira, R.W. Evaluation of different treatments for oroantral/oronasal communications: Experience of 112 cases. *Int. J. Oral Maxillofac. Surg.* **2006**, *35*, 155–158. [CrossRef] [PubMed]
12. Ahmed, M.V.; Kaul, D.; Naz, F.; Tambuwala, A.; Chand, M. Repair of iatrogenic oronasal fistula after periapical surgery. *Univers. Res. J. Dent.* **2013**, *2*, 83. [CrossRef]
13. Majid, O.W. Persistent oronasal fistula after primary management of facial gunshot injuries. *Br. J. Oral Maxillofac. Surg.* **2008**, *46*, 50–52. [CrossRef]
14. Sahoo, N.K.; Desai, A.P.; Roy, I.D.; Kulkarni, V. Oro-Nasal Communication. *J. Craniofacial Surg.* **2016**, *27*, e529–e533. [CrossRef] [PubMed]
15. Tartaro, G.; Rauso, R.; Bux, A.; Santagata, M.; Colella, G. An unusual oronasal fistula induced by prolonged cocaine snort. Case report and literature review. *Minerva Stomatol.* **2008**, *57*, 203–210.
16. Garg, R.; Shah, S.; Uppal, S.; Mittal, R.K. A statistical analysis of incidence, etiology, and management of palatal fistula. *Natl. J. Maxillofac. Surg.* **2019**, *10*, 43–46. [CrossRef]
17. Mahajan, R.K.; Kaur, A.; Singh, S.M.; Kumar, P. A retrospective analysis of incidence and management of palatal fistula. *Indian J. Plast. Surg.* **2018**, *51*, 298–305. [CrossRef]
18. Yuan, N.; Dorafshar, A.H.; Follmar, K.E.; Pendleton, C.; Ferguson, K.; Redett, R.J., 3rd. Effects of Cleft Width and Veau Type on Incidence of Palatal Fistula and Velopharyngeal Insufficiency after Cleft Palate Repair. *Ann. Plast. Surg.* **2016**, *76*, 406–410. [CrossRef]
19. Tse, R.W.; Siebold, B. Cleft Palate Repair: Description of an Approach, Its Evolution, and Analysis of Postoperative Fistulas. *Plast. Reconstr. Surg.* **2018**, *141*, 1201–1214. [CrossRef]
20. Shankar, V.A.; Snyder-Warwick, A.; Skolnick, G.B.; Woo, A.S.; Patel, K.B. Incidence of Palatal Fistula at Time of Secondary Alveolar Cleft Reconstruction. *Cleft Palate-Craniofacial J.* **2018**, *55*, 999–1005. [CrossRef]
21. Xu, Y.; Huang, H.; Wu, M.; Tian, Y.; Wan, Q.; Shi, B.; Hu, T.; Spintzyk, S. Rapid Additive Manufacturing of a Superlight Obturator for Large Oronasal Fistula in Pediatric Patient. *Laryngoscope*, 2022; *online ahead of print*. [CrossRef] [PubMed]
22. Brandão, T.B.; Vechiato Filho, A.J.; de Souza Batista, V.E.; de Oliveira, M.C.Q.; Santos-Silva, A.R. Obturator prostheses versus free tissue transfers: A systematic review of the optimal approach to improving the quality of life for patients with maxillary defects. *J. Prosthet. Dent.* **2016**, *115*, 247–253.e244. [CrossRef] [PubMed]
23. Murthy, J. Descriptive study of management of palatal fistula in one hundred and ninety-four cleft individuals. *Indian J. Plast. Surg.* **2011**, *44*, 41–46. [CrossRef] [PubMed]
24. Li, H.; Yin, N.; Song, T. Oronasal fistula repair using the alveolar ridge approach. *Int. J. Pediatr. Otorhinolaryngol.* **2015**, *79*, 161–164. [CrossRef] [PubMed]
25. Goiato, M.C.; dos Santos, D.M.; Moreno, A.; Santiago, J.F.J.; Haddad, M.F.; Pesqueira, A.A.; Miyahara, G.I. Prosthetic Treatments for Patients with Oronasal Communication. *J. Craniofacial Surg.* **2011**, *22*, 1445–1447. [CrossRef] [PubMed]
26. Gümüş, H.O.; Tuna, S.H. An alternative method for constructing an obturator prosthesis for a patient with a bilateral cleft lip and palate: A clinical report. *J. Esthet. Restor. Dent.* **2009**, *21*, 89–94. [CrossRef]
27. Bartellas, M.; Tibbo, J.; Angel, D.; Rideout, A.; Gillis, J. Three-Dimensional Printing: A Novel Approach to the Creation of Obturator Prostheses Following Palatal Resection for Malignant Palate Tumors. *J. Craniofacial Surg.* **2018**, *29*, e12–e15. [CrossRef]
28. Rodney, J.; Chicchon, I. Digital Design and Fabrication of Surgical Obturators Based Only on Preoperative Computed Tomography Data. *Int. J. Prosthodont.* **2017**, *30*, 111–112. [CrossRef]
29. Ackerman, A.J. Maxillofacial prosthesis. *Oral Surg. Oral Med. Oral Pathol.* **1953**, *6*, 176–200. [CrossRef]
30. Ackerman, A.J. The prosthetic management of oral and facial defects following cancer surgery. *J. Prosthet. Dent.* **1955**, *5*, 413–432. [CrossRef]
31. Boucher, L.J.; Heupel, E.M. Prosthetic restoration of a maxilla and associated structures. *J. Prosthet. Dent.* **1966**, *16*, 154–168. [CrossRef]
32. Curtis, T.A. Treatment planning for intraoral maxillofacial prosthetics for cancer patients. *J. Prosthet. Dent.* **1967**, *18*, 70–76. [CrossRef]

33. Zarb, G.A. The maxillary resection and its prosthetic replacement. *J. Prosthet. Dent.* **1967**, *18*, 268–281. [CrossRef]
34. Jacobson, B.N.; Rosenstein, S.W. Early maxillary orthopedics for the newborn cleft lip and palate patient. An impression and an appliance. *Angle Orthod.* **1984**, *54*, 247–263. [CrossRef] [PubMed]
35. Pielou, W.D. Non-surgical management of Pierre Robin syndrome. *Arch. Dis. Child.* **1967**, *42*, 20–23. [CrossRef] [PubMed]
36. Sullivan, P.G. Early Pre-Surgical Treatment of the Cleft Palate Patient. *J. R. Soc. Med.* **1990**, *83*, 90–93. [CrossRef]
37. Dholam, K.P.; Sadashiva, K.M.; Bhirangi, P.P. Rehabilitation of large maxillary defect with two-piece maxillary obturators. *J. Cancer Res. Ther.* **2015**, *11*, 664. [CrossRef]
38. Murakami, M.; Nishi, Y.; Shimizu, T.; Nishimura, M. A retainer-free obturator prosthesis in a fully dentulous patient with palatal defects. *J. Oral Sci.* **2020**, *62*, 122–124. [CrossRef]
39. Białożyt-Bujak, E.; Wyszyńska, M.; Chladek, G.; Czelakowska, A.; Gala, A.; Orczykowska, M.; Białożyt, A.; Kasperski, J.; Skucha-Nowak, M. Analysis of the Hardness of Soft Relining Materials for Removable Dentures. *Int. J. Environ. Res. Public Health* **2021**, *18*, 9491. [CrossRef]
40. Buurman, D.J.M.; Speksnijder, C.M.; Engelen, B.H.B.T.; Kessler, P. Masticatory performance and oral health-related quality of life in edentulous maxillectomy patients: A cross-sectional study to compare implant-supported obturators and conventional obturators. *Clin. Oral Implant. Res.* **2020**, *31*, 405–416. [CrossRef]
41. Ayad, T.; Xie, L. Facial artery musculomucosal flap in head and neck reconstruction: A systematic review. *Head Neck* **2015**, *37*, 1375–1386. [CrossRef] [PubMed]
42. Sakran, K.A.; Wu, M.; Alkebsi, K.; Mashrah, M.A.; Al-Rokhami, R.K.; Wang, Y.; Mohamed, A.A.; Elayah, S.A.; Al-Sharani, H.M.; Huang, H.; et al. The Sommerlad-Furlow Modified Palatoplasty Technique: Postoperative Complications and Implicating Factors. *Laryngoscope*, 2022; *online ahead of print*. [CrossRef] [PubMed]
43. Chen, N.; Shi, B.; Huang, H. Velopharyngeal Inadequacy-Related Quality of Life Assessment: The Instrument Development and Application Review. *Front. Surg.* **2022**, *9*, 796941. [CrossRef]
44. Xie, Z.; Yang, C.; Zhao, Y.; Yang, Y.; Xia, W.; Zong, Y.; Chi, T.; Shi, B.; Huang, H.; Gong, C. Anxiety in Chinese Patients with Cleft Lip and/or Palate: A Preliminary Study. *Front. Pediatr.* **2022**, *10*, 842470. [CrossRef] [PubMed]
45. Sakran, K.A.; Al-Rokhami, R.K.; Wu, M.; Chen, N.; Yin, H.; Guo, C.; Wang, Y.; Alkebsi, K.; Abotaleb, B.M.; Mohamed, A.A.; et al. Correlation of the Chinese velopharyngeal insufficiency-related quality of life instrument and speech in subjects with cleft palate. *Laryngoscope Investig. Otolaryngol.* **2022**, *7*, 180–189. [CrossRef]
46. Huang, H.; Chen, N.; Yin, H.; Skirko, J.R.; Guo, C.; Ha, P.; Li, J.; Shi, B. Validation of the Chinese Velopharyngeal Insufficiency Effects on Life Outcomes Instrument. *Laryngoscope* **2019**, *129*, E395–E401. [CrossRef]
47. Ysunza, P.A.; Repetto, G.M.; Pamplona, M.C.; Calderon, J.F.; Shaheen, K.; Chaiyasate, K.; Rontal, M. Current Controversies in Diagnosis and Management of Cleft Palate and Velopharyngeal Insufficiency. *Biomed. Res. Int.* **2015**, *2015*, 196240. [CrossRef]
48. Wang, X.; Guo, C.L.; Shi, B.; Yin, H. Velopharyngeal closure pattern and speech characteristics of patients congenital velopharyngeal insufficiency. *Hua Xi Kou Qiang Yi Xue Za Zhi* **2020**, *38*, 662–666. [CrossRef]
49. Tache, A.; Maryn, Y.; Mommaerts, M.Y. Need for velopharyngeal surgery after primary palatoplasty in cleft patients. A retrospective cohort study and review of literature. *Ann. Med. Surg.* **2021**, *69*, 102707. [CrossRef]
50. Smyth, A.G.; Wu, J. Cleft Palate Outcomes and Prognostic Impact of Palatal Fistula on Subsequent Velopharyngeal Function—A Retrospective Cohort Study. *Cleft Palate-Craniofacial J.* **2019**, *56*, 1008–1012. [CrossRef]
51. Gustafsson, C.; Heliövaara, A.; Leikola, J. Long-Term Follow-up of Unilateral Cleft lip and Palate: Incidence of Speech-Correcting Surgeries and Fistula Formation. *Cleft Palate-Craniofacial J.* **2021**, *59*, 1537–1545. [CrossRef] [PubMed]
52. Narayanraopeta, S.; Vemisetty, H.K.; Marri, T.; Konda, P. Rehabilitation of a Unilateral Cleft Palate with Endosseous Implants in an Edentulous Elderly Patient. *Contemp. Clin. Dent.* **2020**, *11*, 285–289. [CrossRef] [PubMed]
53. Bhat, A.M. Prosthetic rehabilitation of a completely edentulous patient with palatal insufficiency. *Indian J. Dent. Res.* **2007**, *18*, 35–37. [CrossRef] [PubMed]
54. Vamsi Krishna, C.H.; Babu, J.K.; Fathima, T.; Reddy, G.V.K. Fabrication of a hollow bulb prosthesis for the rehabilitation of an acquired total maxillectomy defect. *Case Rep.* **2014**, *2014*, bcr2013201400-b. [CrossRef] [PubMed]
55. Oh, W.S.; Roumanas, E.D. Optimization of Maxillary Obturator Thickness Using a Double-Processing Technique. *J. Prosthodont.* **2007**, *17*, 60–63. [CrossRef] [PubMed]
56. Bagis, B.; Aydoğan, E.; Hasanreisoğlu, U. Rehabilitation of a congenital palatal defect with a modified technique: A case report. *Cases J.* **2008**, *1*, 39. [CrossRef]
57. Bhasin, A.S.; Singh, V.; Mantri, S.S. Rehabilitation of Patient with Acquired Maxillary Defect, using a Closed Hollow Bulb Obturator. *Indian J. Palliat. Care* **2011**, *17*, 70–73. [CrossRef]
58. Johns, D.F.; Rohrich, R.J.; Awada, M. Velopharyngeal Incompetence:: A Guide for Clinical Evaluation. *Plast. Reconstr. Surg.* **2003**, *112*, 1890–1898. [CrossRef]
59. Smith, B.; Guyette, T.W. Evaluation of cleft palate speech. *Clin. Plast. Surg.* **2004**, *31*, 251–260. [CrossRef]
60. Woo, A. Velopharyngeal dysfunction. *Semin. Plast. Surg.* **2012**, *26*, 170–177. [CrossRef]
61. Blakeley, R.W. The complementary use of speech prostheses and pharyngeal flaps in palatal insufficiency. *Cleft Palate J.* **1964**, *12*, 194–198.
62. Mazaheri, M.; Millard, R.T. Changes in nasal resonance related to differences in location and dimension of speech bulbs. *Cleft Palate J.* **1965**, *31*, 167–175. [PubMed]

63. Shelton, R.L.; Lindquist, A.F.; Chisum, L.; Arndt, W.B.; Youngstrom, K.A.; Stick, S.L. Effect of prosthetic speech bulb reduction on articulation. *Cleft Palate J.* **1968**, *5*, 195–204. [PubMed]
64. Shelton, R.L.; Lindquist, A.F.; Arndt, W.B.; Elbert, M.; Youngstrom, K.A. Effect of speech bulb reduction on movement of the posterior wall of the pharynx and posture of the tongue. *Cleft Palate J.* **1971**, *8*, 10–17. [PubMed]
65. LaVelle, W.E.; Hardy, J.C. Palatal lift prostheses for treatment of palatopharyngeal incompetence. *J. Prosthet. Dent.* **1979**, *42*, 308–315. [CrossRef]
66. Israel, J.M.; Cook, T.A.; Blakeley, R.W. The use of a temporary oral prosthesis to treat speech in velopharyngeal incompetence. *Facial Plast. Surg.* **1993**, *9*, 206–212. [CrossRef]
67. Bispo, N.H.M.; Whitaker, M.E.; Aferri, H.C.; Neves, J.D.A.; Dutka, J.D.C.R.; Pegoraro-Krook, M.I. Speech therapy for compensatory articulations and velopharyngeal function: A case report. *J. Appl. Oral Sci.* **2011**, *19*, 679–684. [CrossRef]
68. Elangovan, S.; Loibi, E. Two-piece hollow bulb obturator. *Indian J. Dent. Res. Off. Publ. Indian Soc. Dent. Res.* **2011**, *22*, 486–488. [CrossRef]
69. Lin, F.H.; Wang, T.C. Prosthodontic Rehabilitation for Edentulous Patients with Palatal Defect: Report of Two Cases. *J. Formos. Med. Assoc.* **2011**, *110*, 120–124. [CrossRef]
70. Agrawal, K.K.; Singh, B.P.; Chand, P.; Patel, C.B.S. Impact of delayed prosthetic treatment of velopharyngeal insufficiency on quality of life. *Indian J. Dent. Res. Off. Publ. Indian Soc. Dent. Res.* **2011**, *22*, 356–358. [CrossRef]
71. Mack, D.; Becker, P.; Chatterjee, I.; Dobinsky, S.; Knobloch, J.K.; Peters, G.; Rohde, H.; Herrmann, M. Mechanisms of biofilm formation in *Staphylococcus epidermidis* and *Staphylococcus aureus*: Functional molecules, regulatory circuits, and adaptive responses. *Int. J. Med. Microbiol.* **2004**, *294*, 203–212. [CrossRef] [PubMed]
72. Mack, D.; Davies, A.P.; Harris, L.G.; Rohde, H.; Horstkotte, M.A.; Knobloch, J.K. Microbial interactions in Staphylococcus epidermidis biofilms. *Anal Bioanal. Chem.* **2007**, *387*, 399–408. [CrossRef] [PubMed]
73. Xu, L.C.; Siedlecki, C.A. Submicron-textured biomaterial surface reduces staphylococcal bacterial adhesion and biofilm formation. *Acta Biomater.* **2012**, *8*, 72–81. [CrossRef] [PubMed]
74. Ramage, G.; Martínez, J.P.; López-Ribot, J.L. Candida biofilms on implanted biomaterials: A clinically significant problem. *FEMS Yeast Res.* **2006**, *6*, 979–986. [CrossRef] [PubMed]
75. Nett, J.; Andes, D. Candida albicans biofilm development, modeling a host-pathogen interaction. *Curr. Opin. Microbiol.* **2006**, *9*, 340–345. [CrossRef]
76. Thein, Z.M.; Seneviratne, C.J.; Samaranayake, Y.H.; Samaranayake, L.P. Community lifestyle of Candida in mixed biofilms: A mini review. *Mycoses* **2009**, *52*, 467–475. [CrossRef]
77. Beumer, J., 3rd; Kurrasch, M.; Kagawa, T. Prosthetic restoration of oral defects secondary to surgical removal of oral neoplasms. *CDA J.* **1982**, *10*, 47–54.
78. Huber, H.; Studer, S.P. Materials and techniques in maxillofacial prosthodontic rehabilitation. *Oral Maxillofac. Surg. Clin.* **2002**, *14*, 73–93. [CrossRef]
79. Goiato, M.C.; Zucolotti, B.C.; Mancuso, D.N.; dos Santos, D.M.; Pellizzer, E.P.; Verri, F.R. Care and cleaning of maxillofacial prostheses. *J. Craniofacial Surg.* **2010**, *21*, 1270–1273. [CrossRef]
80. Wieckiewicz, W.; Baran, E.; Zenczak-Wiechiewicz, D.; Proniexicz, A. Adhesion of Candida to the obturator and oral mucosa as a cause of the presence of inflammation in patients treated surgically for neoplasia. *Rev. Iberoam. Micol.* **2004**, *21*, 187–190.
81. Mattos, B.S.C.; Sousa, A.A.D.; Magalhaes, M.H.C.G.D.; Andre, M.; Brito E Dias, R. Candida albicans in patients with oronasal communication and obturator prostheses. *Braz. Dent. J.* **2009**, *20*, 336–340. [CrossRef] [PubMed]
82. Depprich, R.A.; Handschel, J.G.; Meyer, U.; Meissner, G. Comparison of prevalence of microorganisms on titanium and silicone/polymethyl methacrylate obturators used for rehabilitation of maxillary defects. *J. Prosthet. Dent.* **2008**, *99*, 400–405. [CrossRef] [PubMed]
83. Atay, A.; Piskin, B.; Akin, H.; Sipahi, C.; Karakas, A.; Saracli, M.A. Evaluation of Candida albicans adherence on the surface of various maxillofacial silicone materials. *J. Mycol. Médicale* **2013**, *23*, 27–32. [CrossRef] [PubMed]
84. Zafar, M.S. Prosthodontic Applications of Polymethyl Methacrylate (PMMA): An Update. *Polymers* **2020**, *12*, 2299. [CrossRef]
85. Nikawa, H.; Yamamoto, T.; Hamada, T. Effect of components of resilient denture-lining materials on the growth, acid production and colonization of Candida albicans. *J. Oral Rehabil.* **1995**, *22*, 817–824. [CrossRef]
86. Nikawa, H.; Jin, C.; Hamada, T.; Makihira, S.; Kumagai, H.; Murata, H. Interactions between thermal cycled resilient denture lining materials, salivary and serum pellicles and Candida albicans in vitro. Part II. Effects on fungal colonization. *J. Oral Rehabil.* **2000**, *27*, 124–130. [CrossRef]
87. Casemiro, L.A.; Martins, C.H.G.; Pires-De-Souza, F.D.C.P.; Panzeri, H. Antimicrobial and mechanical properties of acrylic resins with incorporated silver-zinc zeolite—Part I. *Gerodontology* **2008**, *25*, 187–194. [CrossRef]
88. Wen, J.; Yeh, C.K.; Sun, Y. Functionalized Denture Resins as Drug Delivery Biomaterials to Control Fungal Biofilms. *ACS Biomater. Sci. Eng.* **2016**, *2*, 224–230. [CrossRef]
89. Jo, J.K.; El-Fiqi, A.; Lee, J.H.; Kim, D.A.; Kim, H.W.; Lee, H.H. Rechargeable microbial anti-adhesive polymethyl methacrylate incorporating silver sulfadiazine-loaded mesoporous silica nanocarriers. *Dent. Mater.* **2017**, *33*, e361–e372. [CrossRef]
90. He, J.; Söderling, E.; Vallittu, P.K.; Lassila, L.V.J. Investigation of double bond conversion, mechanical properties, and antibacterial activity of dental resins with different alkyl chain length quaternary ammonium methacrylate monomers (QAM). *J. Biomater. Sci. Polym. Ed.* **2013**, *24*, 565–573. [CrossRef]

91. Nikawa, H.; Jin, C.; Hamada, T.; Makihira, S.; Polyzois, G. Candida albicans growth on thermal cycled materials for maxillofacial prostheses in vitro. *J. Oral Rehabil.* **2001**, *28*, 755–765. [CrossRef] [PubMed]
92. Zhou, L.; Tong, Z.; Wu, G.; Feng, Z.; Bai, S.; Dong, Y.; Ni, L.; Zhao, Y. Parylene coating hinders Candida albicans adhesion to silicone elastomers and denture bases resin. *Arch. Oral Biol.* **2010**, *55*, 401–409. [CrossRef] [PubMed]
93. Khalaf, S.; Ariffin, Z.; Husein, A.; Reza, F. Surface Coating of Gypsum-Based Molds for Maxillofacial Prosthetic Silicone Elastomeric Material: Evaluating Different Microbial Adhesion. *J. Prosthodont.* **2017**, *26*, 664–669. [CrossRef]
94. Tschernitschek, H.; Borchers, L.; Geurtsen, W. Nonalloyed titanium as a bioinert metal-a review. *Quintessence Int.* **2005**, *36*, 523–530. [CrossRef]
95. Perez-Jorge, C.; Arenas, M.A.; Conde, A.; Hernández-Lopez, J.M.; de Damborenea, J.J.; Fisher, S.; Hunt, A.M.; Esteban, J.; James, G. Bacterial and fungal biofilm formation on anodized titanium alloys with fluorine. *J. Mater. Sci. Mater. Med.* **2017**, *28*, 8. [CrossRef]
96. Shi, B.; Huang, H. Computational technology for nasal cartilage-related clinical research and application. *Int. J. Oral Sci.* **2020**, *12*, 21. [CrossRef]
97. Pauwels, R.; Araki, K.; Siewerdsen, J.H.; Thongvigitmanee, S.S. Technical aspects of dental CBCT: State of the art. *Dentomaxillofac. Radiol.* **2015**, *44*, 20140224. [CrossRef]
98. Kuijpers, M.A.R.; Chiu, Y.T.; Nada, R.M.; Carels, C.E.L.; Fudalej, P.S. Three-dimensional Imaging Methods for Quantitative Analysis of Facial Soft Tissues and Skeletal Morphology in Patients with Orofacial Clefts: A Systematic Review. *PLoS ONE* **2014**, *9*, e93442. [CrossRef]
99. Kihara, H.; Hatakeyama, W.; Komine, F.; Takafuji, K.; Takahashi, T.; Yokota, J.; Oriso, K.; Kondo, H. Accuracy and practicality of intraoral scanner in dentistry: A literature review. *J. Prosthodont. Res.* **2020**, *64*, 109–113. [CrossRef]
100. An, H.; Langas, E.E.; Gill, A.S. Effect of scanning speed, scanning pattern, and tip size on the accuracy of intraoral digital scans. *J. Prosthet. Dent.* 2022; *in press*. [CrossRef]
101. Auškalnis, L.; Akulauskas, M.; Jegelevičius, D.; Simonaitis, T.; Rutkūnas, V. Error propagation from intraoral scanning to additive manufacturing of complete-arch dentate models: An in vitro study. *J. Dent.* **2022**, *121*, 104136. [CrossRef] [PubMed]
102. Decazes, P.; Hinault, P.; Veresezan, O.; Thureau, S.; Gouel, P.; Vera, P. Trimodality PET/CT/MRI and Radiotherapy: A Mini-Review. *Front. Oncol.* **2020**, *10*, 614008. [CrossRef] [PubMed]
103. Choi, Y.S.; Shin, H.S. Preoperative Planning and Simulation in Patients With Cleft Palate Using Intraoral Three-Dimensional Scanning and Printing. *J. Craniofacial Surg.* **2019**, *30*, 2245–2248. [CrossRef]
104. Krämer Fernandez, P.; Kuscu, E.; Weise, H.; Engel, E.M.; Spintzyk, S. Rapid additive manufacturing of an obturator prosthesis with the use of an intraoral scanner: A dental technique. *J. Prosthet. Dent.* **2022**, *127*, 189–193. [CrossRef]
105. Williams, R.J.; Bibb, R.; Eggbeer, D.; Collis, J. Use of CAD/CAM technology to fabricate a removable partial denture framework. *J. Prosthet. Dent.* **2006**, *96*, 96–99. [CrossRef]
106. Kattadiyil, M.T.; Mursic, Z.; AlRumaih, H.; Goodacre, C.J. Intraoral scanning of hard and soft tissues for partial removable dental prosthesis fabrication. *J. Prosthet. Dent.* **2014**, *112*, 444–448. [CrossRef]
107. Bibb, R.; Brown, R. The application of computer aided product development techniques in medical modelling topic: Rehabilitation and prostheses. *Biomed. Sci. Instrum.* **2000**, *36*, 319–324.
108. Tian, Y.; Chen, C.; Xu, X.; Wang, J.; Hou, X.; Li, K.; Lu, X.; Shi, H.; Lee, E.-S.; Jiang, H.B. A Review of 3D Printing in Dentistry: Technologies, Affecting Factors, and Applications. *Scanning* **2021**, *2021*, 9950131. [CrossRef]
109. Della Bona, A.; Cantelli, V.; Britto, V.T.; Collares, K.F.; Stansbury, J.W. 3D printing restorative materials using a stereolithographic technique: A systematic review. *Dent. Mater.* **2021**, *37*, 336–350. [CrossRef]
110. Park, S.M.; Park, J.M.; Kim, S.K.; Heo, S.J.; Koak, J.Y. Flexural Strength of 3D-Printing Resin Materials for Provisional Fixed Dental Prostheses. *Materials* **2020**, *13*, 3970. [CrossRef]
111. Schönhoff, L.M.; Mayinger, F.; Eichberger, M.; Reznikova, E.; Stawarczyk, B. 3D printing of dental restorations: Mechanical properties of thermoplastic polymer materials. *J. Mech. Behav. Biomed. Mater.* **2021**, *119*, 104544. [CrossRef] [PubMed]
112. Ajaj-Alkordy, N.M.; Alsaadi, M.H. Elastic modulus and flexural strength comparisons of high-impact and traditional denture base acrylic resins. *Saudi Dent. J.* **2014**, *26*, 15–18. [CrossRef] [PubMed]
113. Meng, T.R., Jr.; Latta, M.A. Physical properties of four acrylic denture base resins. *J. Contemp. Dent. Pract.* **2005**, *6*, 93–100. [CrossRef] [PubMed]
114. Gautam, R.; Singh, R.D.; Sharma, V.P.; Siddhartha, R.; Chand, P.; Kumar, R. Biocompatibility of polymethylmethacrylate resins used in dentistry. *J. Biomed. Mater. Res. Part B Appl. Biomater.* **2012**, *100B*, 1444–1450. [CrossRef] [PubMed]
115. Kedjarune, U.; Charoenworaluk, N.; Koontongkaew, S. Release of methyl methacrylate from heat-curved and autopolymerized resins: Cytotoxicity testing related to residual monomer. *Aust. Dent. J.* **1999**, *44*, 25–30. [CrossRef] [PubMed]
116. Raszewski, Z. Influence of polymerization method on the cytotoxicity of three different denture base acrylic resins polymerized in different methods. *Saudi J. Biol. Sci.* **2020**, *27*, 2612–2616. [CrossRef] [PubMed]
117. Mitra, I.; Bose, S.; Dernell, W.S.; Dasgupta, N.; Eckstrand, C.; Herrick, J.; Yaszemski, M.J.; Goodman, S.B.; Bandyopadhyay, A. 3D Printing in alloy design to improve biocompatibility in metallic implants. *Mater. Today* **2021**, *45*, 20–34. [CrossRef]
118. Heinl, P.; Müller, L.; Körner, C.; Singer, R.F.; Müller, F.A. Cellular Ti–6Al–4V structures with interconnected macro porosity for bone implants fabricated by selective electron beam melting. *Acta Biomater.* **2008**, *4*, 1536–1544. [CrossRef]
119. Zhang, S.; Wei, Q.; Cheng, L.; Li, S.; Shi, Y. Effects of scan line spacing on pore characteristics and mechanical properties of porous Ti6Al4V implants fabricated by selective laser melting. *Mater. Des.* **2014**, *63*, 185–193. [CrossRef]

120. Bose, S.; Ke, D.; Sahasrabudhe, H.; Bandyopadhyay, A. Additive manufacturing of biomaterials. *Prog. Mater. Sci.* **2018**, *93*, 45–111. [CrossRef]
121. Xu, Y.; Xu, Y.; Zhang, W.; Li, M.; Wendel, H.P.; Geis-Gerstorfer, J.; Li, P.; Wan, G.; Xu, S.; Hu, T. Biodegradable Zn-Cu-Fe Alloy as a Promising Material for Craniomaxillofacial Implants: An in vitro Investigation into Degradation Behavior, Cytotoxicity, and Hemocompatibility. *Front. Chem.* **2022**, *10*, 860040. [CrossRef] [PubMed]
122. Chen, S.G.; Yang, J.; Jia, Y.G.; Lu, B.; Ren, L. TiO(2) and PEEK Reinforced 3D Printing PMMA Composite Resin for Dental Denture Base Applications. *Nanomaterials* **2019**, *9*, 1049. [CrossRef] [PubMed]
123. Yue, J.; Zhao, P.; Gerasimov, J.Y.; van de Lagemaat, M.; Grotenhuis, A.; Rustema-Abbing, M.; van der Mei, H.C.; Busscher, H.J.; Herrmann, A.; Ren, Y. 3D-Printable Antimicrobial Composite Resins. *Adv. Funct. Mater.* **2015**, *25*, 6756–6767. [CrossRef]
124. Tu, Z.; Zhong, Y.; Hu, H.; Shao, D.; Haag, R.; Schirner, M.; Lee, J.; Sullenger, B.; Leong, K.W. Design of therapeutic biomaterials to control inflammation. *Nat. Rev. Mater.* **2022**, *7*, 557–574. [CrossRef]
125. Huang, H.; Pan, W.; Wang, Y.; Kim, H.S.; Shao, D.; Huang, B.; Ho, T.C.; Lao, Y.H.; Quek, C.H.; Shi, J.; et al. Nanoparticulate cell-free DNA scavenger for treating inflammatory bone loss in periodontitis. *Nat. Commun.* **2022**, *13*, 5925. [CrossRef]
126. Lee, V.K.; Dai, G. Printing of Three-Dimensional Tissue Analogs for Regenerative Medicine. *Ann. Biomed. Eng.* **2016**, *45*, 115–131. [CrossRef]
127. Buskermolen, J.K.; Reijnders, C.M.A.; Spiekstra, S.W.; Steinberg, T.; Kleverlaan, C.J.; Feilzer, A.J.; Bakker, A.D.; Gibbs, S. Development of a Full-Thickness Human Gingiva Equivalent Constructed from Immortalized Keratinocytes and Fibroblasts. *Tissue Eng. Part C Methods* **2016**, *22*, 781–791. [CrossRef]

Journal of
Functional Biomaterials

MDPI

Article

Adhesive Property of 3D-Printed PEEK Abutments: Effects of Surface Treatment and Temporary Crown Material on Shear Bond Strength

Dingjie Wang [1], Xingting Han [2], Feng Luo [3], Florian M. Thieringer [4,5], Yichen Xu [3,*], Guomin Ou [1,*] and Sebastian Spintzyk [6]

1 State Key Laboratory of Oral Diseases & National Clinical Research Center for Oral Diseases, Department of Oral Implantology, West China Hospital of Stomatology, Sichuan University, Chengdu 610041, China
2 Department of Prosthodontics, Peking University School and Hospital of Stomatology, National Center of Stomatology, National Clinical Research Center for Oral Diseases, National Engineering Laboratory for Digital and Material Technology of Stomatology, Beijing Key Laboratory of Digital Stomatology, NHC Key Laboratory of Digital Technology of Stomatology, 22 Zhongguancun Avenue South, Beijing 100081, China
3 State Key Laboratory of Oral Diseases & National Clinical Research Center for Oral Diseases, Department of Oral Prosthodontics, West China Hospital of Stomatology, Sichuan University, Chengdu 610041, China
4 Medical Additive Manufacturing Research Group, Hightech Research Center, Department of Biomedical Engineering, University of Basel, 4123 Allschwil, Switzerland
5 Department of Oral and Cranio-Maxillofacial Surgery, University Hospital Basel, 4031 Basel, Switzerland
6 ADMiRE Research Center—Additive Manufacturing, Intelligent Robotics, Sensors and Engineering, School of Engineering and IT, Carinthia University of Applied Sciences, 9524 Villach, Austria
* Correspondence: yichen.xu@scu.edu.cn (Y.X.); ouguom@scu.edu.cn (G.O.)

Citation: Wang, D.; Han, X.; Luo, F.; Thieringer, F.M.; Xu, Y.; Ou, G.; Spintzyk, S. Adhesive Property of 3D-Printed PEEK Abutments: Effects of Surface Treatment and Temporary Crown Material on Shear Bond Strength. *J. Funct. Biomater.* **2022**, *13*, 288. https://doi.org/10.3390/jfb13040288

Academic Editor: Josette Camilleri

Received: 11 November 2022
Accepted: 7 December 2022
Published: 9 December 2022

Abstract: Three-dimensionally printed polyetheretherketone (PEEK) materials are promising for fabricating customized dental abutments. This study aimed to investigate the adhesive property of a 3D-printed PEEK material. The effects of surface treatment and temporary crown materials on shear bond strength were evaluated. A total of 108 PEEK discs were 3D printed by fused-filament fabrication. Surface treatments, including sandblasting, abrasive paper grinding, and CO_2 laser ablation, were applied to the PEEK discs, with the untreated specimens set as the control. Afterward, the surface topographies of each group were investigated by scanning electron microscopy (SEM, $n = 1$) and roughness measurements ($n = 7$). After preparing the bonding specimens with three temporary crown materials (Artificial teeth resin (ATR), 3M™ Filtek™ Supreme Flowable Restorative (FR), and Cool Temp NATURAL (CTN)), the shear bond strength was measured ($n = 6$), and the failure modes were analyzed by microscopy and SEM. The results showed that ATR exhibited a significantly higher shear bond strength compared to FR and CTN ($p < 0.01$), and the PEEK surfaces treated by sandblasting and abrasive paper grinding showed a statistically higher shear bond strength compared to the control ($p < 0.05$). For clinical application, the ATR material and subtractive surface treatments are recommended for 3D-printed PEEK abutments.

Keywords: fused filament fabrication; polyether ether ketone; additive manufacturing; shear bond strength; temporary abutment; dental implantology; rapid manufacturing; implant supported restoration

1. Introduction

Implant abutments are sophisticated parts used for connecting the prosthesis and the implant. Nowadays, immediate and early restoration after implant surgery using temporary abutment and prosthesis is encouraged for its positive effect in guiding soft tissue healing and maintaining aesthetic function [1–3]. Although premade temporary abutments might be provided in different dimensions by manufacturers [4], these abutments can hardly fulfill the various prosthodontic requirements in a tooth defect area. Dentists and dental technicians have to spend extra time adjusting the temporary abutments, which significantly decreases medical efficiency [5,6].

With the rapid development of digital technologies in dentistry, computer-aided design (CAD) and additive manufacturing (AM, also known as 3D printing) may provide dentistry with customized solutions [7,8]. According to the operation plan, custom temporary abutments could be designed and fabricated in advance and placed right after surgery [9,10]. Fused filament fabrication (FFF) is one of the most popular 3D printing technologies that show prospects in abutment fabrication due to its advantages in rapid manufacturing and low costs [11]. By melting and extruding, FFF printers can deposit thermoplastic materials onto a build platform, constructing objects layer by layer [12].

Although there might be plenty of material choices for FFF, such as polylactic acid (PLA), acrylonitrile butadiene styrene (ABS), polyethylene terephthalate glycol (PETG), etc., polyetheretherketone (PEEK) seems to be a better choice for biomedical use due to its outstanding chemical resistance, excellent mechanical strength, and superb biocompatibility [13–15]. However, for temporary abutment applications, another aspect that must not be neglected is the material's adhesive property because even a small detachment between the temporary abutment and crown may lead to prosthetic failure, which is not acceptable for both patients and dentists.

Previous studies have indicated that it is difficult to achieve firm adhesion with PEEK due to its hydrophobic nature and low surface energy [16]. Accordingly, various protocols have been investigated to enhance the adhesive ability of PEEK. Surface modifications, including sandblasting, sulfuric acid etching, and atmospheric plasma treatment, were reported to be effective [17–19]. However, less is known about the adhesive property of the 3D-printed PEEK material.

Therefore, the present study aimed to investigate the adhesive property of 3D-printed PEEK used for temporary abutment fabrication. The effects of surface treatment (sandblasting, abrasive paper grinding, and CO_2 laser ablation) and the temporary crown material (Artificial teeth resin, 3M™ Filtek™ Supreme Flowable Restorative, and Cool Temp NATURAL) on the shear bond strength were evaluated. The null hypotheses were set as (i) the surface treatments do not affect the shear bond strength, and (ii) the temporary crown materials have no effect on the shear bond strength.

2. Materials and Methods

2.1. Specimen Preparation

The test specimen was designed as a disk with a diameter of 14 mm and a thickness of 2 mm using open-source CAD software (OpenSCAD, 2021.01, http://www.openscad.org/, accessed on 1 March 2022). A total of 108 specimens were 3D printed horizontally with a layer thickness of 200 μm by a FFF printer (Apium P220, Apium Additive Technologies GmbH, Karlsruhe, Germany) using a PEEK filament (Evonik VESTAKEEP®i4 G resin, Evonik Industries AG, Essen, Germany) [20]. After printing, the support structures were removed, and specimens were collected in zipper storage bags before testing.

2.2. Surface Treatment

Three surface treatments (sandblasting, abrasive paper grinding, and CO_2 laser ablation) were included in this study ($n = 27$). The untreated PEEK specimens were set as the control. For the sandblasting group, the specimens were blasted with 100 μm aluminum oxide abrasive (Al_2O_3, DenTal DENTURE, Chengdu, China) by a sandblasting device (R-603, Zaofeng Technology Co., Ltd., Zhongshan, China) at a distance of 10 mm with a pressure of 0.2 MPa for 10 s. For the abrasive paper group, the specimens were ground with P600 silicon carbide abrasive paper (Suisun Company Limited, Hong Kong, China) under running cooling water for 60 rounds. Regarding the CO_2 laser group, the PEEK specimens were ablated by a CO_2 laser device (JZ-2, Beijing Health Medical Technology Co., Ltd., Beijing, China). The wavelength, frequency, rated input power, and point output power were set as 10.6 ± 0.1 μm, 50 Hz, 450 W, and 30 W, respectively. The diameter of each dot spaced 500 μm apart in the CO_2 laser array was 100 μm. After surface treatment, all specimens were cleaned with compressed air for 20 s to remove material residues.

2.3. *Surface Characterization*

After surface treatment, the PEEK samples (n = 2) of each group were pictured under a stereoscopic microscope (SZX16, Olympus, Tokyo, Japan) at 1× and 4× magnifications and observed by a scanning electron microscope (SEM, JSM-IT500, JEOL, Tokyo, Japan) at 100×, 1000×, and 5000× magnifications, with an acceleration voltage of 10 kV.

In addition, the surface topographies (n = 7 per group) were measured by a confocal white light interferometer (UP-Lambda, RTEC instrument, San Jose, CA, USA). An area of 1.6 mm × 1.2 mm on each sample surface was optically scanned. Afterward, the obtained data were transferred to surface analysis software (MountainsMap Universal 9, Digital Surf, Besancon, France). A Gaussian Filter (ISO 16610-61) with a cut-off value of 0.32 mm was applied to differentiate roughness and waviness. Roughness parameters, including arithmetic mean height (Sa) and dales void volume (Vvv), were calculated based on ISO 25178-2:2012. Finally, the surfaces were 3D reconstructed for visualization.

2.4. *Shear Bond Strength Test*

2.4.1. Fabrication of Bonding Mold

A mold used for bonding the PEEK discs and temporary crown materials was designed by OpenSCAD and Materialise Magics (25.0, Materialise, Leuven, Belgium). The mold consists of two parts (Figure 1a), which can be accurately assembled through a lap joint (Figure 1b). The larger cylindrical space used for placing the PEEK disk was 7.3 mm in radius and 2 mm in height, and the smaller cylindrical space used for adding temporary crown material was 2.5 mm in radius and 2 mm in height. After CAD design, the two parts of the bonding mold were 3D printed by a stereolithography (SLA) printer (Form 3B, Formlabs, Somerville, MA, USA) with a 25 μm-thick layer using model resin (V2, Formlabs). To ensure printing accuracy, the lap joint was placed upward for 3D printing (Figure 1c). After printing, the parts were rinsed with isopropyl alcohol (IPA, Chron chemicals, Chengdu, China) for 10 min in a postcleaning device (Form Wash, Formlabs) and then light cured with 405 nm blue light at 60 °C for 60 min using a postcuring device (Form Cure, Formlabs). Next, the support structures were removed by the finishing kit (Formlabs), and the residues of the support structures were trimmed by a sharp scalpel (Figure 1d–h).

2.4.2. Preparation of the Bonding Specimen

Prior to bonding, a thin layer of liquid separating agent (Xin Shi Ji, Shanghai, China) was smeared uniformly on the lateral surface of the smaller cylindrical channel of the bonding mold. Afterward, the PEEK discs were placed in the larger cylindrical space, and three temporary crown resins (n = 6 per surface treatment group) were applied for bonding (Figure 2a). All temporary crown materials were cured according to manufacturer's instructions at 20–22 °C room temperature and 50% relative atmospheric humidity. Table 1 gives detailed information on the temporary crown resins used in the present study. For short, Artificial teeth resin, 3M™ Filtek™ Supreme Flowable Restorative, and Cool Temp NATURAL were denoted as ATR, FR, and CTN, respectively. After curing, the bonding mold was carefully removed (Figure 2b). The fabricated bonding specimens (Figure 2c) were incubated at 37 °C for 24 h in distilled water to simulate the intraoral aqueous environment.

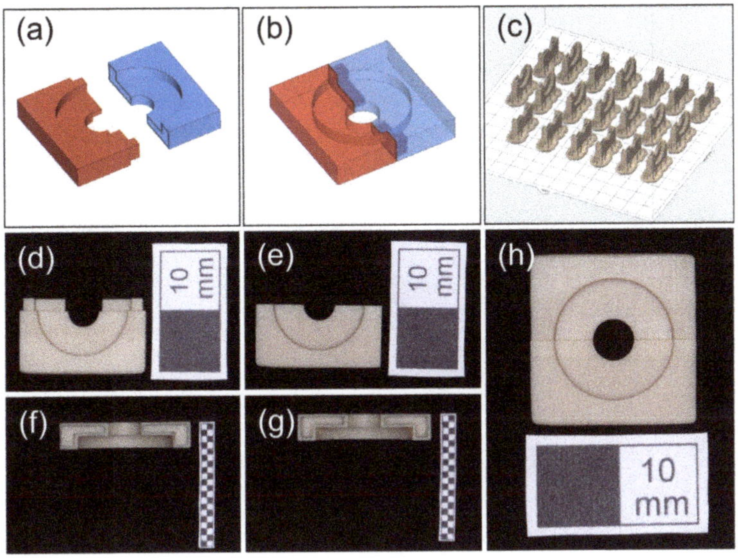

Figure 1. The fabrication of the bonding mold (**a**,**b**). The bonding mold was designed as two parts that can be accurately assembled through a lap joint; (**c**) the direction and layout for 3D printing; (**d**–**h**) the 3D-printed parts could be accurately assembled by lap joint (white scale: 10 mm).

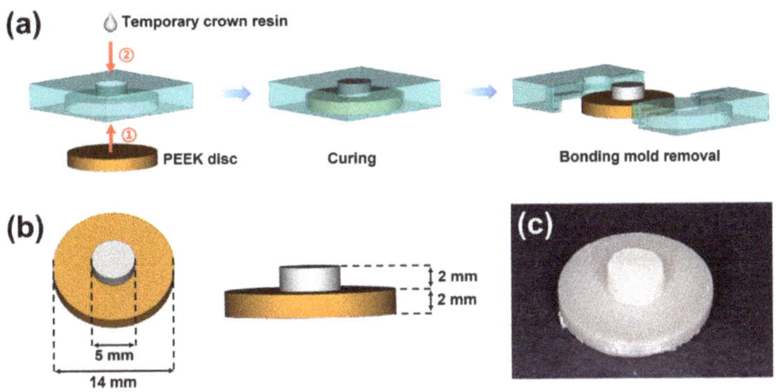

Figure 2. The preparation of the bonding specimen; (**a**) the workflow for fabricating the bonding specimens; (**b**) the dimensions of the bonding specimen after mold removal; (**c**) the fabricated bonding specimen.

Table 1. The temporary crown resins used for shear bond strength testing.

Product Name	Manufacturer, City, Country	Lot No.	Main Component
Artificial teeth resin	Xin Shi Ji, Shanghai, China	20211118	Methyl methacrylate
3M™ Filtek™ Supreme Flowable Restorative	3M, St. Paul, MN, USA	NF20341	Dimethacrylates
Cool Temp NATURAL	Coltène/Whaledent AG, Altstätten, Switzerland	L25817	Trimethylolpropane trimethacrylate, dimethacrylates

2.4.3. Shear Bond Strength Measurement

Before testing, each bonding specimen was fixed in a 30 mm × 30 mm × 10 mm cuboid block made of acrylic resin (Figure 3a). The test block was then immobilized on a universal testing machine (5565, INSTRON, Norwood, MA, USA). A shear load was applied through a blade onto the temporary crown material at a crosshead speed of 1.0 mm/min until debonding (Figure 3b). During measuring, the force and distance were dynamically recorded for plotting the force–distance curves. The shear bond strength was calculated by Equation (1), where S, F, and A represent the shear bond strength (MPa), maximum fracture load (N), and bonding area (mm^2), respectively.

$$S = F/A \quad (1)$$

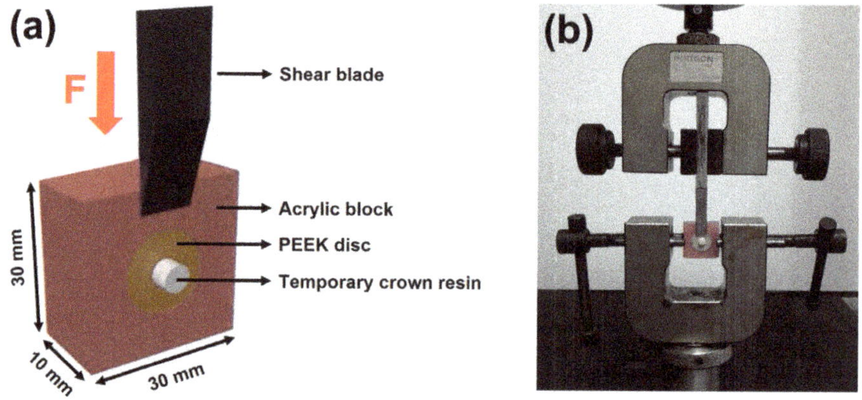

Figure 3. Experimental setup of shear bond strength testing. (**a**) Schematic illustration of the test block; (**b**) shear bond strength test using a universal testing machine.

After measuring, the specimen surfaces were analyzed by a stereoscopic microscope. For further investigation, the debonded samples of the CO_2 laser group were observed by SEM. Failure modes were determined as prefailure: debonding occurred before the shear bond strength test; adhesive failure: less than 33% of the temporary crown material remained at the bonding interface; mixed failure: more than 33% but less than 66% of the temporary crown material remained at the bonding interface, and cohesive failure: more than 66% of the temporary crown material remained at the bonding interface [21].

2.5. Statistical Analysis

All experimental variables were checked for normal distribution by Shapiro–Wilk testing. The roughness data were analyzed by one-way analysis of variance (ANOVA), and Tukey's test was conducted for the subsequent multiple comparisons. A two-way ANOVA was performed to evaluate the effect of surface treatment and temporary crown material on shear bond strength. In the following main effect analysis, Tukey's test was used for the posthoc comparisons. GraphPad Prism (9.4.0, GraphPad Software, San Diego, CA, USA) was utilized for all statistical analyses with an α set to 0.05.

3. Results

3.1. Surface Morphology

The surface morphologies of the 3D-printed PEEK specimens treated with different methods are shown in Figure 4. The sample surfaces in the control group showed clear textures which were generated from the deposition road of PEEK during FFF printing. In contrast, due to the subtractive surface treatment, these textures became blurred in the sandblasting group and disappeared in the abrasive paper group. CO_2 lasering did not affect the surface texture but created orderly arranged pores with a diameter of about 200 μm on the PEEK surface (Figure 4, red arrow).

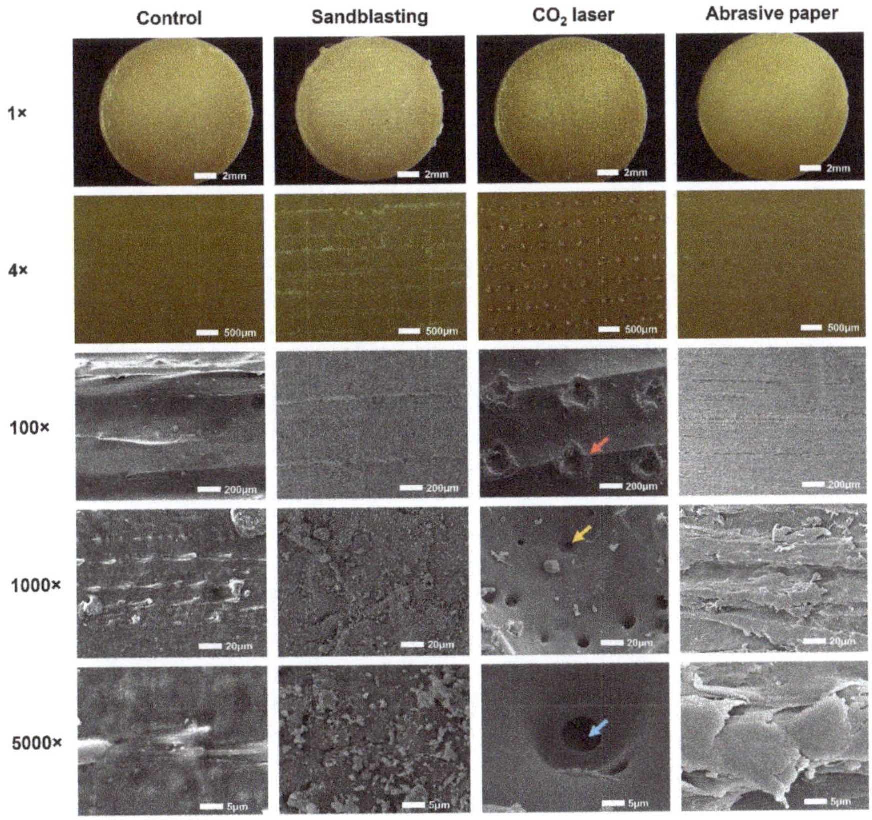

Figure 4. Representative surface morphologies of the 3D-printed PEEK specimens treated with different methods. Images were captured by a stereoscopic microscope (1× and 4×) and a scanning electron microscope (100×, 1000×, and 5000×). The red, yellow, and blue arrows represent the primary, secondary, and tertiary pores created by CO_2 laser ablation, respectively.

More information was obtained from the highly magnified SEM images (1000× and 5000×). The sample surfaces in the sandblasting group showed collisional traces of Al_2O_3 particles, and those in the abrasive paper group exhibited abrasive features generated from silicon carbide grinding. Interestingly, secondary pores with a diameter of approximately 8 μm were found at the bottom and sidewall of the primary pores created by the CO_2 laser (Figure 4, yellow arrow). In some of the secondary pores, we even detected smaller tertiary pores (Figure 4, blue arrow).

3.2. Roughness Measurement

Figure 5a depicts the reconstructed 3D view of the PEEK surfaces with different surface treatments. The untreated surface (control group) exhibited clear surface textures with a vertical drop of 60 μm. After sandblasting and abrasive paper grinding, the surface textures became blurred or disappeared, which further confirmed the finding in the SEM analysis. The CO_2 laser created an array of pores on the 3Diprinted PEEK surface. The edge of the pores has risen, which increased the vertical drop to about 80 μm. As shown in Figure 5b,c, the roughness parameter values of Sa and Vvv in the CO_2 laser group were significantly higher than those of the sandblasting, abrasive paper, and control groups ($p < 0.0001$).

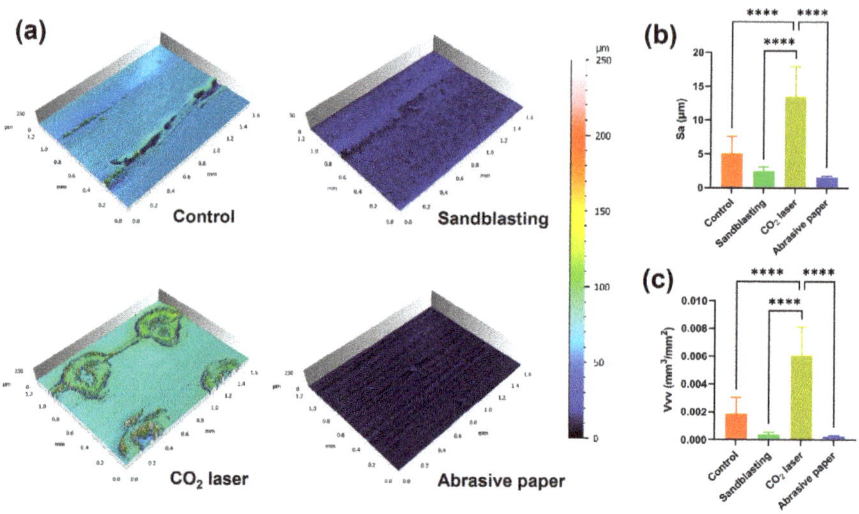

Figure 5. Surface topographies of the 3D-printed PEEK surfaces with various surface treatments; (**a**) the 3D reconstructed surfaces; (**b**) mean (standard deviation) arithmetic mean height (Sa) of each group; (**c**) mean (standard deviation) dales void volume (Vvv) of each group (n = 7). Asterisks indicate statistically significant differences (significance level **** $p < 0.0001$).

3.3. Shear Bond Strength

The means and standard deviations of the recorded shear bond strengths are displayed in Figure 6. The results of the two-way ANOVA indicated both surface treatment (F (3, 59) = 8.132, $p < 0.0001$) and temporary crown material (F (2, 59) = 7.885, $p = 0.0009$) had statistically significant effects on shear bond strength, and no obvious interaction (F (6, 59) = 2.166, $p = 0.0591$) could be detected. In the main effect analyses, multiple posthoc comparisons revealed that the shear bond strengths of the sandblasting group and abrasive paper group were significantly higher than that of the control group ($p < 0.05$), and the shear bond strength generated by ATR was significantly higher than those produced by FR and CTN ($p < 0.01$).

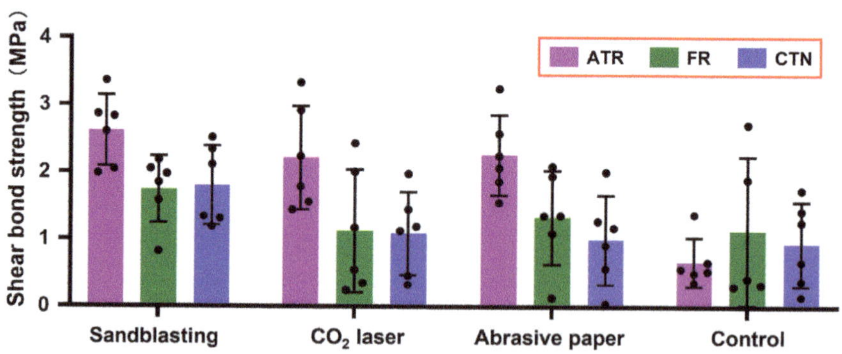

Figure 6. Mean (standard deviation) shear bond strength of each group (n = 6). Black dots represent shear bond strength data for each measurement. ATR, FR, and CTN represent Artificial teeth resin, 3M™ Filtek™ Supreme Flowable Restorative, and Cool Temp NATURAL, respectively.

Figure 7 shows the representative force-distance curves of each group. Generally, the curves started to ascend rapidly once the shear blade contacted the bonding specimen. After the shear blade moved about 0.2–0.3 mm, the curves reached the peak. From this point, the detachment between the temporary crown materials and PEEK discs occurred, and the curves dropped quickly. When compared with the sharp peaks in the control, sandblasting, and abrasive paper groups, the peaks in the CO_2 laser group were round and blunt.

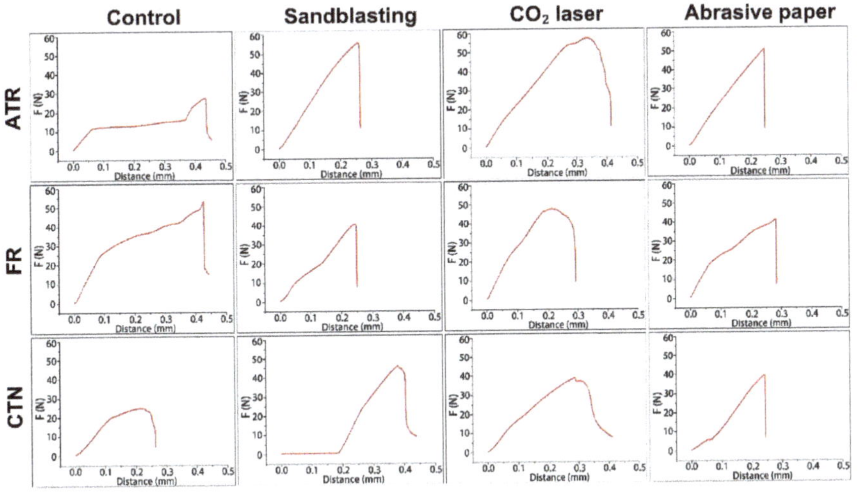

Figure 7. Representative force-distance curves of each group in shear bond strength testing. ATR, FR, and CTN represent Artificial teeth resin, 3M™ Filtek™ Supreme Flowable Restorative, and Cool Temp NATURAL, respectively.

3.4. Failure Mode Analysis

Figure 8 illustrates the composition of the failure mode of each group after the shear bond strength test. Adhesive failures were mostly observed except for the ATR-CO_2 laser group, whose failure mode was dominated by mixed failure (83.33%). In addition, mixed failure could also be found in the FR-CO_2 laser group (33.3%) and the FR-Control group (16.67%). One bonding specimen in the FR-control group exhibited prefailure before shear bond testing. The FR material debonded from the untreated 3D-printed PEEK surface during incubation in distilled water.

Figure 9 shows the sample surfaces after shear bond testing. In terms of adhesive failure, the rapture site was located at the interface between PEEK and the temporary crown material. The residual resin could hardly be observed on the PEEK surfaces.

With regard to mixed failure, a large piece of FR material was found on the untreated PEEK surface. In contrast, residual resin pieces could not be found on the sample surfaces treated by CO_2 laser. The ATR and FR material remained in the pores created by CO_2 laser ablation, which could be further proved by SEM observation.

The results of the SEM exhibit that the residual ATR and FR material remained in the pores created by the CO_2 laser (Figure 10, mixed failure). Interestingly, the material residuals were not solid. Voids were detected at the periphery and interior of the material residuals (Figure 10, red arrow).

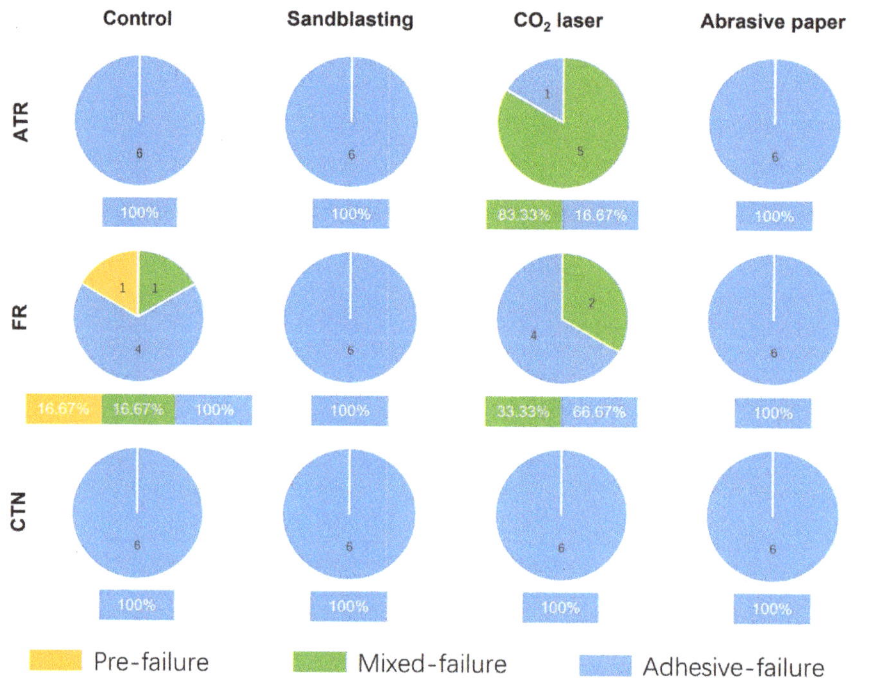

Figure 8. The failure mode for each experimental group after shear bond testing ($n = 6$). ATR, FR, and CTN represent Artificial teeth resin, 3M™ Filtek™ Supreme Flowable Restorative, and Cool Temp NATURAL, respectively.

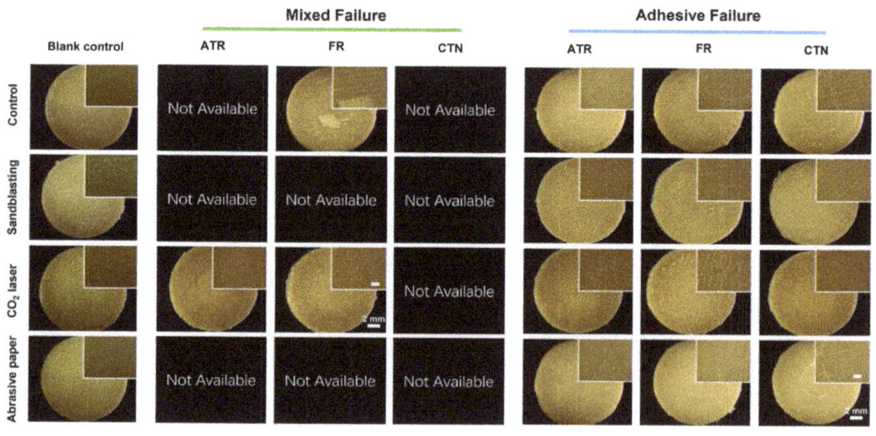

Figure 9. Representative sample surfaces after shear bond strength testing. Images were collected by a stereoscopic microscope at 1× and 4× magnifications. The scale bar in the zoomed figure is 1 mm. ATR, FR, and CTN represent Artificial teeth resin, 3M™ Filtek™ Supreme Flowable Restorative, and Cool Temp NATURAL, respectively.

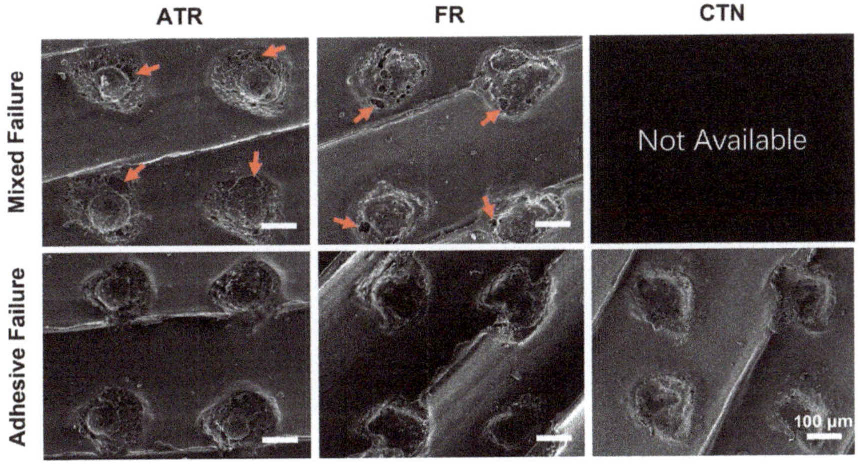

Figure 10. Representative SEM images of the samples treated by CO_2 laser after shear bond strength testing (130×). Red arrows indicate the voids in the residual temporary crown resins. ATR, FR, and CTN represent Artificial teeth resin, 3M™ Filtek™ Supreme Flowable Restorative, and Cool Temp NATURAL, respectively.

4. Discussion

Since its development in 1978 by British chemists, PEEK has attracted increasing attention in the biomedical field due to its excellent mechanical properties [22]. Being chemically inert, PEEK is also deemed a cost-effective alternative to metallic materials used in dentistry [7,23]. The recent development of 3D printing technologies facilitates the fabrication of customized dental prostheses [24]. However, for temporary abutment application, less is known for the adhesive property of 3D-printed PEEK material, especially its bonding with temporary crown materials. According to the shear bond strength test results, both surface treatment and temporary crown material had statistically significant effects on shear bond strength. Therefore, the null hypotheses were rejected.

The results of the SEM (Figure 4) and 3D reconstruction (Figure 5a) indicated that the surfaces of the 3D-printed PEEK samples were not smooth. Our previous study reported that roughness could be generated by the unfilled area between the layers deposited by FFF printers [11]. However, in this study, the roughness was found within a layer on the top surface, so the mechanism behind roughness generation is not the same as before. FFF 3D printers work by extruding molten materials out of a heated nozzle [25]. During the printing of the last layer (top surface), the nozzle repeatedly moves along paths in one direction to deposit materials until the entire surface is covered. In this process, unfilled voids can be found among the paths, generating the surface texture and roughness observed in this study [26]. In clinical practice, sandblasting and abrasive paper are commonly used to roughen surfaces before bonding. However, for a sufficiently rough surface, subtractive surface treatments may only smooth it [11]. This could be verified by the blurred or disappeared surface textures in the sandblasting and abrasive paper group (Figures 4 and 5a). The CO_2 laser generator could emit high-energy beams of photons at a wavelength of 10600 nm from the medium of carbon dioxide gas [27]. Focused CO_2 laser beams produce ultra-high energy density at a convergence point [28]. In the present study, the CO_2 laser array was composed of numerous focused laser beams. These laser beams ablated the PEEK material, creating an array of pores on the surface (Figures 4 and 5a). The second and tertiary pores might be generated by the instant vaporization effect. However, this speculation needs further research for clarification.

This study selected two areal roughness parameters that might be related to bonding for investigation. Sa is the most frequently used areal roughness parameter that characterizes surface height, and Vvv is a 3D functional parameter used for evaluating the void volume at the valley zone. The results of the roughness measurements indicated that Sa and Vvv had similar changes after surface treatments (Figure 5b,c). Sandblasting and abrasive paper grinding slightly reduced the mean value of Sa and Vvv but did not produce statistically significant differences. This is consistent with the results of the SEM analysis. In contrast, CO_2 laser ablation substantially enhanced surface roughness, which could be attributed to the pores generated on the surface.

Currently, there are no acknowledged clinical success criteria for the bond strength between temporary abutment and crown because implant prostheses may undergo complex stresses that are generated from mastication. Nevertheless, the bond strength should be optimized to avoid clinical failure. The results of the shear bond strength tests in this study indicated that ATR had a statistically higher shear bond strength with 3D-printed PEEK surfaces when compared to FR and CTN. This might be explained by the fact that the main components of FR and CTN are multifunctional acrylates (Table 1), which may induce higher volume shrinkage during polymerization [29]. As mentioned above, the PEEK specimens treated with the CO_2 laser seemed to possess surfaces that were better for bonding. Interestingly, the shear bond strength test in this study produced the opposite results, indicating that surface roughness may not be the only decisive factor in shear bond strength. The possible explanation could be related to the decreased wettability. The correlation between roughness and wettability conforms the Wenzel Equation (2) defined in 1936 [30]:

$$\cos\theta_1 = r\cos\theta_2 \tag{2}$$

where θ_1, θ_2, and r represent the apparent contact angle, Young's contact angle, and roughness ratio, respectively. For a hydrophobic PEEK surface, an increase in surface roughness can make the surface even more hydrophobic. Previous studies found that laser treatments could significantly increase the water contact angle on PEEK surfaces to about 110°, turning PEEK surfaces into a highly hydrophobic state [31–34]. In addition to the increased surface roughness, Akkan et al. attributed the increased hydrophobicity to the modification in surface topology [32]. Similarly, Riveiro et al. indicated that the microstructures created by laser ablation could significantly decrease the wettability of PEEK surfaces [34]. It should be noted that the liquid resins used in this study are different from the ultrapure water used for the standard wettability tests in previous studies. Due to the curing from monomer to polymer, the flowability of liquid resins may decrease rapidly during wetting. This phenomenon could further inhibit liquid resins from spreading out on the laser-treated PEEK surfaces, creating unfilled voids at the bonding interface. The above speculations are demonstrated by the results of the failure mode analysis (Figure 10). Void areas were mainly detected at the resin–PEEK interface. In order to further improve the bonding between the 3D-printed PEEK and the temporary crown materials, attempts could be focused on simultaneously improving surface roughness and wettability. Effective ways to enhance surface energy, such as atmospheric plasma [35], deserve further research for a full re-evaluation.

In previous studies, the fabrication of bonding specimens was considered to be time-consuming. The conventional method utilized a specially made metallic mold for sample preparation [8,36]. However, not all labs have access to such equipment, and researchers may fail to fabricate standardized specimens due to poor laboratory conditions. In addition, in order to use the conventional metallic mold, base materials (the material for investigation) must be accurately tailored into a specific dimension, which further increases the difficulties in experiments. More importantly, after bonding, the removal of the metallic mold was along the longitudinal axis of the bonded cylinder material. In this process, the bonded material might be debonded from the base material, and if not, microdetachment may also occur at the bonding interface, affecting shear bond strength. With the help of CAD and AM, this study fabricated customized split molds for sample preparation. This method is not limited by the conventional metallic mold or the sample size. After bonding, the removal of the split mold occurs from the lateral side of the bonded cylinder material, which minimizes the influence of mold removal on the shear bond strength measurement.

At present, different adhesives are being developed to improve the bonding performance of PEEK [37]. The authors assume that the additional use of adhesives may produce higher bond strengths. However, for experimental standardization, we did not include these adhesive systems, which deserves further research for clarification. Another limitation of the present study is that the effect of polymer aging in an oral-aqueous environment on shear bond strength was not considered. For future studies, conducting thermal cycling to simulate the intraoral aging process [38] might be more effective in predicting the long-term adhesive property of additively manufactured PEEK materials.

5. Conclusions

This study investigated the effects of surface treatment and temporary crown material on the shear bond strength of a 3D-printed PEEK material. Our results indicated that surface roughness might not be the only factor in determining the adhesive property of additively manufactured PEEK. For clinical application, the ATR material and subtractive surface treatments (sandblasting and abrasive paper grinding) are recommended for 3D-printed PEEK abutment to obtain a higher shear bond strength.

Author Contributions: Conceptualization, G.O. and Y.X.; methodology, Y.X. and S.S.; validation, G.O.; formal analysis, D.W.; investigation, D.W.; resources, X.H. and F.M.T.; data curation, X.H. and F.L.; writing—original draft preparation, D.W.; writing—review and editing, G.O., S.S. and Y.X.; visualization, G.O.; supervision, G.O.; project administration, G.O. All authors have read and agreed to the published version of the manuscript.

Funding: This research was funded by Sichuan Science and Technology Program (No. 2021YFS0085), the Research and Development Program of West China Hospital of Stomatology Sichuan University (No. RD-02-202114), Research Funding from West China School/Hospital of Stomatology Sichuan University (No. RCDWJS2021-13), National Natural Science Foundation of China (No. 82101073), and China Postdoctoral Science Foundation (No. 2022M710253).

Institutional Review Board Statement: Not applicable.

Informed Consent Statement: Not applicable.

Data Availability Statement: Not applicable.

Acknowledgments: The authors thank Qiang Wei for his technical assistance in the roughness measurements and Neha Sharma for her kind help with specimen preparation.

Conflicts of Interest: The authors declare no conflict of interest.

References

1. Cosyn, J.; Eghbali, A.; De Bruyn, H.; Collys, K.; Cleymaet, R.; De Rouck, T. Immediate Single-Tooth Implants in the Anterior Maxilla: 3-Year Results of a Case Series on Hard and Soft Tissue Response and Aesthetics: Immediate Single-Tooth Implants. *J. Clin. Periodontol.* **2011**, *38*, 746–753. [CrossRef] [PubMed]
2. den Hartog, L.; Huddleston Slater, J.J.R.; Vissink, A.; Meijer, H.J.A.; Raghoebar, G.M. Treatment Outcome of Immediate, Early and Conventional Single-Tooth Implants in the Aesthetic Zone: A Systematic Review to Survival, Bone Level, Soft-Tissue, Aesthetics and Patient Satisfaction. *J. Clin. Periodontol.* **2008**, *35*, 1073–1086. [CrossRef] [PubMed]
3. Chee, W.W.L. Provisional Restorations in Soft Tissue Management around Dental Implants: Chee. *Periodontology 2000* **2001**, *27*, 139–147. [CrossRef] [PubMed]
4. Abichandani, S.; Nadiger, R.; Kavlekar, A. Abutment Selection, Designing, and Its Influence on the Emergence Profile: A Comprehensive Review. *Eur. J. Prosthodont.* **2013**, *1*, 1. [CrossRef]
5. Mirza, J.; Robertson, G. Vital Guide to Dental Implants. *Vital* **2008**, *6*, 19–22. [CrossRef]
6. Hartlev, J.; Kohberg, P.; Ahlmann, S.; Andersen, N.T.; Schou, S.; Isidor, F. Patient Satisfaction and Esthetic Outcome after Immediate Placement and Provisionalization of Single-Tooth Implants Involving a Definitive Individual Abutment. *Clin. Oral Implant. Res.* **2014**, *25*, 1245–1250. [CrossRef]
7. Xu, Y.; Huang, H.; Wu, M.; Tian, Y.; Wan, Q.; Shi, B.; Hu, T.; Spintzyk, S. Rapid Additive Manufacturing of a Superlight Obturator for Large Oronasal Fistula in Pediatric Patient. *Laryngoscope* **2022**, lary.30352. [CrossRef]
8. Li, P.; Krämer-Fernandez, P.; Klink, A.; Xu, Y.; Spintzyk, S. Repairability of a 3D Printed Denture Base Polymer: Effects of Surface Treatment and Artificial Aging on the Shear Bond Strength. *J. Mech. Behav. Biomed. Mater.* **2021**, *114*, 104227. [CrossRef]

9. Rosentritt, M.; Raab, P.; Hahnel, S.; Stöckle, M.; Preis, V. In-Vitro Performance of CAD/CAM-Fabricated Implant-Supported Temporary Crowns. *Clin. Oral Invest.* **2017**, *21*, 2581–2587. [CrossRef]
10. Rayyan, M.M.; Aboushelib, M.; Sayed, N.M.; Ibrahim, A.; Jimbo, R. Comparison of Interim Restorations Fabricated by CAD/CAM with Those Fabricated Manually. *J. Prosthet. Dent.* **2015**, *114*, 414–419. [CrossRef]
11. Xu, Y.; Unkovskiy, A.; Klaue, F.; Rupp, F.; Geis-Gerstorfer, J.; Spintzyk, S. Compatibility of a Silicone Impression/Adhesive System to FDM-Printed Tray Materials—A Laboratory Peel-off Study. *Materials* **2018**, *11*, 1905. [CrossRef] [PubMed]
12. van Noort, R. The Future of Dental Devices Is Digital. *Dent. Mater.* **2012**, *28*, 3–12. [CrossRef] [PubMed]
13. Kurtz, S.M.; Devine, J.N. PEEK Biomaterials in Trauma, Orthopedic, and Spinal Implants. *Biomaterials* **2007**, *28*, 4845–4869. [CrossRef] [PubMed]
14. Skirbutis, G.; Dzingutė, A.; Masiliūnaitė, V.; Šulcaitė, G.; Žilinskas, J. PEEK Polymer's Properties and Its Use in Prosthodontics. A Review. *Stomatologija* **2018**, *20*, 54–58.
15. Katzer, A.; Marquardt, H.; Westendorf, J.; Wening, J.V.; von Foerster, G. Polyetheretherketone–Cytotoxicity and Mutagenicity in Vitro. *Biomaterials* **2002**, *23*, 1749–1759. [CrossRef]
16. Escobar, M.; Henriques, B.; Fredel, M.C.; Silva, F.S.; Özcan, M.; Souza, J.C.M. Adhesion of PEEK to Resin-Matrix Composites Used in Dentistry: A Short Review on Surface Modification and Bond Strength. *J. Adhes. Sci. Technol.* **2019**, *34*, 1241–1252. [CrossRef]
17. Zhou, L.; Qian, Y.; Zhu, Y.; Liu, H.; Gan, K.; Guo, J. The Effect of Different Surface Treatments on the Bond Strength of PEEK Composite Materials. *Dent. Mater.* **2014**, *30*, e209–e215. [CrossRef]
18. Stawarczyk, B.; Jordan, P.; Schmidlin, P.R.; Roos, M.; Eichberger, M.; Gernet, W.; Keul, C. PEEK Surface Treatment Effects on Tensile Bond Strength to Veneering Resins. *J. Prosthet. Dent.* **2014**, *112*, 1278–1288. [CrossRef]
19. Gama, L.T.; Duque, T.M.; Özcan, M.; Philippi, A.G.; Mezzomo, L.A.M.; Gonçalves, T.M.S.V. Adhesion to High-Performance Polymers Applied in Dentistry: A Systematic Review. *Dent. Mater.* **2020**, *36*, e93–e108. [CrossRef]
20. Han, X.; Sharma, N.; Spintzyk, S.; Zhou, Y.; Xu, Z.; Thieringer, F.M.; Rupp, F. Tailoring the Biologic Responses of 3D Printed PEEK Medical Implants by Plasma Functionalization. *Dent. Mater.* **2022**, *38*, 1083–1098. [CrossRef]
21. Spintzyk, S.; Yamaguchi, K.; Sawada, T.; Schille, C.; Schweizer, E.; Ozeki, M.; Geis-Gerstorfer, J. Influence of the Conditioning Method for Pre-Sintered Zirconia on the Shear Bond Strength of Bilayered Porcelain/Zirconia. *Materials* **2016**, *9*, 765. [CrossRef] [PubMed]
22. Wang, Z.; Runzi, M.; Gilchrist, M.; Gong, H. *Mechanical Properties of High-Performance Plastic Polyether-Ether-Ketone (PEEK) Printed by Fused Deposition Modeling*; University of Texas at Austin: Austin, TX, USA, 2021. [CrossRef]
23. Xu, Y.; Xu, Y.; Zhang, W.; Li, M.; Wendel, H.-P.; Geis-Gerstorfer, J.; Li, P.; Wan, G.; Xu, S.; Hu, T. Biodegradable Zn-Cu-Fe Alloy as a Promising Material for Craniomaxillofacial Implants: An in Vitro Investigation into Degradation Behavior, Cytotoxicity, and Hemocompatibility. *Front. Chem.* **2022**, *10*, 860040. [CrossRef] [PubMed]
24. Chen, J.; Yang, R.; Shi, B.; Xu, Y.; Huang, H. Obturator Manufacturing for Oronasal Fistula after Cleft Palate Repair: A Review from Handicraft to the Application of Digital Techniques. *JFB* **2022**, *13*, 251. [CrossRef] [PubMed]
25. Stansbury, J.W.; Idacavage, M.J. 3D Printing with Polymers: Challenges among Expanding Options and Opportunities. *Dent. Mater.* **2016**, *32*, 54–64. [CrossRef] [PubMed]
26. Hohenstein, S.; Bergweiler, G.; Lukas, G.; Krömer, V.; Otten, T. Decision Basis for Multi-Directional Path Planning for Post-Processing Reduction in Material Extrusion. *Prod. Eng.* **2021**, *15*, 457–466. [CrossRef]
27. Witteman, W.J. *The CO_2 Laser*; Softcover Reprint of the Original 1st ed. 1987; Springer: Berlin, Germany, 2013; ISBN 978-3-662-13617-1.
28. Omi, T.; Numano, K. The Role of the CO_2 Laser and Fractional CO_2 Laser in Dermatology. *Laser Ther.* **2014**, *23*, 49–60. [CrossRef] [PubMed]
29. Świderska, J.; Czech, Z.; Kowalczyk, A. Polymerization Shrinkage by Investigation of Uv Curable Dental Restorative Composites Containing Multifunctional Methacrylates. *Pol. J. Chem. Technol.* **2013**, *15*, 81–85. [CrossRef]
30. Wenzel, R.N. Resistance of solid surfaces to wetting by water. *Ind. Eng. Chem.* **1936**, *28*, 988–994. [CrossRef]
31. Çulhaoğlu, A.K.; Özkır, S.E.; Şahin, V.; Yılmaz, B.; Kılıçarslan, M.A. Effect of Various Treatment Modalities on Surface Characteristics and Shear Bond Strengths of Polyetheretherketone-Based Core Materials. *J. Prosthodont.* **2020**, *29*, 136–141. [CrossRef]
32. Akkan, C.K.; Hammadeh, M.; Brück, S.; Park, H.W.; Veith, M.; Abdul-Khaliq, H.; Aktas, C. Plasma and Short Pulse Laser Treatment of Medical Grade PEEK Surfaces for Controlled Wetting. *Mater. Lett.* **2013**, *109*, 261–264. [CrossRef]
33. Gheisarifar, M.; Thompson, G.A.; Drago, C.; Tabatabaei, F.; Rasoulianboroujeni, M. In Vitro Study of Surface Alterations to Polyetheretherketone and Titanium and Their Effect upon Human Gingival Fibroblasts. *J. Prosthet. Dent.* **2021**, *125*, 155–164. [CrossRef] [PubMed]
34. Riveiro, A.; Soto, R.; Comesaña, R.; Boutinguiza, M.; del Val, J.; Quintero, F.; Lusquiños, F.; Pou, J. Laser Surface Modification of PEEK. *Appl. Surf. Sci.* **2012**, *258*, 9437–9442. [CrossRef]
35. Dupuis, A.; Ho, T.H.; Fahs, A.; Lafabrier, A.; Louarn, G.; Bacharouche, J.; Airoudj, A.; Aragon, E.; Chailan, J.-F. Improving Adhesion of Powder Coating on PEEK Composite: Influence of Atmospheric Plasma Parameters. *Appl. Surf. Sci.* **2015**, *357*, 1196–1204. [CrossRef]
36. Kuscu, E.; Klink, A.; Spintzyk, S.; Kraemer Fernandez, P.; Huettig, F. Bonding Interface and Repairability of 3D-Printed Intraoral Splints: Shear Bond Strength to Current Polymers, with and without Ageing. *Materials* **2021**, *14*, 3935. [CrossRef] [PubMed]

37. Stawarczyk, B.; Keul, C.; Beuer, F.; Roos, M.; Schmidlin, P.R. Tensile Bond Strength of Veneering Resins to PEEK: Impact of Different Adhesives. *Dent. Mater. J.* **2013**, *32*, 441–448. [CrossRef]
38. Mair, L.; Padipatvuthikul, P. Variables Related to Materials and Preparing for Bond Strength Testing Irrespective of the Test Protocol. *Dent. Mater.* **2010**, *26*, e17–e23. [CrossRef]

Journal of *Functional Biomaterials*

MDPI

Systematic Review

Four-Dimensional Superimposition Techniques to Compose Dental Dynamic Virtual Patients: A Systematic Review

Ying Yuan [1], Qian Liu [2], Shuo Yang [2,*] and Wulin He [1,*]

1 Department of Orthodontics, Stomatological Hospital, Southern Medical University, Guangzhou 510280, China
2 Center of Oral Implantology, Stomatological Hospital, Southern Medical University, Guangzhou 510280, China
* Correspondence: alex2005191007@163.com (S.Y.); hewulin@i.smu.edu.cn (W.H.); Tel.: +86-15-92033-2690 (S.Y.); +86-15-52109-6103 (W.H.)

Abstract: Four-dimensional virtual patient is a simulation model integrating multiple dynamic data. This study aimed to review the techniques in virtual four-dimensional dental patients. Searches up to November 2022 were performed using the PubMed, Web of Science, and Cochrane Library databases. The studies included were based on the superimposition of two or more digital information types involving at least one dynamic technique. Methodological assessment of the risk of bias was performed according to the Joanna Briggs Institute Critical Appraisal Checklist. Methods, programs, information, registration techniques, applications, outcomes, and limitations of the virtual patients were analyzed. Twenty-seven full texts were reviewed, including 17 case reports, 10 non-randomized controlled experimental studies, 75 patients, and 3 phantoms. Few studies showed a low risk of bias. Dynamic data included real-time jaw motion, simulated jaw position, and dynamic facial information. Three to five types of information were integrated to create virtual patients based on diverse superimposition methods. Thirteen studies showed acceptable dynamic techniques/models/registration accuracy, whereas 14 studies only introduced the feasibility. The superimposition of stomatognathic data from different information collection devices is feasible for creating dynamic virtual patients. Further studies should focus on analyzing the accuracy of four-dimensional virtual patients and developing a comprehensive system.

Keywords: digital dentistry; computer-assisted design (CAD); patient simulation; image fusion; diagnosis; oral; evidence-based dentistry

Citation: Yuan, Y.; Liu, Q.; Yang, S.; He, W. Four-Dimensional Superimposition Techniques to Compose Dental Dynamic Virtual Patients: A Systematic Review. *J. Funct. Biomater.* **2023**, *14*, 33. https://doi.org/10.3390/jfb14010033

Academic Editor: Håvard J. Haugen

Received: 30 November 2022
Revised: 29 December 2022
Accepted: 4 January 2023
Published: 6 January 2023

1. Introduction

Digital workflows are becoming more accurate in dental medicine because of technological innovations. Transferring intraoral and extraoral data to a virtual environment is the first step in digital treatment. Currently, digital information can be captured in different ways, including desktop scanners (DS), intraoral scanners (IOS), facial scanners (FS), cone beam computed tomography (CBCT), computed tomography (CT), cephalometry, and photography. These methods can produce different file formats, such as standard tessellation language (STL), object code (OBJ), polygon (PLY), and digital imaging and communications in medicine (DICOM). A three-dimensional (3D) virtual patient can be created after the alignment and fusion of various data formats, including information about a real patient's teeth, soft tissues, and bones [1]. Thus, if real patients are indisposed, dental treatment plans could still be realized in virtual patients reducing chair time and patients' appointments.

Currently, investigators focus on static virtual patients, with improved gains through new materials, automation, and quality control [2,3]. However, static simulated patients cannot reflect real-time changes. The stomatognathic system comprises the skull, maxilla, mandible, temporomandibular joint (TMJ), teeth, and muscles, and changes in one part will cause synergistic changes in the others [4]. Therefore, integrating TMJ and mandible movement, occlusal dynamic, and soft tissue dynamic information (such as muscle movement and facial expression) to construct four-dimensional (4D) virtual patients is required in the future [5]. Four dimensions use time to express action; thus, four-dimensional patients with temporal information help understand the dynamic interactions of anatomical components under functional activities such as chewing, speech, and swallowing.

The first step in creating a 4D virtual patient is digitalizing the motion data. The virtual facebow (VF) and virtual articulator (VA) combination can facilitate positional relationship replication between the skull and jaws, simulating mandible movements [6]. In addition, a jaw motion analyzer (JMA) moves the digitized dentitions along paths in the computer, helping to visualize kinematic occlusion collisions and the condyle trajectory [7,8]. Information such as the smile line, lip movements, and facial expressions is essential to ensure functional outcomes and aesthetic performance and to construct a pleasant smile [9]. Finally, in various computer-assisted design and computer-aided manufacturing (CAD/CAM) programs, 4D simulated patients with complex movements are built. Currently, 4D virtual patient types vary according to clinical needs. Few studies have comprehensively analyzed the existing 4D dental virtual model construction techniques.

Therefore, this systematic review aims to summarize the current scientific knowledge in the dental dynamic virtual patient field to guide subsequent related research.

2. Materials and Methods

2.1. Eligibility Criteria

This systematic review followed the Preferred Reporting Items for Systematic Reviews and Meta-Analyses (PRISMA) guidelines [10–12]. (Table S1) The present review was not registered because it belongs to the literature reviews that use a systematic search, which PROSPERO does not accept. The focus was on the technique, accuracy, and application of dynamic multi-modal data fusion to create four-dimensional virtual patients in dentistry. The criteria for study selection were (1) the possibility of creating a 4D dental dynamic virtual patient analyzing at least one patient or phantom; (2) possible integration of two or more digital methods, at least one of which captures dynamic information; (3) availability of the methods and devices used; (4) review articles, opinion articles, interviews, charts, and non-English articles were excluded from this systematic review. The PICOS terms were defined as population (P), four-dimensional virtual patient; intervention (I), dynamic digitization technology; comparison © was omitted because the current review was not expected to have randomized controlled trials or relevant controlled trials; outcome (O): dental applications or accuracy analysis; setting (S): multi-modal data fusion.

2.2. Information Sources

The literature search was conducted by reviewing three online databases for eligible studies: PubMed, Medline (Web of Science), and Cochrane Library. The references of the full-text articles were additionally screened manually for other relevant studies. A four round "snowball procedure" was carried out to identify other published articles that met the review's eligibility criteria. (Figure S1) The "snowball procedure" is a multi-round forward screening, after the full-text screening, to search the eligible papers from the reference lists of the included papers. Once a new study/reference is included, its references are called snowball papers, which will undergo a new round of snowball screening. This procedure ends only when no snowball papers can be included in the last round [13].

2.3. Search Strategy

The first search in the database was performed on 5 August 2022. The search strategy was assembled using Medical Subject Headings (MeSH) terms and free-text words. Search terms were grouped according to the PICOS principle (Table 1). Weekly literature tracking was then conducted separately in the three databases using the above search terms to obtain the latest relevant literature.

Table 1. Overview of Electronic Search Strategy.

Subjects	Answers
Database	PubMed, Medline (Web of Science), and Cochrane library
#1 Population	"Patient simulation" [MeSH] or "computer simulation" [MeSH] or "Patient Simulations" or "computer simulations" or "computerized model *" or "Computer Model *" or "Virtual patient *" or "digital patient *" or "simulation patient *" or (("4D" OR "4-D" or "4 dimension *" or "4-dimension *" or "four D" or "four-D" or "four dimension *" or "four-dimension *") and "patient *")
#2 Intervention	"Dental Articulators" [MeSH] or "dynamic" or "Dental Articulator" or "Articulator, Dental" or "Articulator *" or "Condylar movement" or " condylar position" or "computer aided diagnosis axiograph" or "Diagnosis, Computer Assisted" or "Computer-Assisted Diagnosis" or "Computer Assisted Diagnosis" or "Computer-Assisted Diagnosis" or "Diagnoses, Computer-Assisted" or "Jaw motion" or "jaw movement" or "mandibular movement" or "Face bow" or "facebow" or "Electromyography" [MeSH] or "Electromyographies" or "Surface Electromyograph *" or "Electromyography, Surface" or "Electromyogram *" or "EMG" or "EMCP" or "photogrammetry" [MeSH] or "Stereophotogrammetr *" or "Radiostereometric Analysis" or " facial scanning" or "Mastication" [MeSH] or "chewing" or "Dental Occlusion" [MeSH] or "Occlusion?, Dental" or "Dental Occlusions" or "Occlusal Plane *" or "Plane?, Occlusal" or "Canine Guidance" or "Guidance, Canine" or "Occlusal Guidance *" or "Guidance, Occlusal."
#3 Outcome	"Dimensional Measurement Accuracy" [MeSH] or "Dimensional Measurement Accurac *" or "Measurement Accuracy, Dimensional" or "accuracy" or "precision" or "reliability" or "Validity" or "stomatology" or "tooth" or "oral" or "dental" or "Dentistry" or "orthodontics *" or "prosthodontics *" or "implant dentistry*" or "orthognathic *" or "Maxillodental *" or "orthognathic* " or "maxillofacial surgery *" or "plastic surgery *"
#4 Setting	"multi-modal" or "multi-mode" or "multi-modality" or "multimodal" or "multiple-modal" or "multiple-mode" or "multiple-modality" or "multi-source" or "multisource" or "fusion" or "integration" or "superimposition" or "merging" or "registration" or "alignment" or "calibration"
	#1 and #2 and #3 and #4

2.4. Study Records

After the first duplicate check in NoteExpress, the articles were imported into the Rayyan website [14] for the second duplicate check. Titles and abstracts were screened independently by two reviewers (YY and QL) on Rayyan [15], a tool to filter titles and abstracts effectively and to collaborate on the same review. Disagreements were resolved by discussion. For controversial articles labeled as maybe, reviewers discussed including them after reading the full text. After screening, full-text articles of selected titles and abstracts were acquired and read intensively by two reviewers to determine eligible articles. All authors discussed the remaining controversial articles to obtain a consensus. Snowball articles were included from the full-text references and were selected following the same principle.

2.5. Data Extraction

The following parameters were extracted from the selected full-text articles after the selection process: Author(s) and year of publication, study design, sample size, methods (including file format), manufacturer software programs, information, type of superimposition, scope, outcomes, and limitations.

2.6. Evaluation of Quality

The Joanna Briggs Institute (JBI) Critical Appraisal Checklist for quasi-experimental studies [16] and case reports [17] was used for the non-randomized controlled experimental studies/case reports. Two reviewers (YY and QL) independently assessed the methodological quality of the included studies. For every question in the checklist, except Q3 in the Checklist for Quasi-Experimental Studies, a yes answer means the question is low risk. Studies that met 80–100%/60–79%/40–59%/0–39% of the criteria were considered to have a low/moderate/substantial/high risk of bias, respectively [13]. In cases of disagreement, the decision was made by discussion among all authors.

3. Results

3.1. Search

The systematic search was completed on 30 November 2022. The QUORUM diagram details are shown in Figure 1. The snowball procedure is illustrated in Figure S1. The search yielded 1776 titles, 173 titles and abstracts were identified. Subsequently, 19 full texts were selected by two reviewers, of which 550 references underwent four rounds of citation checks using the snowball procedure, yielding 14 articles. Six controversial articles were excluded from the 19 full-text articles based on the outcomes of the discussion between all authors. The reasons for exclusion were as follows: single dynamic data ($n = 1$), non-dynamic patients ($n = 4$), and no virtual patient built ($n = 1$). Finally, 27 articles were included in the systematic review; the reasons for excluding other papers are shown in Figure 1.

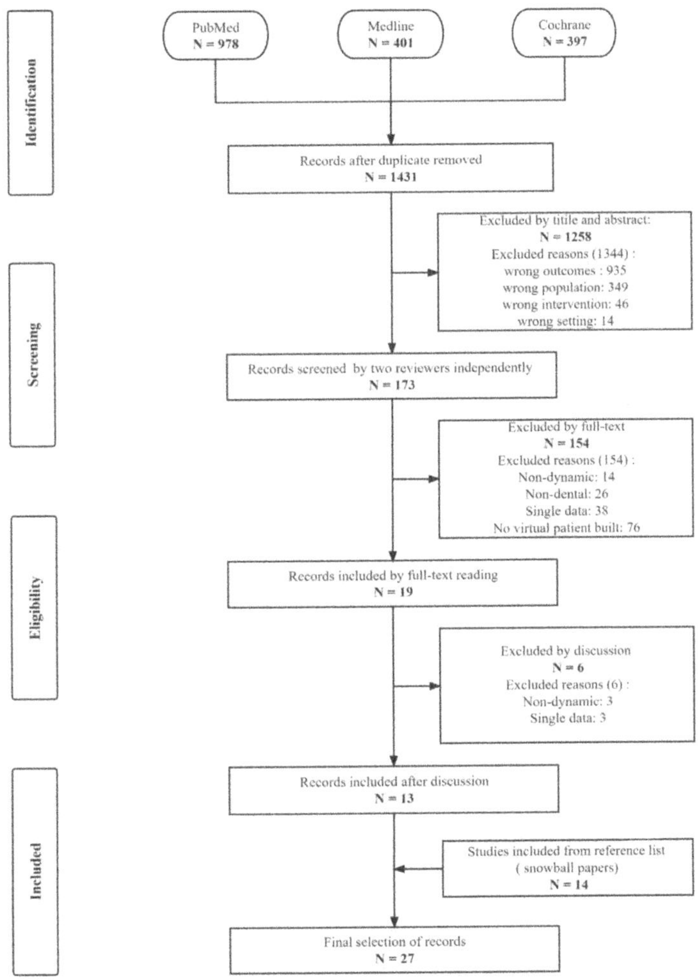

Figure key: The number of excluded reasons may be larger than that of excluded articles, in the same box, due to more than one excluded reason for some papers.

Figure 1. PRISMA flow chart. This diagram describes the identification, screening, exclusion reasons, and included procedures of 27 included articles. The number of excluded reasons exceeds that of excluded articles only at the screening phase, where the total number of reasons excluded was 1344, but 1258 articles were excluded.

3.2. Description of Studies

The characteristics of the included studies are presented in Table 2. The included publications were dated from 2007 to 2022, without intentional time restriction, since the 4D virtual patient is a new technology recently proposed. The review included 17 case reports and 10 nonrandomized controlled experimental studies. No randomized controlled trials (RCT) were found. Most of the 27 studies had only one subject/phantom, except for 2 [18,19]. In total, 75 patients and three phantoms were included in creating 4D virtual patient models.

Two or more methods, including static information and dynamic information collection devices, can acquire different formats of 3D data. The present article focuses on dynamic data (Table 3); 10 studies acquired real-time jaw-motion data, 4 analyzed the dynamic facial information, 13 simulated the jaw position, and 2 examined the coordinated movement of the masticatory system. Three to five types of information acquired from the above data were integrated to create 4D virtual patients. Additionally, the included studies focused on different clinical scopes: prosthetic dentistry (n = 19) (including implant dentistry (n = 4)), maxillofacial surgery (n = 8), and orthodontics (n = 5).

3.3. Risk of Bias in Included Studies

Table S2 describes the risk of bias assessment in the 10 non-randomized controlled experimental studies. All had a low risk of bias for Q1, Q4, Q7, and Q8 of the JBI Critical Appraisal Checklist. For Q2, Q3, and Q9, 80%, 90%, and 60% of the studies had a low risk of bias, respectively. For the overall risk, 90% of the studies showed moderate risk and 10% showed substantial risk (Figure 2a). Table S3 describes the risk of bias assessment of the 17 case reports. All these had a low risk of bias for Q5, Q6, Q7, and Q9 of the JBI Critical Appraisal Checklist. Except for four studies that indicated low risk, Q4 was not applicable in most studies where specific diseases were not crucial during the construction of the simulated model. However, for Q1, Q2, and Q3, 76.5%, 69.5%, and 52.9% of the case reports, respectively, showed a high risk of bias. For the overall risk, 52.9%, 29.4%, 11.8%, and 5.9% showed high, moderate, substantial, and low risk, respectively (Figure 2b).

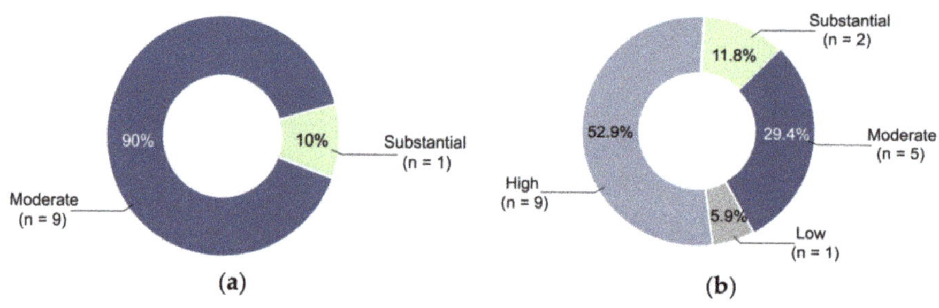

Figure 2. Percentage of different risk levels: (**a**) For the overall risk in non-randomized controlled experimental studies, 90% showed moderate risk and 10% showed substantial risk. (**b**) For the overall risk in case reports, 52.9%, 29.4%, 11.8%, and 5.9% showed high, moderate, substantial, and low risk, respectively.

Table 2. Information on the 27 included studies to compose dynamic simulated dental patient.

Author Ref./Year	Study Designs (Humans/ Phantom)	Sample Size (n)	Methods (+File Format)	Manufacturer Software Programs	Information	Types of Registration/ Superimposition	Scopes	Outcomes	Limitations
Zambrana N [20]/2022	Case report: method description (Humans)	Unknown	1. CBCT (DICOM) 2. IOS(STL) 3. Target tracking camera (MP4) 4. DS 5. Open-source CAD software	1. (Promax 3D Mid, Planmeca OY, Helsinki, Finland) 2. (TRIOS3; 3Shape A/S, Copenhagen, Denmark) 3. Camera of a mobile phone (iPhone 7; Apple Inc., Cupertino, CA, USA) 4. (Swing; DOF Inc., Seoul, Republic of Korea) 5. (Blender 3D; Blender Foundation, Amsterdam, Netherlands)	1. TMJs, maxilla, and mandible 2. Dentitions and the maxillomandibular relationship 3. Mandibular movements 4. Maxillomandibular relationship, including the marker boards 5. 4D virtual patient with mandibular kinematic path and TMJ kinematic path	Markerboards + marker board from occlusal registration (IOS + DS): point-based	Prosthetic Dentistry	Feasibility	1. Lack of a rigorous validation 2. Lack of comparison with other jaw-tracking systems and software programs
Kim JE [21]/2019	Case report: method description (Humans)	Unknown	1. DS 2. FS + targets 3. Image registration software 4. CAD software	1. (Identica hybrid, Medit Inc., Seoul, Republic of Korea) 2. (Rexcan CS2, Medit Inc., Seoul, Republic of Korea) 3. (EzScan8, Medit Inc., Seoul, Republic of Korea) 4. (Exocad; exocad GmbH, Darmstadt, Germany)	1. Virtual casts 2. Anterior part of dentition, face, and target position 3. Four types of antagonist mesh + models superimposing dental cast data on facial scan data 4. Occlusal contacts in the MICP and occlusal interference during eccentric movement	FS + DS: point-based(teeth) + horn alignment algorithm + ICP algorithm	Prosthetic dentistry	Feasibility	1. Time-consuming 2. The need to maintain the retractor 3. Lack of validation of accuracy
Revilla-Leon M [22]/2022	Case report: method description (Humans)	Unknown	1. FS(PLY) 2. IOS(STL) 3. VA 4. DS(PLY) + scan body system 5. CAD software	1. (InstaRisa Facial Scanner; InstaRisa, Clovis, CA, USA) 2. (TRIOS4, wireless, v21.2.0; 3Shape A/S, Copenhagen, Denmark) 3. (Panadent Articulator; Panadent, Colton, CA, USA) 4. (Medit T500; Medit Inc., Seoul, Republic of Korea) + (Kois Scan Body System; Kois Center, LLC, Seattle, WA, USA) 5. (DentalCAD 3.0, Galway; exocad GmbH, Darmstadt, Germany)	1. Static and smiling face 2. The dentitions and a maxillomandibular registration 3. Jaw-correction function 4. NHP + registration tool 5. Virtual patient with a digitized scan body, maxillary digital scan, and reference facial scan	1. Maxillary cast + the scan body (DS + IOS): point-based + best-fit alignment (teeth as common information) 2.Scan body + reference face (DS + FS): point-based (scan-body as common information) 3. Reference face + the smiling face (FS + FS): point-based (facial markers as common information)	Prosthetic dentistry	Feasibility	1. Lack of validation of accuracy 2. Factors affecting accuracy: 1). the eye closure due to the intense light 2) color and surface texture of the scan body system

Table 2. *Cont.*

Author Ref./Year	Study Designs (Humans/ Phantom)	Sample Size (n)	Methods (+File Format)	Manufacturer Software Programs	Information	Types of Registration/ Superimposition	Scopes	Outcomes	Limitations
Li J [19]/2022	Non-randomized controlled experimental study (Humans)	Two healthy subjects	1. FS (STL/PLY) 2. IOS (STL) 3. DS + virtual facebow fork 4. CBCT + implant planning software (STL) 5. Open-source software 6. CAD software	1. Application (Hege 3D scanner;) in (iPhone 11 Pro; Apple Inc., Cupertino, CA, USA) 2. (TRIOS Color Pod; 3Shape, Copenhagen, Denmark) 3. (D2000; 3Shape, Copenhagen, Denmark) 4. (3D Accuitomo 170; J Morita) + (Blue Sky Plan; Blue Sky Bio, LLC, Libertyville, IL, USA) 5. (Blender 3D; Blender Foundation, Amsterdam, Netherlands) 6. (Exocad version 2.2; exocad GmbH, Darmstadt, Germany)	1. Facial information 2. Maxillary dentition 3. Alignment of dentition and the face 4. Model with face soft tissue and maxillary dentition 5 + 6. A composite model including the face, facebow fork, and the maxillary dentition	CBCT + FS + IOS: Surface-based + the best-fitting algorithm	Prosthetic dentistry: CAD/CAM	1. Feasibility 2. Accuracy of the VF: (1) High trueness: 1.14 ± 0.40 mm (2). High precision: 1.08 ± 0.52 mm (the difference of eight measurements was small)	The factors influencing accuracy were not explored.
Revilla-Leon M [23]/2022	Case report: method description (Humans)	One subject	1. IOS(STL) 2. FS (PLY+STL) 3. Extra-oral scan body system 4. DS (STL) 5. Open-source software 6. 3D modeling software 7. Avatar generator software (OBJ) 8. Motion Engine software program + a facial tracking app in a smartphone	1. (TRIOS 4, wireless, v. 21.2.0; 3Shape A/S, Copenhagen, Denmark) 2. (Instarisa facial scanner; Instarisa, Clovis, CA, USA) 3. (AFT Scan Body Teeth; AFT Dental System) 4. (T710 Scanner; Medit Inc., Seoul, Republic of Korea) 5. (Blender 3D; Blender Foundation, Amsterdam, Netherlands) 6. (Zbrush 2022; Maxon, Friedrichsdorf, Germany) 7. (Character Creator, v.3.44; Reallusion, SFO, CA, USA) 8. (Iclone, v.7.93; Reallusion) + (Live Face, v.1.08; Reallusion) in (iPhone 12 Pro; Apple Inc., Cupertino, CA, USA)	1. Dentitions and a maxillo-mandibular registration 2. Facial scan with and without scan body system 3. Registration tool 4. Digitizing scan body 5. All data converted to the OBJ format 6. Avatar without the hair and eyes 7. Avatar with the hair and eyes, dentition, and scan body 8. Virtual patient with lip dynamics	IOS + DS: fiducial markers based (extra-oral scan body system)	1. Prosthetic dentistry 2. Maxillofacial surgery	Feasibility	1. Lack of validation of accuracy 2. Complicated and time-consuming

Table 2. *Cont.*

Author Ref./Year	Study Designs (Humans/Phantom)	Sample Size (n)	Methods (+File Format)	Manufacturer Software Programs	Information	Types of Registration/Superimposition	Scopes	Outcomes	Limitations
Lepidi L [24]/2021	Case report: method description (Humans)	One subject	1. IOS(STL) 2. FS(OBJ) 3. DS(STL) 4. CBCT (DICOM) 5. Dental CAD software (STL)	1. (Cs 3600, Carestream Health, Rochester, NY, USA) 2. (Bellus 3D; Bellus 3D Inc., Campbell, CA, USA) 3. (Neway, Faro Technologies Inc., Brescia, Italy) 4. (Cs 9600 3D, Carestream Health, Rochester, NY, USA) 5. (Exocad; exocad GmbH, Darmstadt, Germany)	1. Maxillary and mandibular fixed prosthesis in the desired CR and VDO 2. Facial scan with and without the fork 3. Cast with implant analogs 4. CBCT image contained the restorations 5. Virtual patient mounting to the VA The occlusal discrepancy between MICP/CR	1. Cast with implant analogs, the casts with prostheses, and FS (DS + IOS + FS): fiducial markers based 2. FS with and without fork (FS + FS): point-based	Prosthetic dentistry	Feasibility	1. Lack of validation of accuracy 2. Complicated and time-consuming
Kim SH [25]/2020	Non-randomized controlled experimental study (phantom)	A skull phantom	1. MDCT + occlusal splint with a registration body 2. DS 3. The flat-panel display monitor with a 3D depth camera 4. EM tracking system	1. (SOMATOM Sensation 10, Siemens, Munich, Germany) 2. (Maestro 3D, Maestro, Pisa, Italy) 3. (QCT130, One Inc., Seoul, Republic of Korea) 4. (Aurora, Northern Digital Inc., Waterloo, ON, Canada)	1. The maxillary and skeletal models 2. The maxillary dentition 3. 3D virtual objects superimposed on the real patient image 4. 3D tracking positions of the bone segment and the reference in physical space	1. EM tracking + CT image spaces + AR camera spaces: fiducial markers based (markers on the registration body) 2. Dentition + maxillary model (MDCT + DS): ICP algorithm	Orthognathic surgery: AR-assisted free-hand orthognathic surgery	High accuracy: the MADs of the difference between actual and measured positions exhibited no significant differences between the SRT (0.20, 0.34, 0.29) and BRT (0.23, 0.37, 0.30)	Lack of validation of usability and accuracy in real patients
Lee SJ [26]/2019	Non-randomized controlled experimental study (phantom and humans)	A skull phantom + a patient	1. MDCT + Occlusal splint attached with a registration body 2. DS 3. EM tracking system	1. (SOMATOM Sensation 10, Siemens, Munich, Germany) 2. (Maestro 3D, Maestro, Pisa, Italy) 3. (Aurora, Northern Digital Inc., Waterloo, ON, Canada)	1. MPS and other skeletal models 2. Artifact-free model of the dentitions + final occlusal model 3. 3D tracking positions of MPS in a virtual maxillomandibular complex	1. CT image + physical spaces: fiducial markers based (registration body) 2. Dentition + maxillary model (MDCT + DS): ICP algorithm	Orthognathic surgery: (MPS repositioning; model-guided surgery)	1. Convenient 2. Accuracy: (The RMS differences between the simulated and intraoperative MPS models and between the simulated and postoperative CT models were 1.71 ± 0.63 mm and 1.89 ± 0.22 mm, respectively.)	Further development is needed to increase accuracy by reducing technical errors in the tracking devices, imaging errors from the modalities, registration errors, application errors, and human error

Table 2. *Cont.*

Author Ref./Year	Study Designs (Humans/ Phantom)	Sample Size (n)	Methods (+File Format)	Manufacturer Software Programs	Information	Types of Registration/ Superimposition	Scopes	Outcomes	Limitations
Li J [27]/2020	Case report: case Reports (humans)	One subject	1. IOS(STL) + a wax rim 2. CBCT (DICOM) + gothic arch 3. FS (STL) 4. Implant planning software 5. Dental CAD software + free CAD software.	1. (TRIOS; 3Shape, Copenhagen, Denmark) 2. (3D Accuitomo 170; J. Morita, Osaka, Japan) 3. (3dMDtrio System; 3dMD, Atlanta, GA, USA) 4. (Blue Sky Plan; Blue Sky Bio, LLC Libertyville, IL, USA). 5. (Exocad version 2.2; exocad GmbH, Darmstadt, Germany) + (Meshmixer; Autodesk, San Rafael, CA, USA)	1. Arches in approximate CR and VDO 2. (1) The maxilla, mandible, infraorbital points, and external acoustic meatus (2) The CR and VDO records 3. Smiling face 3D reconstruction of the face 4. Two-piece surgical templates 5. (1) Custom bases for a gothic arch tracer (2) The 3D images of FS, IOS, and skull mounted on a VA	Unknown	Dental implant surgery Pros-thetic dentistry	Feasibility: successful prosthetic outcome	1. Obtaining an occlusal record with increased VDO using the gothic arch tracer is difficult for the patient with remaining teeth. 2. Lack of a quantitative validation of accuracy
Li J [28]/2021	Case report: method description (Humans)	Unknown	1. IOS (STL) 2. Dental CAD software 3. CBCT (DICOM) 4. FS (PLY) 5. Implant planning software (STL) 6. Open-source 3D software program	1. (TRIOS 3; 3Shape A/S, Copenhagen, Denmark) 2. (exocad; exocad GmbH, Darmstadt, Germany) 3. not mentioned 4. (Hege 3D scanner) in (iPhone 11 Pro; Apple Inc., Cupertino, CA, USA) 5. (BlueSkyPlan v4.70; Blue Sky Bio LLC, Libertyville, IL, USA) 6. (Blender 3D; Blender Foundation, Amsterdam, Netherlands)	1. Arches in approximate CR and VDO 2. Design of the tooth-supported template 3. The maxilla, mandible, infraorbital points, and external acoustic meatus 4. Facial scan 5. 3D bone model and face model 6. Tooth-supported gothic arch tracer Alignment of the facebow and the skull Virtual patient with FS, IOS, CBCT	Unknown	Dental implant surgery Pros-thetic dentistry: (complex implant-supported Prosthe-ses)	1. Predictability 2. Feasibility	1. The need for a CBCT scan 2. Lack of a quantitative validation of accuracy

Table 2. *Cont.*

Author Ref./Year	Study Designs (Humans/Phantom)	Sample Size (n)	Methods (+File Format)	Manufacturer Software Programs	Information	Types of Registration/Superimposition	Scopes	Outcomes	Limitations
Olszewski R [29]/2008	Non-randomized controlled experimental study (Humans)	Unknown	1. CT (DICOM) + MRI 2. DS 3. 3D tracking device 4. System for planning and assisting orthognathic surgery	Data integration module of the system: MedicalStudio1 platform 1. ACRO 3D: 3D CT based craniofacial cephalometric Analysis 2. ACROTooth: virtual occlusion 3. TMJSim: TMJ movement Simulation 4. ACROSim: virtual surgery planning 5. ACROGuide: intra-operative AR assistance	1. Maxilla, mandible, skull, and skin 2. Dental casts. 3. Mandibular movement: (1) translation and the rotation (2) the centric position for the condyles (3) The joint's degrees of freedom 4. Virtual models	The physical space + the digital world: 1. algorithm (minimizing the mean square distance between the points) 2. fiducial markers (attached to the tracked surgical tools) based	Maxillofacial orthognathic surgery	Complete Accuracy: 1. ACRO 3D module: validated 2. TMJSim: partly validated; ACROGuide: partly validated.	Before clinical application: 1. accuracy needs to be further validated. 2. technology and algorithms need to be further improved.
Fushima K [18]/2007	Non-randomized controlled experimental study (Humans)	More than 50 cases	1. Lateral and PA cephalograms 2. 3D motion tracking 3. Facebow-transfer 4. Surgical simulation system: mandibular motion tracking system (ManMoS)	1. Scanner (ES-2200, Epson Co., Owa, Suwa, Nagano, Japan) 2. (Polhemus, Colchester, VT, USA) 3. Unknown 4. (FASTRAK, Virtual Realities, LLC., League City, TX, USA)+2	1. Skeletal in a virtual space 2. Mandibular motion tracking 3. A record of how the upper dentition relates to the TMJ 4. Pilot surgical prediction and real-time surgical simulation: model of the craniofacial skeleton with the centric stops of the dental arches	Dentition + skeleton + motion tracking in a virtual space: fiducial markers based	Maxillofacial orthognathic surgery	1. Feasibility 2. Trueness: sufficient (small SD) 3. Precision: SD in the 40 recordings was less than 0.1 mm	Complicated and time-consuming
Kois JC [30]/2022	Case report: method description (Humans)	Unknown	1. IOS 2. KFRG 3. Photography 4. CAD software 5. VA module	1. (TRIOS 3; 3Shape A/S, Copenhagen, Denmark) 2. (Kois Center, LLC, Seattle, WA, USA) 3. Digital single-lens reflex camera (D850; Nikon Inc., Tokyo, Japan) 4. (DentalCAD; exocad GmbH, Darmstadt, Germany)	1.Maxillary and mandibular arches 2. NHP 3. Photograph of dentition and face 4. Virtual Orientation: the project scene 5. Facially generated virtual mounting	IOS + photograph: "Align Mesh" tool	Prosthetic dentistry	Feasibility	Lack of the validation of accuracy

Table 2. *Cont.*

Author Ref./Year	Study Designs (Humans/ Phantom)	Sample Size (n)	Methods (+File Format)	Manufacturer Software Programs	Information	Types of Registration/ Superimposition	Scopes	Outcomes	Limitations
Kwon JH [31]/2019	Case report: method description (Humans)	One subject	1. CBCT (DICOM) 2. Image processing software (STL) 3. DS 4. FS + targets 5. Image registration tool	1. (PaXZenith3D, Vatech Co., Ltd., Hwaseong, Republic of Korea;) 2. (OnDemand3D, Cybermed Co., Ltd., Seoul, Republic of Korea.) 3. (Identica hybrid, Medit Inc., Seoul, Republic of Korea.) 4. (Rexcan CS2, Medit Inc., Seoul, Republic of Korea.) 5. (Ezscan8, Medit Inc., Seoul, Republic of Korea)	1. Maxilla and mandible 2. 3D skull and jaws 3. digital casts 4. The oral cavity and face in MICP 5. A 3D model with a CBCT scan Real-time mandibular motions	CBCT + DS: fiducial markers based (to generate the transformation matrix by comparing reference points)	Prosthetic dentistry	1. Convenient 2. Stability: 36 mm in the mandible 30.78 mm/ 37.74 mm in the left/right condyle 3. Accuracy: high (4.1–6.9 mm)	The sample size needs to be expanded for further validation.
Lam WYH [32]/2016	Non-randomized controlled experimental study (Humans)	Unknown	1. FS with and without facebow 2. DS 3. IOS 4. Open-source software 5. CBCT scan+ the occlusal wafer + radiopaque markers	1. (3dMDface; 3dMD Inc, Atlanta, GA, USA) 2. (Handyscan 3D; Creaform) 3. (True Definition; 3M ESPE, Iverson Dental Labs, MARB, CA, USA) 4.(MeshLab v1.3.3; Visual Computing Lab of the ISTICNR, Pisa, Italy) 5. Image analysis software (3D Slicer9 version 4.3; Slicer community, Boston, MA, USA)	1.3D face with/without facebow 2. Facebow 3. (1) Dentitions and the maxillomandibular relationship; (2) Buccal surface of the maxillary dentition and the occlusal wafer 4. Virtual patients transferred to the VA 5. The facial skin, teeth, and radiographic markers	1. FS + DS + IOS: point-based + ICP algorithm 2. Face-bow (DS) + CBCT: fiducial markers based + point-based algorithm	Prosthetic dentistry	1. Feasibility 2. Accuracy (1) error in tooth registration: less than 1 mm (2) the facial alignment (the mean distances of the nasion, alares, and tragions: 0.83 mm, 0.77 mm, and 1.70 mm)	1. The registration accuracy needs to be improved. 2. Distortion needs to be avoided. 3. Time-consuming
Lam WYH [33]/2018	Non-randomized controlled experimental study (Humans)	One subject	1. FS 2. IOS 3. DS 4. Open-source software 5. CBCT 6. CAD software	1. (3dMDface; 3dMD Inc., Atlanta, GA, USA) 2. (True definition scanner; 3M ESPE, Iverson Dental Labs, MARB, CA, USA) 3. (DAVID SLS-3; Hewlett-Packard, Palo Alto, CA, USA) 4. (MeshLab v1.3.3; Visual Computing Lab of the ISTI-CNR, Pisa, Italy) 5. (ProMax 3D Mid, Planmeca OY, Helsinki, Finland) 6. (Exocad; Exocad GmbH, Darmstadt, Germany)	1. 3D face in NHP position or with a VF 2. (1) Dentitions and the maxillomandibular relationship; (2) Buccal relationship of the maxillary teeth and VF 3. VF 4. The dentition and the 3D facial photographs in NHP 5. The dentition, jaws, and 3D facial photograph in NHP 6. Virtual patients transferred to the VA	FS + DS + IOS: point-based + ICP algorithm Face-bow (DS) + CBCT: fiducial markers based + point-based algorithm	Prosthetic dentistry	Good precision of the SP NHP technique: positional differences of less than 1 degree and 1 mm in five repeated measurements in one patient	The sample size needs to be expanded for further validation.

Table 2. *Cont.*

Author Ref./Year	Study Designs (Humans/Phantom)	Sample Size (n)	Methods (+File Format)	Manufacturer Software Programs	Information	Types of Registration/Superimposition	Scopes	Outcomes	Limitations
Shao J [34]/2019	Case report: clinical report	One subject	1. Multi-slice spiral CT 2. Facial photograph 3. Imaging software 4. DS 5. CAD software 6. CAD software + 3D printer	1. (Philips MX16 EVO CT; Koninklijke Philips N.V., Amsterdam, NL) 2. (3dMDface System; 3dMD, Atlanta, GA, USA) 3. (Dolphin Imaging & Management Solutions; Patterson Dental, Chatsworth, CA, USA) 4. (3ShapeA/S, Copenhagen, Denmark) 5. (Exocad GmbH, Darmstadt, Germany) 6. (Formlabs Form 2; Formlabs, Boston, MA, USA)	1. Bone and dentition 2. Virtual face model with bone and soft tissue in real-time in NHP 3. Simulation of the lateral facial profile influenced by the retrusion of anterior maxillary teeth Dentition casts scan preserving their articulator-mounted relationship 5. The restorations with incisal edges were retruded for 5.0 mm 6. Prosthetically driven planning	Unknown	Dental implant surgery Prosthetic dentistry	Feasibility	Lack of the validation of accuracy
He S [35] /2016	Non-randomized controlled experimental study: (Humans)	One subject	1. CBCT + GALAXIS 3D software (DICOM) 2. SICAT JMT + system	1. (Sirona Galileos, Bensheim, Germany) 2. (SICAT Function; SICAT, Bonn, Germany)	1. CT images: condylar status 2. (1) Jaw movements + incisor ranges (2) Models integrating CBCT and JMT data: movement of the mandible (including the translation of the condyles)	CBCT + JMT data: fiducial markers based (radiopaque markers on bite tray)	Maxillofacial orthognathic surgery Digital dentistry	1. Reliable accuracy: the same positions between the simulated condylar position with that in the second CBCT 2. High precision	Further studies are needed to validate its accuracy.
Park JH [36] /2021	Case report: clinical report (Humans)	One subject	1. IOS + software program (STL) 2. CBCT 3. VA program (STL)	1. (TRIOS; 3Shape, Copenhagen, Denmark) + (Ortho Analyzer; 3Shape, Copenhagen, Denmark) 2. (Alphard Vega; Asahi Roentgen, Kyoto, Japan) 3. (R2GATE 2.0.0; Megagen, Seoul, Republic of Korea)	1. Cast in CRO and MICP position 2. CBCT model in CRO 3. (1) The maxillary cast registered on the CBCT model (2) VA (3) Models Superimposing the mandible position in CRO and MICP	IOS + CBCT: point-based + ICP algorithm	Prosthetic dentistry	Feasibility	Lack of the validation of accuracy

Table 2. *Cont.*

Author Ref./Year	Study Designs (Humans/Phantom)	Sample Size (n)	Methods (+File Format)	Manufacturer Software Programs	Information	Types of Registration/Superimposition	Scopes	Outcomes	Limitations
Amezua X [37]/2021	Non-randomized controlled experimental study (phantom)	A skull phantom	1. CAD software 2. DS by Industrial reference scanner (STL) 3. FS by the reference scanner (STL)/SWL scanner/SFM scanner methods (OBJ) 4. RE software	1. (Solid Edge ST10; Siemens, Munich Ger) 2. (ATOS Compact Scan 5M scanner with ATOS Professional V7.5 software; GOM, GmbH, ZEISS, BS, Ger) 3. SFM: (PENTAX K-S1; Ricoh Imaging Co, Ltd., Tokyo, Japan) + (Agisoft Metashape Professional; Agisoft, SPB, Russia) SWL: (Go! SCAN20 scanner with VX element 6.3 SR1 software; Creaform, Inc., Lévis, CAN) 4. (Geomagic Studio 2013; Geomagic, Inc., RTP, NC, USA)	1. IOTE 2 + 4. The maxillary model without regions not correspond to the teeth 3. (1) FS with IOTE (2) IOTE-free FS (3) FS with the mouth open 4. Models aligning IOTE-free FS and the maxillary scan	FS + DS/transferring maxillary digital scan to standard virtual patient: point-based + ICP algorithm	Prosthetic dentistry	Reliable accuracy: (below 1 mm): 0.182 mm for the RE group, 0.241 mm for the SWL group, and 0.739 mm for the SFM group	1. Further studies are needed to validate its accuracy. 2. In vitro experiments may underestimate the scanning error.
Dai F [38] /2016	Case report	One subject	1. DS (STL) 2. Spiral CT (DICOM) + 3D software (STL) 3. Ultrasonic axiograph Arcus Digma system (TXT file) 4. RE software 5. Mathematical software MATLAB 7.0 + Amira software FE modeling 6. Analysis software Ansys 15.0	1. (Roland DG., Hamamatsu, Japan) 2. (PHILIPS Inc., Andover, MA, USA) + Amira5.2.2 (Visage Imaging Inc., SD, CA, USA) 3. (KaVo, Biberach, Germany) 4. Rapidform 2006 (Inus Technology Inc., Seoul, Republic of Korea) 5. MATLAB 7.0 (Math Works Inc., Natick, MA, USA) 6. (ANSYS Inc., Canonsburg, PA, USA)	1. The upper cast with the bite fork and the occlusion 2. 3D bone, muscle, and teeth 3. Mandibular movement 4. Static model of the masticatory system 5. Dynamic model with a simulation of mandibular movement 6. The FE masticatory system model	1. Casts + casts made in occlusion (IOS + IOS): regional registration method 2. Registration of the different coordinate systems: based on the global coordinates + the (bite fork)	Digital dentistry Orthodontics	1. Feasibility 2. Accuracy: (1) the static masticatory system model: small difference (0.32 ± 0.25 mm) indicated good accuracy (2) the FE model showed accuracy similarity to that of the T-Scan (3) The accuracy of the 3D Arcus Digma system: 0.1 mm and 1.5°	The sample size needs to be expanded for further validation.

Table 2. *Cont.*

Author Ref./Year	Study Designs (Humans/Phantom)	Sample Size (n)	Methods (+File Format)	Manufacturer Software Programs	Information	Types of Registration/Superimposition	Scopes	Outcomes	Limitations
Savoldelli C [39]/2012	Non-randomized controlled experimental study: (Humans)	One subject	1. Multislice CT with a splint 2. MRI with a splint 3. 3D image segmentation software 4. FE analysis software	1. (General Electric Medical System, UWM, WI, USA) 2. Gyroscan Intera 1.5-T MR system (Philips Medical Systems, Best, NL) 3. (AMIRA®) (Visage Imaging, Inc., SD, CA, USA) 4. FORGE (Transvalor, Glpre 2005, Antibes, France)	1. Bone components of skull and mandible, dental arches when the jaw was opened 10 mm. 2. Soft tissues such as joint discs, temporomandibular capsules, and ligaments when the jaw was opened 10 mm. 3. Virtual models with surface and volume meshes of the above components 4. (1) Boundary conditions for closing jaw simulations by different jaw muscles (2) Stress distribution in both joint discs	MRI + CT: based on the anatomical structures (Hounsfield unit values + manual identification)	Digital dentistry Orthodontics	1. Feasibility 2. Accuracy (high): stress levels (5.1 MPa) were within the range of reported stress (0.85–9.9 MPa)	1. The material behavior of the articular discs was a linear elastic model and not a non-linear material model. 2. the sample size needs to be expanded for further validation.
Terajima M [4]/2008	Case report: method description (Humans)	One subject	1. CT + image processing software + visualization software 2. DS (VIVID format) 3. Jaw-movement analyzer 4. Image measurement software	1. CT scanner (Aquilion, Toshiba Medical, Tokyo, Japan) + (Mimics version 7.0, CDI, Tokyo, Japan) + (Magics, CDI, Tokyo, Japan) 2. (VIVID 900, Minolta, Tokyo, Japan) 3. (TRI-MET, Tokyo-Shizaisha, Tokyo, Japan) 4. (3D-Rugle, Medic Engineering, Kyoto, Japan)	1. Reconstruction of images integrating the CT, the 3D dental surface, ceramic spheres 2. Dental surface + ceramic spheres 3. Mandibular movement 4. Condyle position relative to the condylar fossa + contact areas during jaw movements	1. CT +DS: fiducial markers based (ceramic balls) 2. Registration of 3D maxillofacial-dental images and that in the TRI-MET system: the least squares method	Digital dentistry Orthodontics Orthognathic surgery	Feasibility	Further studies are needed to validate its accuracy.

Table 2. *Cont.*

Author Ref./Year	Study Designs (Humans/ Phantom)	Sample Size (n)	Methods (+File Format)	Manufacturer Software Programs	Information	Types of Registration/ Superimposition	Scopes	Outcomes	Limitations
Perez-Giugovaz MG [40] /2021	Case report: method description (Humans)	One subject	1. IOS (STL) 2. CAD software (STL) 3. Printer software 4. FS + a facebow record 5. DS (STL) 6. CAD software	1. (Cs 3600, Carestream Health, Rochester, NY, USA) 2. (MeshMixer; Autodesk, SR, CA, USA) 3. (ChiTuBox V1.7.0; ChiTuBox, Shenzhen, China) 4. (Bellus Face Camera Pro; Bellus3D Inc., Campbell, CA, USA) 5. (Open Technologies Small; Faro, Lake Mary, FL, USA) 6. (Dental CAD Plovdiv; exocad GmbH, Darmstadt, Germany)	1. Maxillary and mandibular casts 2. Virtual design of: (1) custom tray and mandibular occlusion rim with gothic arch tracer (2) scan body 3. Manufacture of the above devices 4. Facial scan with occlusion rim and scan body 5. Casts with occlusion rims and the scan body 6. The virtual patient with the casts mounted on the VA	1. FS + scan body (DS): fiducial markers based 2. FS with occlusion rim + FS with scan body: facial point-based	Digital dentistry Prosthetic dentistry: CAD/CAM	Feasibility	Further studies are needed to validate its accuracy.
Solaberrieta E [41]/2015	Case report: method description (Humans)	Unknown	1. IOS 2. Camera + reverse engineering software + target 3. Reverse engineering software	1. (3Shape TRIOS; 3Shape A/S, Copenhagen, Denmark) 2. (Nikon D3200; Nikon Inc., Tokyo, Japan) + (Agisoft Photoscan; Agisoft LLC, SPB, Russia) 3. Rapidform 2006 (Inus Technology Inc., Seoul, Republic of Korea)	1. (1) Maxillary and mandibular casts (2) The casts + facebow fork (3) Casts in the VA in MICP 2. 3D face with targets on the facebow fork 3. (1) Alignment of the face and facebow fork, the maxillary cast, and the facebow fork (2) Casts transferred to VA	1. The maxillary cast + facebow fork + 3D face-facebow fork (IOS + FS): best-fit command 2. Alignment of cranial coordinate system: facial point-based	Digital dentistry Prosthetic dentistry: CAD/CAM	Feasibility	Additional studies need to validate the accuracy of the new systems.

Table 2. *Cont.*

Author Ref./Year	Study Designs (Humans/Phantom)	Sample Size (n)	Methods (+File Format)	Manufacturer Software Programs	Information	Types of Registration/Superimposition	Scopes	Outcomes	Limitations
Granata S [42]/2020	Case report (Humans)	One subject	1. Geometric occlusal registration prototype device 2. FS (OBJ) 3. IOS (PLY/STL) 4. CBCT with DGB1 (DCM) 5. DS 6. 3D-guided surgery planning software + CAD design software	1. (DGB) (Digitalbite; Digitalsmile srl, Pietracamela, Italy) 2. (Bellus3D; Bellus3D Inc., Campbell, CA, USA) 3. (Cs 3600, Carestream Health, Rochester, NY, USA) 4. (Cs 9300, Carestream Health, Rochester, NY, USA) 5. (InEosXs; Dentsply Sirona, Charlotte, NC, USA) 6. (DDS-Pro; Dentalica Spa, Milano, Italy) + (Exocad; Exocad GmbH, Darmstadt, Germany)	1. Auxiliary equipment for registration 2. (1) Face with a maximum smile, face with DGB1, and face with mouth open and MICP (2) Face with DGB2 in three poses 3. Maxillary and mandibular dental arches 4. Bone and dental arches after placing DGB1 5. DGB1 devices 6. Virtual patient and virtual prosthetic planning	FS in three poses + IOS: fiducial markers + geometric reference-based (DGB with radiopaque landmark) + best-fitting algorithm	Digital dentistry Prosthetic dentistry: CAD/CAM Dental implant surgery	Feasibility Dynamic Inexpensive	1. The distortion caused by the processing of the original files and the matching method 2. Further studies are needed to validate its accuracy.
Noguchi N [43]/2007	Case report (Humans)	One subject	1. DS + FS 2. Cephalometry + digital radiograph system 3. A digitizer 4. 3D shape analysis software	1. (SURFLACER 3D-VMS250/300, UNISN Inc., Osaka, Japan) 2. (FCR, Fuji Film Co., Ltd., Tokyo, Japan) 3. (KW4610, Graphtec, Yokohama, Japan) 4. (SURFLACER 3D-VMP300 (UNISN Inc., Osaka, Japan) 5. (Imageware Surfacer, Metrix Software Solutions Ltd., Montreal, Canada)	1. (1) Dentition and occlusal impression; (2) facial soft-tissue 2. Data for the mandible 3. The traced bone, teeth, and soft tissue 4. (1) Virtual models integrating the data above (2) Movement displayed using a color map	Projection-matching technique: based on the contour line of the projection image	Orthodontics Orthognathic surgery	1. Feasibility 2. Accuracy: registration error was the same as that in conventional Cephalometry.	Further studies are needed to validate its accuracy.

Abbreviations: cone beam computed tomography (CBCT); intraoral scanners (IOS); desktop scanners (DS); facial scanners (FS); virtual facebows (VF); virtual articulator (VA); standard tessellation language (STL); the object code (OBJ); polygon (PLY); digital imaging and communications in medicine (DICOM); data communication module (DCM); moving picture experts group 4 (MP4); text file (TXT); the iterative closest point (ICP) algorithm; computer-assisted design (CAD); maximal intercuspation position (MICP); centric relation (CR); vertical dimension of occlusion (VDO); natural head position (NHP); Kois Facial Reference Glasses (KFRG); electromagnetic (EM); mean absolute deviations (MADs); the root mean square (RMS); standard deviations (SD); mandibular proximal segment (MPS); intraoral transfer element (IOTE); reverse engineering (RE) software; finite element (FE) software.

Table 3. Summary of the dynamic data.

Type of Method	Reference	Ways to Acquire Dynamic Data	Type of Dynamic Data
Target tracking video	[20]	Target tracking video-camera of a mobile phone with 4000-pixel (4K) resolution (iPhone 7; Apple Inc., Cupertino, CA, USA)	Mandibular movements: mandibular kinematic path and TMJ kinematic path
FS + targets	[21,31]	FS (Rexcan CS2, Medit Inc., Seoul, Republic of Korea) + lip and cheek retractor + nonreflective targets attached to incisors	Real-time mandibular motions
JMT (Jaw motion tracker)	[25,26]	EM tracking system (Aurora, Northern Digital Inc., Waterloo, ON, Canada) + skin-attached dynamic reference frame	3D tracking of the positions of the bone segment
	[18]	EM tracking (Polhemus, Burlington, VT, USA) Facebow-transfer Three rectangular coordinate systems (Cartesian)	Mandibular motion tracking Real-time surgical simulation
	[4]	Optoelectronic analysis system with 6 degrees of freedom (TRI-MET, Tokyo-Shizaisha, Tokyo, Japan) Image measurement software (3D-Rugle, Medic Engineering, Kyoto, Japan)	4D display of mandibular movement Condyle position relative to the condylar fossa Contact areas during jaw movements
	[29]	Ultrasonic tracking device with six degrees of freedom (TMJSim: TMJ movement) in (MedicalStudio1 platform)	Mandibular movement: translation and the rotation The centric position for both of the condyles Joint's degrees of freedom
	[35]	SICAT JMT + system (SICAT Function; SICAT, Bonn, Germany) (with ultrasonic tracking device)	Mandibular movements: opening, right and left lateral movement, and protrusion Incisor ranges movement of the mandible, including the translation of the condyles
	[38]	Ultrasonic axiograph Arcus Digma system (KaVo, Biberach, Biberach, Germany)	Mandibular movements
Facial tracking system	[23]	Facial tracking app (Live Face, v.1.08; Reallusion) in a smartphone (iPhone 12 Pro; Apple Inc., Cupertino, CA, USA)	Lip dynamics (including rest/"m" sound/smile/speech)
	[42]	FS (Bellus3D; Bellus3D Inc, Campbell, CA, USA) Geometric occlusal registration prototype device (DGB) (Digitalbite; Digitalsmile srl, Pietracamela, Italy)	Face with a maximum smile, with mouth open and MICP
	[34]	3D facial photograph Dolphin 3D Imaging	Lateral facial profile influenced by retrusion of anterior maxillary teeth.
	[43]	FS (SURFLACER 3D-VMS300, UNISN Inc., Osaka, Japan) Frontal and lateral cephalometry (FCR, Fuji Film Co. Ltd., Tokyo, Japan) Digitizer (KW4610, Graphtec, Yokohama, Japan)	The traced bone, teeth, and outline of the soft tissue

Table 3. *Cont.*

Type of Method	Reference	Ways to Acquire Dynamic Data	Type of Dynamic Data
Method to acquire positional relationships of jaws, skull, and TMJ	[22,30]	FS (InstaRisa Facial Scanner; InstaRisa, Clovis, CA, USA)/photograph Scan body system (Kois Scan Body System; Kois Center, LLC, Seattle, WA, USA) VA (Panadent Articulator; Panadent, Colton, CA, USA) CAD software	Dynamic facial information NHP Maxillomandibular registration and jaw-correction function
	[41]	Photograph + reverse engineering software (Agisoft Photoscan; Agisoft LLC, SPB, Russia) + targets VF + VA Reverse engineering software	Mandibular position in MICP 3D face Maxillary and mandibular dentiton transfered on VA
	[19,42]	FS: smartphone (iPhone 11 Pro; Apple Inc., Cupertino, CA, USA) with a 3D scan application (Hege 3D scanner)/((Bellus3D; Bellus 3D Inc., Campbell, CA, USA) DGB) VF fork CAD software	Facial information Occlusion + position of the maxilla
	[24]	FS (Bellus 3D; Bellus 3D Inc., Campbell, CA, USA) VF CAD software	Position of the maxilla CR position with the joint axis of the VA Occlusal discrepancy
	[32,33]	FS (3dMDface; 3dMD Inc, Atlanta, GA, USA)/(DAVID SLS-3; Hewlett-Packard, Palo Alto, CA, USA) VF CAD software	The digital teeth to the 3D facial photographs in NHP Maxillomandibular relationship with maxilla
	[37]	FS by the reference scanner, SWL scanner, and SFM scanner RE software program	Maxillomandibular relationship with the maxilla Face with the mouth open
	[40]	FS (Bellus3D; Bellus3D Inc., Campbell, CA, USA) Facebow record CAD software program	The definitive casts mounted on the VA to simulate the jaw position
	[27,28]	FS: (3dMDtrio System; 3dMD, Atlanta, GA, USA)/application (Hege 3D scanner) in (iPhone 11 Pro; Apple Inc., Cupertino, CA, USA) CBCT + gothic arch tracing + articulator CAD software, Implant planning software	Maxillary and the mandible arches aligned in a proximal CR and VDO 3D face
	[36]	CBCT VF VA (R2GATE 2.0.0; Megagen, Seoul, Republic of Korea)	Mandible position in both CRO and MICP
FE analysis system	[38]	FE modeling and analysis software Ansys 15.0 (ANSYS Inc., Canonsburg, PA, USA) Reverse engineering software (Rapidform 2006 (Inus Technology Inc., Seoul, Republic of Korea))	A dynamic model of the individualized masticatory system including the cranio-maxilla, the mandible, masticatory muscles, and 28 complete teeth The FE masticatory system model
	[39]	FE analysis software (FORGE (Transvalor, Glpre 2005, Antibes, France)) 3D image segmentation software (AMIRA®) (Visage Imaging, In, SD, CA, USA)	Boundary conditions for closing jaw simulations by different load directions of jaws muscles The stress distribution in both joint discs during closing conditions

Abbreviations are the same as that of Table 2.

3.4. Outcomes

3.4.1. Dynamic Data Collection Methods of 4D Virtual Patients

1. Mandibular movement (jaw motion)

Ten studies acquired real-time jaw-motion data. Only two mentioned the dynamic file format (Moving Picture Experts Group 4 (MP4) [20]/text file (TXT) [20,38]). One study captured videos of mandibular movements using a target-tracking camera [20]. Two studies introduced FS and targets to record jaw movements and the accumulated movement paths [21,31]. The kinematics of the occlusion/condyle can be simulated by combining jaw tracking with DS/CBCT. The accuracy of the FS + targets method, evaluated by the distance between the targets, showed a value of 4.1–6.9 mm (a minor error compared with that of laboratory scanners [78 mm]). Three studies used an electromagnetic (EM) system to track the positions of the maxillary bone segment (MBS) and the reference in physical space [25] or the mandibular proximal segment (MPS) and condyle position [26], or mandibular repositioning and occlusal correction [18]. One study used an optical analyzer to track the position of light-emitting diodes mounted on facebow and jaw movements [4]. Three studies acquired mandibular movements [29,35,38], condylar motions [29,38], and collision detection [38] by the ultrasonic system. The ultrasound Arcus Digma system has an accuracy of 0.1 mm and 1.5°, and its pulse running time was converted to 3D coordinate values and saved as TXT files [38].

2. Dynamic faces

One study [23] used a smartphone facial tracking app to track lip dynamics. Three studies acquired semi-dynamic faces using FS and image processing software. Semi-dynamic facial information refers to dynamic bits of facial details [42] or faces at different times [34,43] instead of real-time facial changes.

3. Positional relationships of the dentition, jaws, skull, and TMJ (jaw position)

Thirteen studies obtained positional relationships to help simulate jaw movements on the VA. First, the facial reference system helped to locate the natural head position (NHP), a reliable plane to align the skull and VA hinge axis [22,30]. The common VF techniques involve transferring the maxillary dentition to the FS, guided by an Intraoral transfer element (IOTE) (i.e., facebow fork) or a scan body, and to the FS without an IOTE/scan body [19,22,24,32,33,37,40,42]. A modified IOTE combined with LEGO blocks, trays, and impressions enabled a convenient transfer procedure [32,33]. The VF showed average trueness of 1.14 mm and precision of 1.09 mm based on the FS of a smartphone [19]. There is a difference between the virtual transferred maxillary position and its real position (a trueness of 0.138 mm/0.416 mm and a precision of 0.022 mm/0.095 mm were obtained using the structured white light (SWL)/structure-from-motion (SFM) scanning method). This difference is mainly caused by the registration error, which may be reduced by different alignment methods and IOTE [37]. Special VF transfers the relationship by anatomical points/marker planes, reducing the errors by omitting the traditional facebow transfer [36]. The FS can be replaced by photographs [30,41]/CBCT [27,28]; however, the accuracy was not calculated. Once VF transfer is completed, VA systems are available for jaw motion simulation.

4. Coordinated movement of the masticatory system

Two studies used finite element (FE) software to analyze mandibular movements, masticatory muscle performance, occlusal force [38], and stress distribution of TMJ [39].

3.4.2. 4D Superimposition Techniques of Virtual Patients

The main components of stomatognathic information include static information such as the skeletal components (SK) of the skull, jaws, and TMJ, dentition (DENT), dynamic information such as mandibular movement (MM), jaw position (JP), occlusal analyses (OA), and motion of TMJ soft tissues and muscles. Moreover, soft tissues of the face (SF) can be either in a dynamic or static form. Three to five of these components were combined to create virtual models based on 4D superimposition techniques. These models are created using various software systems.

1. Five types of information superimpositions

- SK + DENT + MM + SF + TMJ soft tissues/SK + DENT + MM + OA + muscles

Two studies created models [29,38] including the skeletal components from CT (DICOM format), dentitions from DS (STL format), and jaw movement from the Jaw motion tracker (JMT) (TXT format [38]). Lszewski et al. [29] added MRI's facial and TMJ soft tissues to the above data. The virtual model's superimposition and real scenes were based on an algorithm and fiducial markers. The module's accuracy requires further validation before clinical application. Dai et al. [38] included masticatory muscles and occlusal analyses in their model. The regional registration method was used to superimpose the dentition in the occlusion. Their FE model showed an accuracy ($\leq$0.5 mm) similar to that of the T-Scan.

2. Four types of information superimpositions

- SF + DENT + OA + MM/JP:

Two studies integrated facial tissues from FS, dentition from DS [21]/IOS [19], occlusal contact analyzed by the CAD software, and mandibular movement tracking from FS + targets [21]/jaw position from VF + VA [19]. The alignment of the maxillary cast, scan body, and facial tissues was based on the points of the teeth [21]. The superimposition of FS and CBCT was based on the facial surface. The deviation of their models was 1 mm in linear distance and 1° in angulation [19].

- SK + SF + DENT + JP:
- Six studies [24,27,28,32–34] constructed virtual patients mounted on the VA with faces from the FS (OBJ [24]/STL [27]/PLY [28] format)/photographs [34], bones from CBCT/CT (DICOM format), and dentition/prosthesis from the IOS (STL format). The models of three studies [24,27,28] were centric relation occlusion (CRO) and vertical dimension of occlusion (VDO). The prosthetic outcomes demonstrated a good fit, occlusion, and esthetics. Three studies created models based on points and fiducial markers [24,32,33]. One study [32] aligned the models with CBCT using the Iterative Closest Point (ICP) algorithm. The errors in tooth registration were less than 1 mm, whereas those of the nasion, alares, and tragions were 0.83 mm, 0.77 mm, and 1.70 mm, respectively. Further research is required to reduce this discrepancy and distortion. Granata et al. [42] created a virtual patient with faces in a smiling, open mouth, and maximum intercuspation (MICP) positions from FS (OBJ format), dental arches from IOS (PLY/STL format), and bones and jaw position from CBCT with the occlusal registration device Digitalbite (DGB). The superimposition was based on fiducial markers and a best-fitting algorithm.
- SK + SF + DENT + MM:

Kwon et al. [31] introduced a virtual patient with skeletons from CBCT, dentitions from DS, and a face combined with mandibular motions using FS + targets. The alignment was based on triangulated mesh points and a transformation matrix. The tracking system stability and reproduction were acceptable compared with routine VF transfer. Noguchi et al. [43] created virtual models integrating the mandible, TMJs, and outline of the soft tissue from cephalometry and dentition from DS and traced their movement. The projection-matching technique is based on the contour line of the projection image, and the registration error is the same as that in conventional cephalometry.

- SK + DENT + MM + TMJ soft tissues:

Savoldelli et al. [39] combined the bone components of jaws and TMJs, dental arches from multi-slice CT, and soft tissues of TMJs when the jaw was opened 10 mm. Then, the joint discs' boundary conditions and stress distribution were analyzed showing a high level of accuracy (stress levels of the model [5.1 MPa] were within the range of reported stress [0.85–9.9 Mpa]).

3. Three types of information superimpositions

- SK + DENT + MM:

Zambrana et al. [20] constructed a virtual patient with jaws and TMJs from CBCT (DICOM format), dentition with maxillomandibular relationship from IOS (STL format), and mandibular movements from a target tracking video (MP4 format). Registration was based on the surface and points. Five studies combined bones from CT [4]/MDCT [25,26]/CBCT [35]/cephalograms [18], dentition from DS [4,25,26]/CBCT [35], and positions of MBS [25]/MPS [26]/mandibular movement [4,18,35] from the JMT. Integrations [4,18,25,26,35] were based on fiducial markers, and some were combined with the least-squares method [4]/ICP algorithm [25,26]. The technique of Fushima et al. [18] showed high accuracy (the minor standard deviations (SD) <0.1 mm). The two methods showed no significant difference between the actual and measured positions [25]. The condylar landmark results showed high accuracy (differences between MPS models and those between CT models were 1.71 ± 0.63 mm and 1.89 ± 0.22 mm, respectively) [26]. He et al. [35] analyzed the condyle position, which was more accurate than CBCT, with records showing high precision over three days.

- SF/SK + DENT + JP:

One study [36] combined bones from CBCT with mandibular position from VA and VF and dentition in CRO/MICP from IOS to simulate a patient. Six studies [22,30,37,40,41] described methods to create models with faces from FS [22,23,37,40]/2D photograph [30,41], dentition models from IOS [22,23,30,41]/DS [37], and mandibular position using VF and VA. Kois et al. [30] used the "Align Mesh" tool to align data. The Alignment of the casts, scan body, facial scan, or alignment of the facebow forks was based on points [22,41]. IOS and CBCT/FS were registered based on points and the ICP algorithm [36,37] or fiducial markers of the scan body [23,40].

4. Software programs to create the virtual patients

An open-source program (Blender 3D; Blender Foundation, Amsterdam, Netherlands) can import MP4 dynamic information directly. The MP4 file can be transferred into TXT format using a direct linear transform (DLT) algorithm, which facilitates the integration [20], marking the reference points to align the models acquired from other CAD software programs using Python [19]. In this program, STL/PLY files can be transferred to the OBJ format to facilitate the fusion of different formats [23].

Exocad (Exocad; exocad GmbH, Darmstadt, Germany) is the most common multifunction CAD software used to build virtual patients. It can align the IOS/DS with the FS, with/without CBCT, guided by the scan body [22,40]/IOTE [19]/gothic arch tracer, wax [27,40]/DGB [42], positions NHP, transfers facebow, and finally integrates the model into the VA. Exocad can also analyze the occlusal discrepancy between MIP and CR [21,24] or design a tooth-supported template [28]. Finally, restorations can be created based on the VDO and the occlusal plane [34]. Other software programs for constructing the 4D virtual patients are presented in Table 4.

FE modeling and analysis software can convert 4D virtual models created by other software into numerical models and analyze dynamic/static components. The FE software commonly used for 4D virtual models is ANSYS (ANSYS Inc., Canonsburg, PA, USA) [38]/FORGE (Transvalor, Glpre 2005, Antibes, France) [39]. Processing software such as AMIRA (Visage Imaging, Inc., SD, CA, USA) are often used to obtain surface and volume meshes. The volumes of the anatomical components were input to FE software and meshed as the element; subsequently, the mesh quality and nodes' quality were verified. The accuracy of a FE model is determined using geometric models.

Table 4. Summary of the software to create 4D patients.

References	Software	Type	Possible Registration	Imported Data
[19,20,23,28]	(Blender 3D; Blender Foundation, Amsterdam, The Netherlands)	Free open-source CAD software	Point-based Direct linear transform (DLT) algorithm	CBCT (DICOM) IOS (STL) JMT video (MP4) FS (PLY + STL)
[19,21,22,24,27,28, 30,34,40,42]	(Exocad; exocad GmbH, Darmstadt, Germany)	CAD software	Point-based Fiducial markers-based Surface-based Best-fitting algorithm	CBCT (DICOM) IOS/DS (STL/PLY) FS (STL/PLY/OBJ) Photograph
[23]	(Zbrush 2022; Maxon, Friedrichsdorf, Ger-many)	3D modeling software	Fiducial markers-based	IOS/DS (OBJ)
[19,21]	(EzScan8, Medit Inc., Seoul, Republic of Korea)	The image registration software	Point-based Horn alignment algorithm ICP algorithm	CBCT (STL transferred from DICOM) DS (STL) FS (STL)
[32,33]	(MeshLab v1.3.3; Visual Computing Lab of the ISTICNR, Pisa, Italy)	Open-source software	Point-based Fiducial markers based ICP algorithm	CBCT, IOS, FS
[34]	(Dolphin Imaging & Management Solutions; Patterson Dental, Chatsworth, CA, USA)	3D Imaging software	Unknown	CT Facial photograph DS
[36]	(R2GATE 2.0.0; Megagen, Seoul, Republic of Korea)	Virtual articulator program	Point-based ICP algorithm	CBCT (DICOM) IOS (STL)
[38]	Rapidform 2006 (Inus Technology Inc., Seoul, Republic of Korea)	Reverse engineering software	Regional registration method Global coordinates-based	CT (DICOM) DS (VIVID) 3D motion tracking (TXT)
[38]	(ANSYS Inc., Canonsburg, PA, USA)	FE analysis software	*	*
[39]	(AMIRA) (Visage Imaging, Inc., SD, CA, USA)	3D Imaging software	Hounsfield-unit values-based Manually identification of anatomical structures	CT, MRI
[39]	FORGE (Transvalor, Glpre 2005, Antibes, France)	FE analysis software	*	*
[41]	Rapidform 2006 (Inus Technology Inc., Seoul, Republic of Korea)	RE software	Best-fit command Facial point-based	IOS 2D photograph
[25,26]	(Aurora, Northern Digital Inc., Waterloo, ON, Canada)	Orthognathic navigation systems	Fiducial markers based ICP algorithm	CBCT/MDCT (DICOM) DS (STL) IOS (STL)
[29]	MedicalStudio1 platform: ACRO	Orthognathic planning and navigation systems	Fiducial markers based Algorithm minimizing the mean square distance between the points	CT (DICOM) MRI DS

Table 4. *Cont.*

References	Software	Type	Possible Registration	Imported Data
[18]	ManMoS: (FASTRAK, Virtual Realities, LLC., League City, TX, USA) + (Polhemus, Colches-ter, VT, USA)	Orthognathic simulation systems	Fiducial markers based	Lateral and posteroanterior cephalograms 3D motion tracking
[35]	(SICAT, Bonn, Germany)	Orthognathic planning systems	Fiducial markers based	CBCT (DICOM) 3D motion tracking
[4]	(TRI-MET, Tokyo-Shizaisha, Tokyo, Japan)	Orthognathic planning systems	Fiducial markers based The least squares method	CT DS 3D motion tracking
[43]	(Imageware Surfacer, Metrix Software Solutions Ltd., Montreal, Canada)	Orthognathic planning systems	Projection-matching technique	Lateral and posteroanterior cephalograms DS FS 3D motion tracking

* The model of the FE software is based on the 3D models from other software, so its types of information and registration are not shown in the table. Abbreviations are the same as that of Table 2.

3.4.3. Clinical Applications of 4D Virtual Patients

The 4D virtual patient is built to apply to different clinical scopes.

1. Application in prosthetic dentistry and dental implant surgery

In this field, the 4D virtual patient mainly involves locating the jaw positions and condylar axes, obtaining functional data, simulating the mandible and condyle trajectories, and analyzing occlusion. Traditional restoration processes focusing only on static occlusion may lead to poor occlusal function and TMJs disorder. However, for digital workflows, VF techniques locate the position of jaws and condyles [19,22–24,27,28,30,32,33,37,40–42], and VA [22,24,27,28,30,36,40] or JMA [20,31,35] helps simulate or record patient-specific mandibular and TMJ kinematics. These procedures obtain the correct MICP, CRO, and VDO for coordinated dental implants and restorations in a stable position [24,27,28,36]. The dynamic occlusal analysis allows the detection of occlusal interference during eccentric movements to design anatomic prostheses [4,19,21]. Additionally, the condylar motion trajectory and mandibular movement pathway can help diagnose and treat TMJ diseases and facilitate occlusal reconstruction [20,31]. Furthermore, adding facial information to the virtual patient is beneficial to harmonizing the prosthesis with the face, ensuring an aesthetic effect [23,30,34,41,42]. CAD systems perform the above digital analysis and design, and finally, dock CAM for guide template and restoration fabrication. The innovative workflows of 4D virtual models are particularly suitable for complex implant rehabilitation [24,27,28] and complete denture restoration [40], resulting in excellent repair results with patient satisfaction.

2. Application in maxillofacial surgery

Traditional orthognathic surgery planning is effective but time-consuming, and many factors, such as the occlusal recording and mounting, affect the accuracy. Computer-assisted orthognathic surgery is an interdisciplinary subject that combines signal engineering, medical imaging, and orthognathic surgery to improve efficiency. In addition to facial esthetics, reducing mandibular spin is essential to obtaining stable skeletal and occlusal outcomes and preventing temporomandibular disorders. Therefore, studies have improved accuracy and minimized spin using 4D virtual models in surgical systems.

This article mainly included planning [4,18,29,35,36,43] and navigation systems [25,26]. The ACRO system integrates modules for planning, assisting surgery, and bringing information from virtual planning to the operating room [29]. The Aurora system (Aurora, Northern Digital Inc., Waterloo, ON, Canada) uses augmented reality (AR) to locate bone segments and condylar positions [25,26]. The 4D analysis system TRI-MET (Tokyo-Shizaisha, Tokyo, Japan) was used to simulate mandibular motion, condylar to articular fossa distance, and occlusal contact [4]. The mandibular motion tracking system (ManMoS), a communication tool for operational trial and error, predicts changes in occlusion and repeatedly determines mandibular position [18]. The SICAT system (SICAT, Bonn, Germany) can show the motion of the incisors and condyles during mandibular movement, thereby avoiding additional radiation exposure [35]. The simulation system based on cephalometry allows the location of 3D bone changes without CBCT, reducing radiation and errors in manual pointing [43].

3. Application in orthodontics

Most of the included papers have cross-disciplinary applications. The VF, VA, and JMA to locate the mandible and condyles are also crucial for orthodontics. This helps reconstruct stable and balanced occlusion in the optimum position and prevents recurrence and TMJ symptoms. The simulated position of the incisors, jaws, and soft tissue provides a visual treatment objective. As mentioned above, analysis of pre- and post-operative tissue changes [34], orthodontic-orthognathic planning [4,43], and dynamic facial information of virtual patients are also applicable to aesthetic orthodontics plans and outcomes.

Traditional orthodontic treatment often focuses on dentition and bone problems in three dimensions but ignores the improvement of mastication efficiency and TMJ health. Quantifying masticatory function is essential for occlusal evaluation and orthodontic tooth movement. The directly mentioned application of the dynamic patient models in orthodontics mainly involves the analysis of masticatory muscles, stress distribution in the articular disc, mandibular movements, and occlusion [38,39]. FE methods can simulate dynamic masticatory models, including muscle forces to the teeth, to determine the magnitude and direction of the bite force. High-resolution FE models analyzed the stress distribution and symmetry of TMJ and the boundary conditions of mandibular during the closure process. Further studies are expected to enable the prediction of various stress loads on the TMJ disc in the context of mandibular trauma, surgery, or dysfunction.

4. Discussion

With the development of computer-aided design/computer-aided manufacturing (CAD/CAM) technology, virtual patient construction, incorporating multi-modal data, has been widely used in multiple fields of dentistry. Building a 4D virtual dental patient using dynamic information is of great interest. The present review revealed the methods, manufacturer software programs, information, registration techniques, scopes, accuracy, and limitations of existing approaches to dynamic virtual patient construction. Data from multiple sources and formats were captured using various methods and programs. Specific alignment methods integrate various types of information to build 4D virtual patients in different clinical settings.

4.1. Dynamic Data Collection Methods

The JMT system, FS + targets, and target tracking camera were used to acquire real-time jaw motion data. The EM JMT system uses a magnetic sensor to track jaw motion, bone segment, and condyle positions and a receiver to detect movement, which is popular in minimally invasive surgeries [18,25,26]. Nevertheless, electromagnetic interference can affect the device's accuracy [44]. An optical JMT system can display condyles and mandibular movement trajectories. The limitation of this method is the strict conditions and motion restriction by a sizeable facebow [4,20]. Ultrasonic JMT systems transfer acoustic signals from the transmitter into spatial information to record movements, which may be vulnerable to environmental conditions [29,35,38]. FS or photographs combined with targets can simultaneously capture jaw motion and the face and show a small error [21,31,41]. The mobile phone's camera, connected to a marker board, captures movements inexpensively and conveniently but without a test of the accuracy [20]. These methods capture much information, such as mandibular movements and kinematics of the condyles, in all degrees of freedom (including excursive movements, maximum mouth opening, protrusion, and lateral excursions). However, few studies have reported real-time jaw motion accuracy and file format.

Integrating kinematic digital VF with VA makes capturing jaw movements with acceptable accuracy possible [19,37]. Most VA assembly procedures include a digital impression of dentition, occlusal recording, VF transfer of the maxilla position to the skull, and mounting the models to VA. Various software [19,22,24,27,28,30,33,34,36,37,40–42] include the VA procedure. Although this approach does not present motion in real-time, it is compatible with file formats and requires more information. The FE method accurately simulates the masticatory system [38,39]. In addition, studies on the fusion of dynamic facial information in 4D virtual patients are lacking.

4.2. 4D Superimposition Techniques

The 4D superimposition technique integrates three to five types of information to create dynamic patients. The technology's superimposition methods, software, and outcome varied among different research.

4.2.1. Superimposition Methods

Image fusion and virtual patient creation are based on selecting the corresponding marker for the superimposition of data from multiple sources. The construction of a simulated model may involve multiple alignment processes. The alignment methods in the included articles were mainly based on points [20–22,24,32,33,36,37,40,41], fiducial markers [4,18,22–27,29,31–33,35,40–42], surface [19,20], and anatomical structure [39,43]. Registration based on additional attached fiducial markers is also a point-based registration method. Some algorithms, such as the ICP [21,26,27,32,33,36,37], best-fit [19,22,29,41,42], and least squares methods, were used to help the registration processes [4]. The specific alignment techniques used in each study are presented in Table 2.

Real-time mandibular movement data integration was mainly based on the fiducial marker [4,18,26,27,29,35], but the specific integration principles were not described. The alignment step affected the final virtual model's accuracy. However, only two studies have evaluated the accuracy of alignment methods [32,34]. Therefore, further studies should introduce registration methods for the dynamic data of 4D patients and quantify the accuracy and optimization methods for every alignment step.

4.2.2. Software Programs

The 4D virtual model ensures that a comprehensive model contains the required data, while the FE analysis software chunks the complex model into simple units connected by nodes, facilitating simple algorithms for the analysis and interpretation of complex data. Although few FE software programs analyzed 4D models, several FE-related programs were used to analyze 3D models. For example, Hypermesh (Altair, Troy, MI, USA) is an important preprocessing software, and Abaqus (Abaqus Inc., Providence, RI, USA) is a common FE analysis software that interprets geometric models [45,46].

Currently, commercial software is available for dynamic dental virtual models; however, the principle is not specified, and the accuracy needs further improvement. Various computer software packages have made it convenient to use diverse clinical information. However, creating patient models by superimposing multi-modal data is still new. Integrating data from diverse file formats may be incompatible and inaccurate. Other limitations are tedious processes, requiring different hardware and software to acquire and analyze data, and expensive fees.

4.2.3. Outcome of the Technology

Most recent studies on 4D patients have included only one patient or phantom for feasibility exploration. Further studies are required to increase the sample size. In that regard, fourteen studies did not evaluate their results' reliability [4,20–24,27,28,30,34,36,40–42], six assessed different dynamic techniques' accuracy [19,29,31,33,35,37], five evaluated the simulated model's accuracy [18,25,26,38,39], and two reported the registration accuracy [32,43]. Owing to the differences in the sources, formats, integration methods, and use of different 4D virtual patients, there is still a need for a unified standard to assess 4D virtual patients' accuracy.

4.3. Clinical Applications

Depending on the clinical needs, 4D virtual patients integrate static virtual patients with dynamic information. The present review focuses on applying dynamic information of the virtual models: the establishment of occlusion in a stable mandibular and condylar position is of great benefit for restoration design, implant planning, orthodontic treatment, and orthognathic surgery. In addition to analysis of masticatory function and occlusal interference in the functional state, real-time jaw and joint movements are also useful for cause analysis and treatment of TMJ disorders and intraoperative navigation in implantology and orthognathic. The dynamic face of the virtual models facilitates the smile design. Overall, dynamic virtual patient models facilitate pre-treatment planning, intraoperative assessment, and stable, healthy, and aesthetic treatment outcomes for actual patients. Virtual patient construction achieves an intuitive presentation, facilitating communication and clinical decision-making in dentistry.

There are some limitations to this systematic review. First, the included studies were either non-randomized experimental studies or case reports. Few studies showed low overall risk. Therefore, the scientific level of clinical evidence is lacking. Appropriate statistical methods are needed to evaluate non-randomized experimental studies. The case reports lacked demographic characteristics, medical histories, and current clinical conditions. Second, the total sample size was 78, which needs further expansion. Third, there was considerable variation in subjects, interventions, outcomes, study design, and statistical methods across the included studies. Due to high heterogeneity, we did not conduct a meta-analysis but only qualitatively discussed the technique and application. Fourth, a few included articles were a series of related studies done by the same authors, which may increase the bias of the results. Although all criteria are met, such cases should be avoided in future studies. Finally, the number of included snowball papers was more than that obtained by electronic searches. There were no suitable articles in the weekly literature tracking. Thus, our search strategy needs to be improved.

High-quality clinical studies such as RCT should be conducted in 4D virtual patients to ensure an appropriate design, sufficient sample size, and less heterogeneity. Registration methods for dynamic data should be introduced and optimized. Future investigators should evaluate the accuracy of the currently available techniques to create 4D dynamic virtual patients. Furthermore, it is better to establish a unified evaluation standard conducive to quantitative analysis. In addition, integrating all of the required information within one system should also be considered.

5. Conclusions

Based on the included articles, the following conclusions were drawn:

- Dynamic data collection methods of 4D virtual patients include the JMT, FS + targets, and target tracking camera to acquire real-time jaw motion, VF and VA to simulate jaw position, facial tracking systems, and FE programs to analyze the coordinated movement of the masticatory system.
- Superimposition of the skeleton, TMJs, soft tissue, dentition, mandibular movement/position, and occlusion from different static/dynamic information collection devices in various file formats is feasible for 4D dental patients.
- Four-dimensional virtual patient models facilitate pre-treatment planning, intraoperative assessment, and stable, healthy, and aesthetic treatment outcomes in different clinical scopes of dentistry.
- There is a lack of well-designed and less heterogeneous studies in the field of 4D virtual patients.

- Further studies should focus on evaluating the accuracy of the existing software, techniques, and final models of dental dynamic virtual patients and developing a comprehensive system that combines all necessary data.

Supplementary Materials: The following supporting information can be downloaded at: https://www.mdpi.com/article/10.3390/jfb14010033/s1, Figure S1: Snowball procedure; Table S1: PRISMA 2020 checklist; Table S2: Joanna Briggs Institute Critical Appraisal Checklist for Quasi-Experimental Studies (non-randomized experimental studies); Table S3: Joanna Briggs Institute Critical Appraisal Checklist for case reports.

Author Contributions: Conceptualization, S.Y. and Y.Y.; methodology, formal analysis, and investigation, Y.Y. and Q.L.; resources and data curation, S.Y. and W.H.; writing—original draft preparation, review visualization, and editing, Y.Y.; supervision, project administration, W.H.; funding acquisition, S.Y. All authors have read and agreed to the published version of the manuscript.

Funding: This research was funded by the Research Initiation Program of Southern Medical University (grant number PY2018N094), Guangdong Medical Research Foundation (grant number A2020458). The APC was funded by Stomatological Hospital, Southern Medical University, the Guangdong Basic and Applied Basic Research Foundation (grant numbers 2021A1515111140 and 2021B1515120059), Research and Cultivation Program of Stomatological Hospital, Southern Medical University (grant number PY2018021).

Institutional Review Board Statement: Not applicable.

Informed Consent Statement: Not applicable.

Data Availability Statement: Not applicable.

Acknowledgments: The authors thank An Li for his great advice on this review.

Conflicts of Interest: The authors declare no conflict of interest.

Abbreviations

desktop scanners (DS); intraoral scanners (IOS); facial scanners (FS); cone beam computed tomography (CBCT); computed tomography (CT); virtual facebow (VF); virtual articulator (VA); jaw motion analyzer (JMA); electromagnetic (EM) system; intraoral transfer element (IOTE); structured white light (SWL); structure-from-motion (SFM); finite element (FE) software; reverse engineering (RE) software; augmented reality (AR); computer-assisted design (CAD); computer-aided manufacturing (CAM); mandibular motion tracking system (ManMoS); standard tessellation language (STL); object code (OBJ); polygon (PLY); digital imaging and communications in medicine (DICOM); moving picture experts group 4 (MP4); text file (TXT); skeletal components (SK); dentition (DENT); mandibular movement (MM); jaw position (JP); occlusal analyses (OA); soft tissues of the face (SF); three-dimensional (3D); four-dimensional (4D); temporomandibular joint (TMJ); maxillary bone segment (MBS); mandibular proximal segment (MPS); centric relation (CR); centric relation occlusion (CRO); vertical dimension of occlusion (VDO); maximum intercuspation (MICP); natural head position (NHP); iterative closest point (ICP) algorithm; direct linear transform (DLT) algorithm; preferred reporting items for systematic reviews and meta-analyses (PRISMA); medical subject headings (MeSH); Joanna Briggs Institute (JBI); standard deviations (SD).

References

1. Joda, T.; Bragger, U.; Gallucci, G. Systematic literature review of digital three-dimensional superimposition techniques to create virtual dental patients. *Int. J. Oral Maxillofac. Implant.* **2015**, *30*, 330–337. [CrossRef]
2. Joda, T.; Gallucci, G.O. The virtual patient in dental medicine. *Clin. Oral Implant. Res.* **2015**, *26*, 725–726. [CrossRef] [PubMed]
3. Perez-Giugovaz, M.G.; Park, S.H.; Revilla-Leon, M. 3D virtual patient representation for guiding a maxillary overdenture fabrication: A dental technique. *J. Prosthodont.* **2021**, *30*, 636–641. [CrossRef]
4. Terajima, M.; Endo, M.; Aoki, Y.; Yuuda, K.; Hayasaki, H.; Goto, T.K.; Tokumori, K.; Nakasima, A. Four-dimensional analysis of stomatognathic function. *Am. J. Orthod. Dentofac.* **2008**, *134*, 276–287. [CrossRef] [PubMed]

5. Mangano, C.; Luongo, F.; Migliario, M.; Mortellaro, C.; Mangano, F.G. Combining intraoral scans, cone beam computed tomography and face scans: The virtual patient. *J. Craniofac. Surg.* **2018**, *29*, 2241–2246. [CrossRef]
6. Lepidi, L.; Galli, M.; Mastrangelo, F.; Venezia, P.; Joda, T.; Wang, H.L.; Li, J. Virtual articulators and virtual mounting procedures: Where do we stand? *J. Prosthodont.* **2021**, *30*, 24–35. [CrossRef]
7. Ruge, S.; Quooss, A.; Kordass, B. Variability of closing movements, dynamic occlusion, and occlusal contact patterns during mastication. *Int. J. Comput. Dent.* **2011**, *14*, 119–127. [PubMed]
8. Mehl, A. A new concept for the integration of dynamic occlusion in the digital construction process. *Int. J. Comput. Dent.* **2012**, *15*, 109.
9. Schendel, S.A.; Duncan, K.S.; Lane, C. Image fusion in preoperative planning. *Facial Plast. Surg. Clin.* **2011**, *19*, 577–590. [CrossRef]
10. Page, M.J.; McKenzie, J.E.; Bossuyt, P.M.; Boutron, I.; Hoffmann, T.C.; Mulrow, C.D.; Shamseer, L.; Tetzlaff, J.M.; Akl, E.A.; Brennan, S.E.; et al. The PRISMA 2020 statement: An updated guideline for reporting systematic reviews. *BMJ Br. Med. J.* **2021**, *372*, n71. [CrossRef]
11. Shamseer, L.; Moher, D.; Clarke, M.; Ghersi, D.; Liberati, A.; Petticrew, M.; Shekelle, P.; Stewart, L.A. Preferred reporting items for systematic review and meta-analysis protocols (PRISMA-P) 2015: Elaboration and explanation. *BMJ Br. Med. J.* **2015**, *354*, i4086. [CrossRef] [PubMed]
12. The Preferred Reporting Items for Systematic Reviews and Meta-Analyses (PRISMA) Website. Available online: http://www.prisma-statement.org (accessed on 26 November 2022).
13. Li, A.; Thomas, R.Z.; van der Sluis, L.; Tjakkes, G.H.; Slot, D.E. Definitions used for a healthy periodontium-A systematic review. *Int. J. Dent. Hyg.* **2020**, *18*, 327–343. [CrossRef] [PubMed]
14. The Rayyan Website. Available online: https://rayyan.ai/reviews (accessed on 26 November 2022).
15. Ouzzani, M.; Hammady, H.; Fedorowicz, Z.; Elmagarmid, A. Rayyan-a web and mobile app for systematic reviews. *Syst. Rev. Lond.* **2016**, *5*, 210. [CrossRef] [PubMed]
16. Tufanaru, C.; Munn, Z.; Aromataris, E.; Campbell, J.; Hopp, L. Chapter 3: Systematic reviews of effectiveness. In *JBI Manual for Evidence Synthesis*; Aromataris, E., Munn, Z., Eds.; JBI: Adelaide, SA, Australia, 2020.
17. Moola, S.; Munn, Z.; Tufanaru, C.; Aromataris, E.; Sears, K.; Sfetcu, R.; Currie, M.; Qureshi, R.; Mattis, P.; Lisy, K.; et al. Chapter 7: Systematic reviews of etiology and risk. In *JBI Manual for Evidence Synthesis*; Aromataris, E., Munn, Z., Eds.; JBI: Adelaide, SA, Australia, 2020.
18. Fushima, K.; Kobayashi, M.; Konishi, H.; Minagichi, K.; Fukuchi, T. Real-time orthognathic surgical simulation using a mandibular motion tracking system. *Comput. Aided Surg.* **2007**, *12*, 91–104. [CrossRef]
19. Li, J.; Chen, Z.; Decker, A.M.; Wang, H.L.; Joda, T.; Mendonca, G.; Lepidi, L. Trueness and precision of economical smartphone-based virtual facebow records. *J. Prosthodont.* **2022**, *31*, 22–29. [CrossRef]
20. Zambrana, N.; Sesma, N.; Fomenko, I.; Dakir, E.I.; Pieralli, S. Jaw tracking integration to the virtual patient: A 4D dynamic approach. *J. Prosthet. Dent.* 2022; *in press*. [CrossRef]
21. Kim, J.-E.; Park, J.-H.; Moon, H.-S.; Shim, J.-S. Complete assessment of occlusal dynamics and establishment of a digital workflow by using target tracking with a three-dimensional facial scanner. *J. Prosthodont. Res.* **2019**, *63*, 120–124. [CrossRef]
22. Revilla-Leon, M.; Zeitler, J.M.; Kois, J.C. Scan body system to translate natural head position and virtual mounting into a 3-dimensional virtual patient: A dental technique. *J. Prosthet. Dent.* 2022; *in press*. [CrossRef]
23. Revilla-Leon, M.; Zeitler, J.M.; Blanco-Fernandez, D.; Kois, J.C.; Att, W. Tracking and recording the lip dynamics for the integration of a dynamic virtual patient: A novel dental technique. *J. Prosthodont.* **2022**, *31*, 728–733. [CrossRef]
24. Lepidi, L.; Galli, M.; Grammatica, A.; Joda, T.; Wang, H.L.; Li, J. Indirect digital workflow for virtual cross-mounting of fixed implant-supported prostheses to create a 3D virtual patient. *J. Prosthodont.* **2021**, *30*, 177–182. [CrossRef]
25. Kim, S.-H.; Lee, S.-J.; Choi, M.-H.; Yang, H.J.; Kim, J.-E.; Huh, K.-H.; Lee, S.-S.; Heo, M.-S.; Hwang, S.J.; Yi, W.-J. Quantitative augmented reality-assisted free-hand orthognathic surgery using electromagnetic tracking and skin-attached dynamic reference. *J. Craniofac. Surg.* **2020**, *31*, 2175–2181. [CrossRef]
26. Lee, S.-J.; Yang, H.J.; Choi, M.-H.; Woo, S.-Y.; Huh, K.-H.; Lee, S.-S.; Heo, M.-S.; Choi, S.-C.; Hwang, S.J.; Yi, W.-J. Real-time augmented model guidance for mandibular proximal segment repositioning in orthognathic surgery, using electromagnetic tracking. *J. Cranio Maxillofac. Surg.* **2019**, *47*, 127–137. [CrossRef]
27. Li, J.; Chen, Z.; Dong, B.; Wang, H.L.; Joda, T.; Yu, H. Registering maxillomandibular relation to create a virtual patient integrated with a virtual articulator for complex implant rehabilitation: A clinical report. *J. Prosthodont.* **2020**, *29*, 553–557. [CrossRef]
28. Li, J.; Att, W.; Chen, Z.; Lepidi, L.; Wang, H.; Joda, T. Prosthetic articulator-based implant rehabilitation virtual patient: A technique bridging implant surgery and reconstructive dentistry. *J. Prosthet. Dent.* 2021; *in press*. [CrossRef]
29. Olszewski, R.; Villamil, M.B.; Trevisan, D.G.; Nedel, L.P.; Freitas, C.M.; Reychler, H.; Macq, B. Towards an integrated system for planning and assisting maxillofacial orthognathic surgery. *Comput. Methods Programs Biomed.* **2008**, *91*, 13–21. [CrossRef]
30. Kois, J.C.; Kois, D.E.; Zeitler, J.M.; Martin, J. Digital to analog facially generated interchangeable facebow transfer: Capturing a standardized reference position. *J. Prosthodont.* **2022**, *31*, 13–22. [CrossRef] [PubMed]
31. Kwon, J.H.; Im, S.; Chang, M.; Kim, J.-E.; Shim, J.-S. A digital approach to dynamic jaw tracking using a target tracking system and a structured-light three-dimensional scanner. *J. Prosthodont. Res.* **2019**, *63*, 115–119. [CrossRef] [PubMed]
32. Lam, W.Y.; Hsung, R.T.; Choi, W.W.; Luk, H.W.; Pow, E.H. A 2-part facebow for CAD-CAM dentistry. *J. Prosthet. Dent.* **2016**, *116*, 843–847. [CrossRef] [PubMed]

33. Lam, W.; Hsung, R.; Choi, W.; Luk, H.; Cheng, L.; Pow, E. A clinical technique for virtual articulator mounting with natural head position by using calibrated stereophotogrammetry. *J. Prosthet. Dent.* **2018**, *119*, 902–908. [CrossRef] [PubMed]
34. Shao, J.; Xue, C.; Zhang, H.; Li, L. Full-arch implant-supported rehabilitation guided by a predicted lateral profile of soft tissue. *J. Prosthodont.* **2019**, *28*, 731–736. [CrossRef]
35. He, S.; Kau, C.H.; Liao, L.; Kinderknecht, K.; Ow, A.; Saleh, T.A. The use of a dynamic real-time jaw tracking device and cone beam computed tomography simulation. *Ann. Maxillofac. Surg.* **2016**, *6*, 113–119. [CrossRef] [PubMed]
36. Park, J.H.; Lee, G.-H.; Moon, D.-N.; Kim, J.-C.; Park, M.; Lee, K.-M. A digital approach to the evaluation of mandibular position by using a virtual articulator. *J. Prosthet. Dent.* **2021**, *125*, 849–853. [CrossRef] [PubMed]
37. Amezua, X.; Iturrate, M.; Garikano, X.; Solaberrieta, E. Analysis of the influence of the facial scanning method on the transfer accuracy of a maxillary digital scan to a 3D face scan for a virtual facebow technique: An in vitro study. *J. Prosthet. Dent.* **2021**, *128*, 1024–1031. [CrossRef]
38. Dai, F.; Wang, L.; Chen, G.; Chen, S.; Xu, T. Three-dimensional modeling of an individualized functional masticatory system and bite force analysis with an orthodontic bite plate. *Int. J. Comput. Assist. Radiol.* **2016**, *11*, 217–229. [CrossRef]
39. Savoldelli, C.; Bouchard, P.-O.; Loudad, R.; Baque, P.; Tillier, Y. Stress distribution in the temporo-mandibular joint discs during jaw closing: A high-resolution three-dimensional finite-element model analysis. *Surg. Radiol. Anat.* **2012**, *34*, 405–413. [CrossRef] [PubMed]
40. Perez-Giugovaz, M.G.; Mostafavi, D.; Revilla-Leon, M. Additively manufactured scan body for transferring a virtual 3-dimensional representation to a digital articulator for completely edentulous patients. *J. Prosthet. Dent.* **2021**, *128*, 1171–1178. [CrossRef] [PubMed]
41. Solaberrieta, E.; Garmendia, A.; Minguez, R.; Brizuela, A.; Pradies, G. Virtual facebow technique. *J. Prosthet. Dent.* **2015**, *114*, 751–755. [CrossRef]
42. Granata, S.; Giberti, L.; Vigolo, P.; Stellini, E.; Di Fiore, A. Incorporating a facial scanner into the digital workflow: A dental technique. *J. Prosthet. Dent.* **2020**, *123*, 781–785. [CrossRef]
43. Noguchi, N.; Tsuji, M.; Shigematsu, M.; Goto, M. An orthognathic simulation system integrating teeth, jaw and face data using 3D cephalometry. *Int. J. Oral Maxillofac. Surg.* **2007**, *36*, 640–645. [CrossRef]
44. Stella, M.; Bernardini, P.; Sigona, F.; Stella, A.; Grimaldi, M.; Fivela, B.G. Numerical instabilities and three-dimensional electromagnetic articulography. *J. Acoust. Soc. Am.* **2012**, *132*, 3941–3949. [CrossRef]
45. Chen, G.; Lu, X.; Yin, N. Finite-element biomechanical-simulated analysis of the nasolabial fold. *J. Craniofac. Surg.* **2020**, *31*, 492–496. [CrossRef] [PubMed]
46. Dai, J.; Dong, Y.; Xin, P.; Hu, G.; Xiao, C.; Shen, S.; Shen, S.G. A novel method to determine the potential rotational axis of the mandible during virtual three-dimensional orthognathic surgery. *J. Craniofac. Surg.* **2013**, *24*, 2014–2017. [CrossRef] [PubMed]

Journal of
Functional
Biomaterials

MDPI

Article

Response of Human Gingival Fibroblasts and *Porphyromonas gingivalis* to UVC-Activated Titanium Surfaces

Yin Wen [1,†], Hao Dong [1,†], Jiating Lin [1], Xianxian Zhuang [1], Ruoting Xian [1], Ping Li [1,*] and Shaobing Li [1,2,3,*]

1 Center of Oral Implantology, Stomatological Hospital, School of Stomatology, Southern Medical University, Guangzhou 510280, China
2 First Clinical Medical College, Xinjiang Medical University, Urumqi 830011, China
3 The First People's Hospital of Kashgar Region, Kashgar 844000, China
* Correspondence: ping_li_88@smu.edu.cn (P.L.); lishaobing@smu.edu.cn (S.L.)
† These authors contributed equally to this work.

Abstract: Ultraviolet (UV) photofunctionalization has been demonstrated to synergistically improve the osteoblast response and reduce biofilm formation on titanium (Ti) surfaces. However, it remains obscure how photofunctionalization affects soft tissue integration and microbial adhesion on the transmucosal part of a dental implant. This study aimed to investigate the effect of UVC (100–280 nm) pretreatment on the response of human gingival fibroblasts (HGFs) and *Porphyromonas gingivalis* (*P. g.*) to Ti-based implant surfaces. The smooth and anodized nano-engineered Ti-based surfaces were triggered by UVC irradiation, respectively. The results showed that both smooth and nano-surfaces acquired super hydrophilicity without structural alteration after UVC photofunctionalization. UVC-activated smooth surfaces enhanced the adhesion and proliferation of HGFs compared to the untreated smooth ones. Regarding the anodized nano-engineered surfaces, UVC pretreatment weakened the fibroblast attachment but had no adverse effects on proliferation and the related gene expression. Additionally, both Ti-based surfaces could effectively inhibit *P. g.* adhesion after UVC irradiation. Therefore, the UVC photofunctionalization could be more potentially favorable to synergistically improve the fibroblast response and inhibit *P. g.* adhesion on the smooth Ti-based surfaces.

Keywords: titanium; dental implants; gingival sealing; biofilm; gingival fibroblasts; *Porphyromonas gingivalis*; ultraviolet; photo-functionalization

Citation: Wen, Y.; Dong, H.; Lin, J.; Zhuang, X.; Xian, R.; Li, P.; Li, S. Response of Human Gingival Fibroblasts and *Porphyromonas gingivalis* to UVC-Activated Titanium Surfaces. *J. Funct. Biomater.* **2023**, *14*, 137. https://doi.org/10.3390/jfb14030137

Academic Editors: Pankaj Vadgama and Jayakumar Rangasamy

Received: 15 December 2022
Revised: 17 February 2023
Accepted: 21 February 2023
Published: 28 February 2023

1. Introduction

Titanium (Ti) has been the preferred choice for the fabrication of dental implants due to its excellence in biocompatibility, corrosion resistance, and mechanical properties [1,2]. In the Ti dental implant system, a stable transmucosal region of the abutments, necks, or soft tissue integration, plays a critical role in achieving long-term success in dental implant restorations [3,4]. Unfortunately, the peri-implant tissue is susceptible to bacterial invasion compared with the periodontal tissues of natural teeth and the bioinert of Ti materials may aggravate this adverse situation [5,6]. Hence, to achieve better soft tissue integration for Ti implants, efforts should be made to enhance gingival sealing and prevent bacterial aggression.

Various surface modification strategies have been applied to improve soft tissue integration, including machining, acid-etching, argon plasma, laser melting, and anodization [7]. However, most methods are insufficient to inhibit bacterial accumulation because either cells or microbes can be affected by the surface properties of the substrates [8,9]. From the aspect of cells, the anodized nano-Ti surface observed a better adhesion of gingival fibroblasts compared with other types of surfaces. Thereby, the anodization has been considered a promising technique for the surface modification of Ti materials [10,11]. On the other hand, the machined smooth Ti surface displayed less biofilm formation and more convenient plaque control than other surface designs. The surface is characterized as an arithmetic mean roughness value of fewer than 0.2 μm, which is currently recommended for abutments [12,13]. Thus, it would be challenging to fabricate a Ti abutment surface that synergistically improves cell adhesion and inhibits bacterial attachment.

Ultraviolet (UV) photofunctionalization is expected to overcome this issue. It refers to a phenomenon of modification of Ti surfaces that occurs after UV treatment, including the alteration of physicochemical properties and the improvement of bioactivity [14]. Current studies revealed that UV-activated Ti implant surfaces enhanced bone regeneration in vitro and in vivo [15,16]. Meanwhile, UV irradiation, which rarely changes the structure of the objects, has been widely used in surface sterilization for centuries, and it has become attractive for its potential application in light-based antimicrobial therapies [17–19]. Interestingly, the effect of UV photofunctionalization may depend on varying factors including the wavelength of UV light, the duration of UV irradiation, the types of materials, the surface properties of the materials, and even the type of organisms. For example, it was reported that Ultraviolet C (UVC; 100–280 nm) pretreatment generated more hydrophilic groups and achieved a higher level of cell adhesion than Ultraviolet A (UVA; 315–400 nm) [20]; the hydrophilic groups increased with a longer duration of UV irradiation [21]; UV pretreatment promoted osteoblastic differentiation of cells on Ti surface with anatase-enriched coating [22]. While many pieces of research have focused on UV irradiation, the effects of UV on Ti with different surface structures, especially the nano-structures, are rarely studied. Moreover, these studies paid little attention to the clinical problems of gingival healing and the accumulation of periodontal pathogens. To the best of our knowledge, it remains unclear how human gingival fibroblasts (HGFs) and *Porphyromonas gingivalis* (*P. g.*) react to the UVC-activated Ti with different surfaces and whether such a response would be beneficial for soft tissue integration.

This study aimed to investigate the effect of UVC photofunctionalization on the biological behaviors of HGFs and *P. g.* on different Ti-based surfaces, which corresponded to two representative types of surface design that are currently applied in commercial abutments. The cellular response of HGFs on different Ti surfaces with related molecular mechanisms concerning the expression of focal adhesion was investigated. Additionally, the response of *P. g.* as well as the co-culture of HGFs with pre-accumulated *P. g.* on different surfaces were also assessed.

2. Materials and Methods

2.1. Specimen Preparation

Ti discs (commercial grade IV) with a diameter of 15 mm and thickness of 1 mm were used for the following surface treatments. The smooth surfaces were obtained by polishing with multiple sandpapers (Hermes-abrasives, Hamburg, Germany) up to P4000 and silicon rubber burs (0147, Qumo, Wuhan, China), then cleaned ultrasonically (KQ-700DE, Shumei, Kunshan, China) in acetone (GCRF, Guangzhou, China), ethanol (GCRF, Guangzhou, China), and ultrapure water (GOKU-B1, EWLL BIO, Guangzhou, China) sequentially, and then dried at 25 °C, labeled S. The nano-surfaces were obtained by anodization (labeled N), as Wu et al. previously reported [23]. Specifically, the polished and clean samples were immersed in an electrolyte of 0.5% hydrofluoric acid (Macklin, Shanghai, China) for 15 min at 20 V under a direct current power supply system (DP3303D, MESTEK, Shenzhen, China). The UVC pretreatment was performed on the smooth or nano-Ti surfaces using a UVC bactericidal lamp (TUV30W, Philips, Amsterdam, Netherland) with an intensity of 107 $\mu W/cm^2$ (k = 253.7 nm) for 24 h at 25 °C, as Li et al. previously reported [24]. Prepared samples were divided into four groups: (1) S group: smooth Ti surfaces; (2) S + UVC group: smooth Ti surfaces with UVC pretreatment; (3) N group: nano-Ti surfaces; and (4) N + UVC group: nano-Ti surfaces with UVC pretreatment.

2.2. *Surface Characterization*

The physicochemical properties of the samples were characterized, as mentioned previously [25,26]. The surface topography was analyzed by field-emission scanning electron microscopy (FE-SEM; Sigma 300, Zeiss, Oberkochen, Germany). The surface roughness was assessed by atomic force microscopy (AFM; Bruker dimension icon, Bruker, Karlsruhe, Germany). The surface chemistry of the samples was investigated by energy-dispersive X-ray spectroscopy (EDS, XFlash 6 | 30, Bruker, Karlsruhe, Germany) and X-ray photoelectron spectroscopy (XPS; K-Alpha, Thermo Fisher Scientific, Waltham, MA, USA). Wettability was measured by static water contact angles (WCA) on the sessile-drop method measuring device (DSA XROLL, Betops, Guangzhou, China).

2.3. *Response of Human Gingival Fibroblasts*

2.3.1. Cell Culture and Seeding

Cell cultivation followed the protocols provided by Gibco®. HGF (CRL-2014; ATCC, Manassas, VA, USA) was obtained commercially and cultured in Dulbecco's modified eagle medium (DMEM; Gibco, Waltham, MA, USA), which was supplemented with 10% fetal bovine serum (FBS; Gibco, Waltham, MA, USA) and 1% penicillin–streptomycin (Gibco, Waltham, MA, USA) and incubated at 37 °C, 5% CO_2. The medium was changed every 2 days. Passages 3–7 were used in the following experiments.

2.3.2. Initial Cell Adhesion and Cell Proliferation

Initial cell adhesion was evaluated by nucleus counting following the protocol provided by Meilun. HGFs were seeded at 1×10^4 cells/cm^2 on the samples in 24-well plates. After 0.5, 1, and 2 h of incubation, samples were rinsed gently with phosphate-buffered saline (PBS; Gibco, Waltham, MA, USA) twice following the fixation in 4% paraformaldehyde (PFA; Meilun, Dalian, China) for 15 min and staining with 1 μM 4′,6-diamidino-2-phenylindole (DAPI; Meilun, Dalian, China) for 10 min at 25 °C. Cells were observed and documented using an inverted fluorescence microscope (DMIL LED, Leica, Wetzlar, Germany).

The tetrazolium-based assay was used to assess the relative cell proliferation, as followed by Dojindo Laboratories. HGFs were seeded at 2×10^4 cells/cm^2 on the samples in 24-well plates. After 1, 4, and 7 d of incubation, the samples were washed in PBS twice and added to 600 μL of fresh medium plus 60 μL of CCK-8 solution (Dojindo, Kumamoto, Japan) per well for 3 h of incubation at 37 °C, 5% CO_2. The acquired fluid samples were collected in a 96-well plate (100 μL per well) and then measured by a microplate reader (SpectraMax Plus384, Molecular Devices, San Jose, CA, USA) at 450 nm. To quantify the relative cell proliferation, the samples without UVC pretreatment were set as the controls.

2.3.3. Cell Spreading with Focal Adhesion

The morphology of cell spreading with the distribution of focal adhesions was visualized by immunofluorescence staining following the experimental protocol provided by Proteintech®. HGFs were seeded at 2×10^4 cells/cm^2 on the samples in 24-well plates. After 2 and 24 h of incubation, cells were fixed with 4% PFA for 15 min, permeabilized with 0.2% Triton X-100 solution for 5 min, blocked with 5% goat serum for 30 min at 37 °C, and then incubated with the anti-focal adhesion kinase (FAK) primary antibody (1:100, Proteintech, Wuhan, China) overnight at 4 °C. After that, samples were incubated with the Alexa Fluor® 488-labeled secondary antibody (1:2000, AAT Bioquest, Pleasanton, CA, USA) and Alexa Fluor® 594-Phalloidin (1:1000, AAT Bioquest, Pleasanton, CA, USA) for 1 h, respectively. Finally, samples were immersed in a mounting medium with DAPI. Images were captured by a confocal laser-scanning microscope (STELLARIS, Leica, Wetzlar, Germany). Biometric analysis was performed based on these images.

2.3.4. Adhesion-Related Gene Expression

The reverse transcription–quantitative polymerase chain reaction (RT-qPCR) assay was performed to investigate the adhesion-related gene expression levels of HGFs, which followed the Minimum Information for Publication of Quantitative Real-Time PCR Experiments (MIQE) guidelines [27]. HGFs were seeded at 2×10^5 cells/cm^2 on the samples in 24-well plates. After 2 and 24 h of incubation, total RNAs were isolated using Trizol reagent (Accurate Biology, Changsha, China), and cDNAs were generated using the Reverse Transcription Mix Kit (Accurate Biology, Changsha, China). Quantitative polymerase chain reaction (qPCR) was conducted using SYBR Green Premix Kit (Accurate Biology, Changsha, China) on a real-time PCR system (LightCycler 96, Roche, Basel, Switzerland). The relative changes in mRNA expression determined from qPCR experiments were calculated by the $2^{-\Delta\Delta CT}$ method. Glyceraldehyde-3-phosphate dehydrogenase (GAPDH) was used as the reference gene. The groups without UVC pretreatment were set as the controls. The primers (Tsingke Biotechnology, Beijing, China) used are shown in Table 1.

Table 1. Primers used for a real-time polymerase chain reaction in this study.

Gene	Forward Primer Sequence (5′-3′)	Reverse Primer Sequence (3′-5′)
FAK	GCTTACCTTGACCCCAACTTG	ACGTTCCATACCAGTACCCAG
Integrin β_1	CCTACTTCTGCACGATGTGATG	CCTTTGCTACGGTTGGTTACATT
Vinculin	CGAATCCCAACCATAAGCAC	CGCACAGTCTCCTTCACAGA
GAPDH	GGAGCGAGATCCCTCCAAAAT	GGCTGTTGTCATACTTCTCATGG

2.4. *Response of Porphyromonas gingivalis*

2.4.1. Preparation of *Porphyromonas gingivalis*

Microbial culture followed the protocols provided by AOBOX. *P. g.* (33277, ATCC, Manassas, VA, USA) and was maintained in the brain heart infusion (BHI) broth (AOBOX, Beijing, China) that was supplemented with 5 µg/mL hemin (AOBOX, Beijing, China) and 5 µg/mL vitamin K (AOBOX, Beijing, China) and incubated in an anaerobic jar (C-31, MGC, Tokyo, Japan) at 37 °C. After incubation, the *P. g.* suspension was adjusted to an optical density $(OD)_{600} = 0.15$ by the microplate reader for counting the colony-forming units (CFUs), as Di Giulio et al. previously described [28].

2.4.2. Bacterial Viability and Biofilm Formation

The viability of *P. g.* was evaluated by live/dead bacterial staining following the protocols provided by AAT Bioquest. *P. g.* strains were seeded at 1×10^8 CFU/mL of 2 mL on the samples in 24-well plates for 24 h. After incubation, samples were gently washed with saline (Kelun, Chengdu, China) and stained using a Live/Dead Bacterial Assay Kit (AAT Bioquest, Pleasanton, CA, USA). Images were recorded by the inverted fluorescence microscope at an emission wavelength of 530 nm for MycoLight™ Green and 660 nm for propidium iodide.

Biofilm formation was measured by crystal violet staining, as Kamble et al. previously reported [29]. Specifically, *P. g.* were seeded at 1×10^8 CFU/mL of 2 mL on the samples in 24-well plates for 24 h. After incubation, samples were fixed with 99% methanol for 15 min and submerged in 0.5% crystal violet solution (Macklin, Shanghai, China) for 20 min, following extensive washing with ultrapure water to remove the residual dye. After drying out, the samples were added with 33% acetic acid (600 µL per well) and gently shaken for 30 min to release the bound crystal violet. Finally, the OD values were measured at 590 nm.

2.5. *Co-Culture of Human Gingival Fibroblasts and Porphyromonas gingivalis*

The adhesion of HGFs under *P. g.* pre-accumulation was explored by a co-culture model using the established protocol provided by Gao et al. [30]. *P. g.* suspended in BHI was priorly added onto the samples at 1×10^8 CFU/mL of 2 mL in 24-well plates and incubated in an anaerobic jar at 37 °C for 60 min. Subsequently, HGFs were seeded at

2×10^4 cells/cm^2 in DMEM supplemented with 10% FBS and 2% BHI without antibiotics at 37 °C, 5% CO_2, for 2 h. After incubation, the samples were assessed by fluorescence staining with Alexa Fluor® 594-Phalloidin and DAPI, which was processed as mentioned above.

2.6. Statistical Analysis

Each experiment was performed in triplicates. Data were analyzed by SPSS 25. 0 (SPSS, Chicago, IL, USA). Statistical difference was analyzed using the student's t test. The level of p-value < 0.05 was considered significant.

3. Results

3.1. Surface Characterization

The physicochemical properties of all the groups are summarized in Table 2.

Table 2. Physiochemical properties of the samples.

Physiochemical Properties		Groups			
		S	S + UVC	N	N + UVC
Nanotube diameter (nm)		-	-	95.66 ± 9.52	94.52 ± 8.06
Surface roughness (nm)	Sa	13.85 ± 5.05	18.00 ± 6.20	27.10 ± 3.91	23.77 ± 7.02
	Sq	18.07 ± 6.85	23.60 ± 8.47	33.80 ± 4.55	30.13 ± 8.58
EDS atomic percentage (%)	O	16.11 ± 1.12	17.99 ± 0.88	47.99 ± 0.76 ^{B}	50.62 ± 0.80 ^{b}
	C	4.35 ± 0.26	3.81 ± 0.46	1.45 ± 0.03	1.55 ± 0.06
	Ti	79.54 ± 1.35	78.20 ± 1.10	50.56 ± 0.78 ^{B}	47.83 ± 0.85 ^{b}
XPS atomic percentage (%)	O 1s	50.94 ± 0.38 ^{A}	57.05 ± 1.19 ^{a}	44.77 ± 0.25	46.23 ± 0.52
	C 1s	33.35 ± 1.63 ^{A}	24.35 ± 0.34 ^{a}	34.15 ± 0.99	33.57 ± 1.85
	Ti 2p	16.76 ± 0.79	18.60 ± 1.34	20.09 ± 0.13	20.14 ± 0.36
Water contact angle (°)		103.20 ± 0.93	≈0	47.75 ± 0.38	≈0

Sa: arithmetic mean height; Sq: root mean square height. The labeled superscript letters within a row indicate a statistical difference in values between smooth Ti (A) and smooth + UVC Ti (a) groups, or between nano-Ti (B) and nano + UVC Ti (b) groups ($p < 0.05$).

The surface roughness of the smooth-based Ti was no more than 20 nm in Sa while the nano-Ti was slightly rougher, which turned out to be 20–30 nm in Sa (Table 2). The smooth-based Ti surfaces exhibited a smooth, flat, and scratch-like pattern with parallel polishing lines in low magnification (2000×) (Figure 1A). Under high magnification (50,000×) (Figure 1C), some irregular pits were distributed among the polishing lines. By contrast, the scratching texture was not observed on the nano-based Ti surfaces in low magnification (Figure 1B). Under high magnification, it displayed a wide range of tubular structures in tight and orderly arrangement (Figure 1D). Each unit of the structures was approximately 90 nm in diameter (Table 2), like empty nano-scale test tubes that opened at the top and closed at the bottom, suggesting the existence of TiO_2 nanotube arrays. No significant changes in topography or roughness were observed on smooth or nano-Ti surfaces after UVC pretreatment.

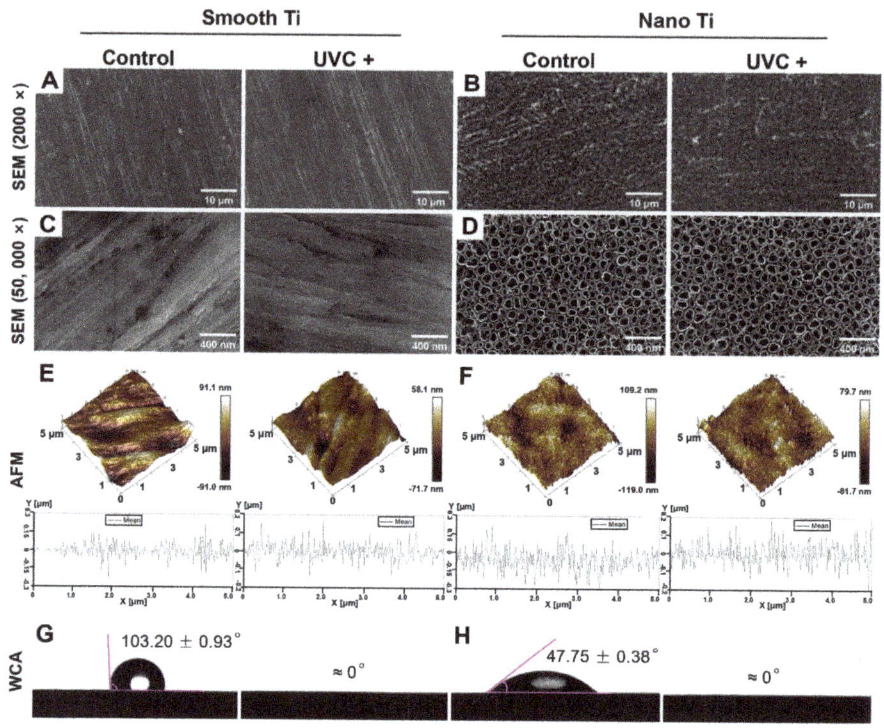

Figure 1. Surface characterization of samples after UVC irradiation: representative SEM images of (**A**) smooth-based Ti (magnification 2000×), (**B**) nano-based Ti (magnification 2000×), (**C**) smooth-based Ti (magnification 50,000×), and (**D**) nano-based Ti (magnification 50,000×); AFM images with the surface roughness profile at a range of 5 μm × 5 μm for (**E**) smooth-based Ti and (**F**) nano-based Ti before and after UVC irradiation; representative images of water contact angle measured on (**G**) smooth-based Ti and (**H**) nano-based Ti, before and after UVC irradiation.

Surface wettability was measured by the static WCA based on the established principle: WCA > 90° was described as hydrophobic; WCA < 90° was considered hydrophilic; and WCA that was very close to 0° ascribed to the superhydrophilicity [31]. As shown in Table 2 and Figure 1G,H, the static WCA was 103.20° (±0.93) on the smooth Ti and 47.75° (±0.38) on the nano-Ti. After UVC pretreatment, both smooth and nano-Ti surfaces displayed overspreading of the water droplets with a static WCA at 0°. Therefore, the smooth Ti surfaces were hydrophobic while the nano-Ti surfaces were hydrophilic. After UVC pretreatment, both surfaces acquired superhydrophilicity.

The main chemical compositions of the samples were all related to the O, C, and Ti elements, which were measured by the EDS and XPS analysis (Table 2). EDS showed that both the S + UVC group (Figure 2A) and the N + UVC group (Figure 2B) exhibited an increased content of oxygen and a decreased content of carbon compared with the control groups. This result was confirmed by XPS. As illustrated in the peak fitting images of XPS, the O 1s peak (Figure 3A,B) was resolved into two individual peaks. The major peak at 530.00 eV corresponded to the TiO_2 phase, and the second peak located at a binding energy of 532.15 eV was attributed to the hydroxyl group. The C 1s peak (Figure 3C,D) at approximately 284.8 eV was ascribed to the carbon contaminants. After UVC pretreatment, both smooth and nano-Ti surfaces displayed a higher peak of a hydroxyl group from O 1s peaks (Figure 3A,B) and lower C 1s peaks (Figure 3C,D) compared with the untreated samples. Therefore, after UVC pretreatment, both surfaces exhibited a higher proportion of hydroxyls and a lower percentage of carbohydrate contaminants compared with the untreated ones.

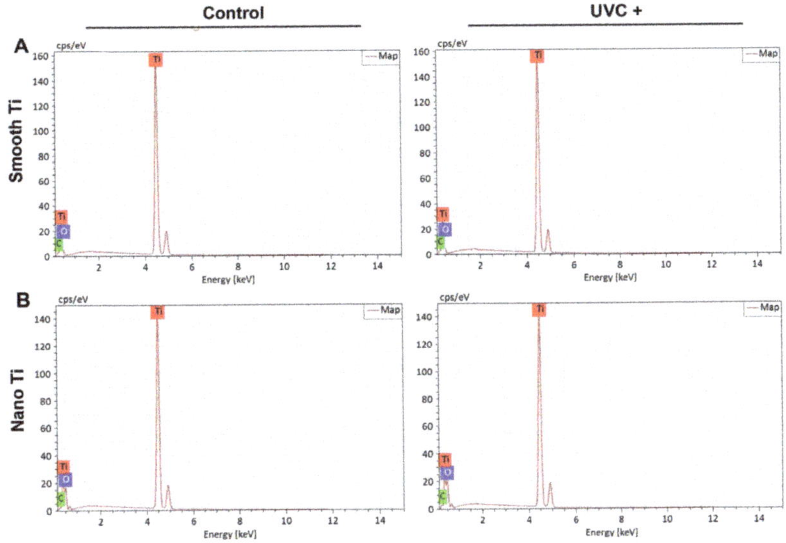

Figure 2. Element peaks of EDS analysis for (**A**) smooth Ti and (**B**) nano-Ti.

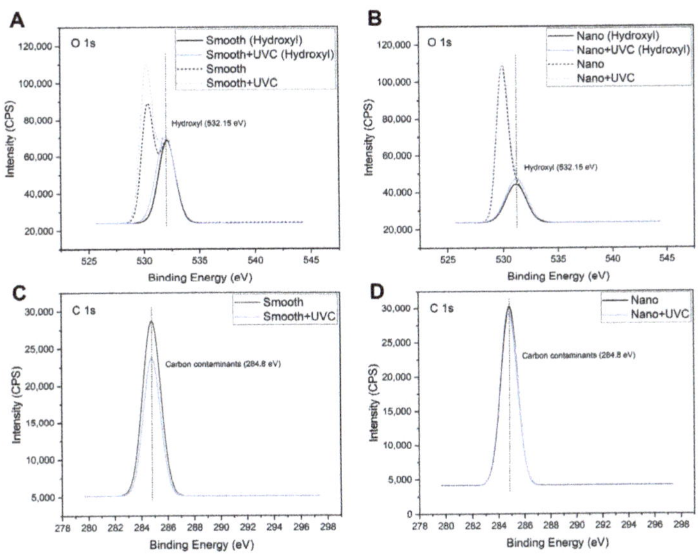

Figure 3. Peak fitting of XPS for the samples: (**A**) O 1s for smooth-based Ti, (**B**) O 1s for nano-based Ti, (**C**) C 1s for smooth-based Ti, and (**D**) C 1s for nano-based Ti.

3.2. Response of Human Gingival Fibroblasts

3.2.1. Initial Cell Adhesion and Cell Proliferation

After 0.5, 1, and 2 h of incubation, the S + UVC group (Figure 4A) was found to have a significantly greater number of adhered HGFs compared with the S group ($p < 0.001$ at 0.5 and 1 h; $p = 0.018$ at 2 h) (Figure 4C). In contrast, the N + UV group (Figure 4B) displayed a lower number of adhered HGFs compared with the N group at 0.5, 1, and 2 h, with significant differences at 0.5 h ($p < 0.001$) and 1 h ($p = 0.005$) (Figure 4D).

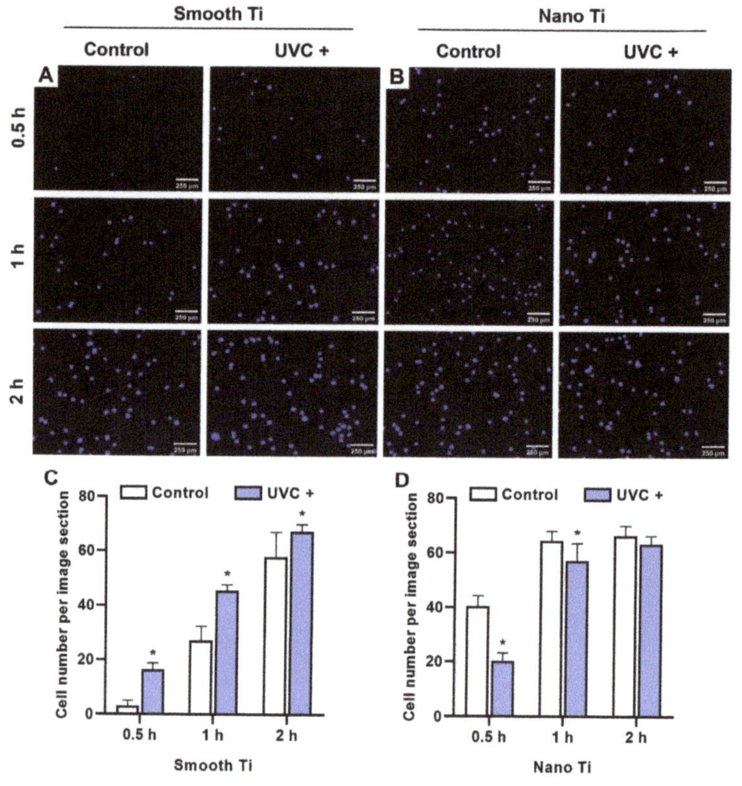

Figure 4. Adhesion of HGFs on the samples. Representative fluorescent images of (**A**) smooth-based Ti and (**B**) nano-based Ti for 0.5, 1, and 2 h, stained by DAPI. Quantitative analysis of adhered cells per image section (2.5 mm^2) of (**C**) smooth-based Ti and (**D**) nano-based Ti for 0.5, 1, and 2 h. * represents $p < 0.05$ when compared to the control.

The relative cell proliferation of HGFs in the S + UVC group (Figure 5A) was significantly higher than the S group for 1 d ($p = 0.011$), 4 d ($p < 0.001$), and 7 d ($p < 0.001$). By contrast, the results showed no significant difference between N + UVC and N groups (Figure 5B) in the relative proliferation rate at 1, 4, and 7 d ($p > 0.05$).

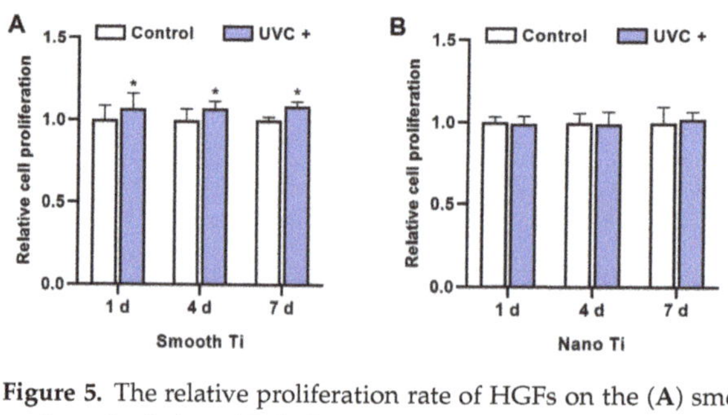

Figure 5. The relative proliferation rate of HGFs on the (**A**) smooth-based and (**B**) nano-based Ti surfaces for 1, 4, and 7 d, determined by CCK-8 assay (λ = 450 nm). * represents $p < 0.05$ when compared to the control.

3.2.2. Cell Spreading with Distribution of Focal Adhesion

After adhering to the Ti surfaces, HGFs spread with a rearrangement of the cytoskeleton (in red) and an expression of the focal adhesive structures (in green). Representative fluorescent images for 2 h of incubation time showed that HGFs in the S + UVC group (Figure 6B) extensively stretched out with thick actin filaments and a peripheral protrusion of filopodia. In contrast, the S group (Figure 6A) exhibited HGFs with poorly elongated shapes. Similarly, HGFs in the N + UVC group (Figure 6D) extended better than in the N group (Figure 6C), in which most of the cells still showed a shrunk and non-elongated morphology. A biometric analysis showed that the cell area of HGFs at 2 h (Figure 6I) was significantly larger in the S + UVC group ($p < 0.001$) or N + UVC group ($p = 0.002$) compared with the untreated groups. After 24 h of incubation, HGFs on all the surfaces (Figure 6E–H) were thoroughly elongated in a spindle shape with more protruded filopodia and a higher expression of focal adhesive structures, especially at the leading edge of the protrusion. No significant differences in cell morphology were observed among these groups at 24 h. Therefore, UVC pretreatment promoted the early spreading of HGFs, both on smooth and nano-Ti surfaces.

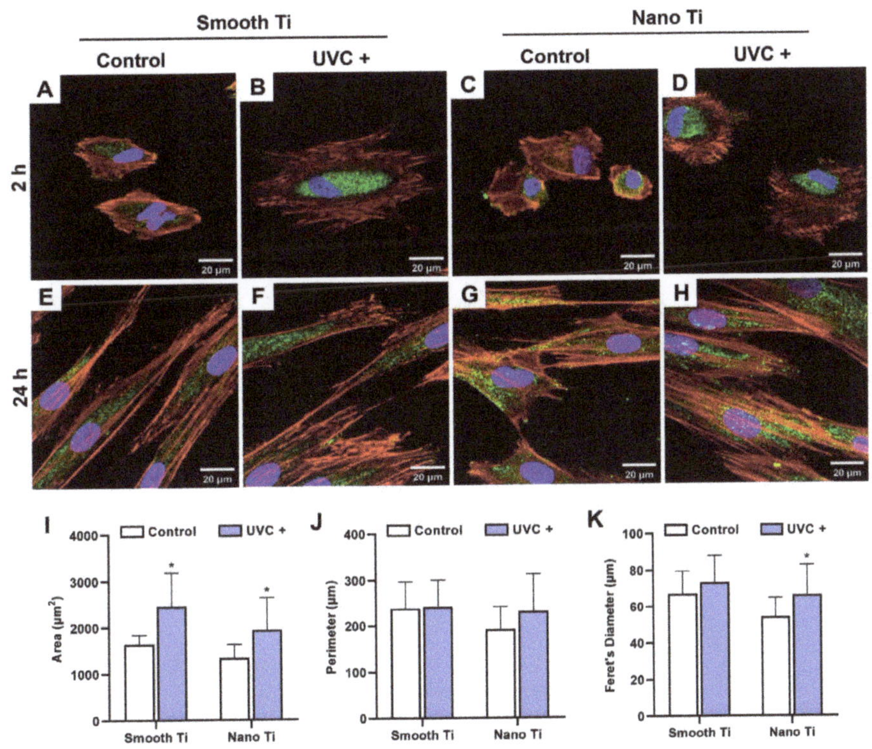

Figure 6. The spreading of HGFs on the samples with immunofluorescent staining for cytoskeleton with phalloidin (red), FAK (green), and nuclei with DAPI (blue): representative confocal microscopic images of the spread of HGFs at 2 h for (**A**) smooth Ti, (**B**) smooth + UVC Ti, (**C**) nano-Ti, and (**D**) nano + UVC Ti, and at 24 h for (**E**) smooth Ti, (**F**) smooth + UVC Ti, (**G**) nano-Ti, and (**H**) nano + UVC Ti; biometric analysis of (**I**) cellular area, (**J**) perimeter, and (**K**) Feret's diameter at 2 h. * represents $p < 0.05$ when compared to the control.

3.2.3. Adhesion-Related Gene Expression

After 2 h of incubation, the S + UVC group (Figure 7A) displayed a higher mRNA expression of FAK, integrin β_1, and vinculin compared with the S group, with significant differences in integrin β_1 ($p = 0.009$) and vinculin ($p < 0.001$). The N + UVC group (Figure 7B) exhibited insignificantly different mRNA levels of FAK and integrin β_1 compared with the N group, although a higher level of vinculin was observed ($p = 0.180$). After 24 h, the S + UVC group (Figure 7C) showed higher mRNA levels of integrin β_1 ($p = 0.133$) and vinculin ($p = 0.105$), with a lower mRNA level of FAK ($p = 0.391$) compared with the S group, while the N + UVC group (Figure 7D) exhibited similar mRNA levels of FAK, integrin β_1, and vinculin ($p = 0.105$). It should be noted that no significant differences were observed in all the groups regarding the mRNA expressions of FAK, integrin β_1, and vinculin at 24 h (Figure 7C,D).

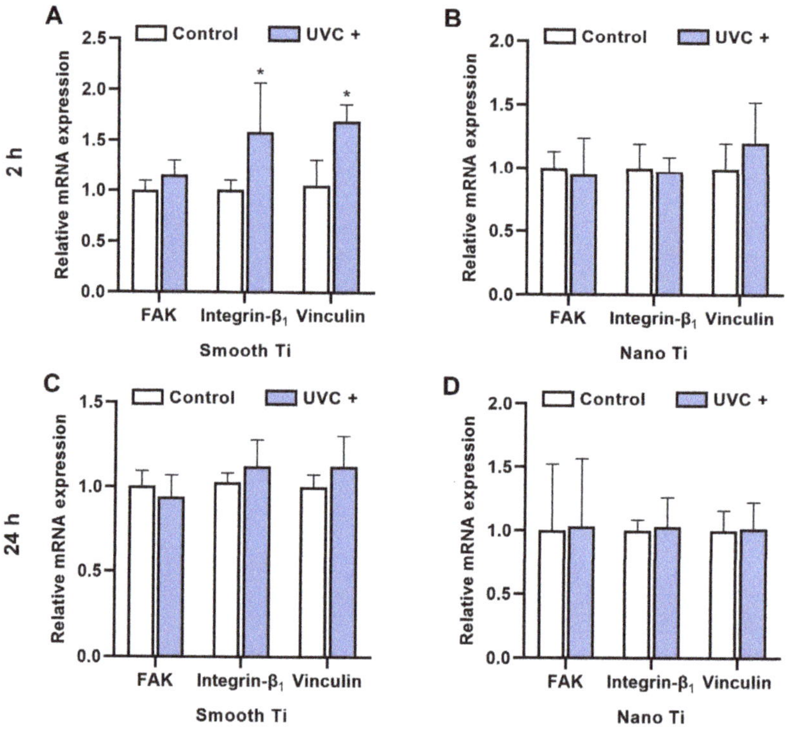

Figure 7. Relative mRNA expression (FAK, integrin β1, and vinculin) of HGFs on the samples of (**A**) smooth-based Ti and (**B**) nano-based Ti for 2 h, (**C**) smooth-based Ti, and (**D**) nano-based Ti for 24 h. * represents $p < 0.05$ when compared to the control.

3.3. Response of Porphyromonas gingivalis

The adhered *P. g.* was stained by live/dead assay after 24 h of incubation (Figure 8A). After UVC pretreatment, both smooth Ti and nano-Ti surfaces (Figure 8B) demonstrated a reduction in the live/dead rate of *P. g.* ($p < 0.05$). Furthermore, the biofilm formation was evaluated by crystal violet (Figure 8C). After UVC pretreatment, both smooth Ti ($p = 0.002$) and nano-Ti ($p = 0.049$) surfaces showed that the OD values of crystal violet decreased, indicating that the adhered *P. g* inhibited.

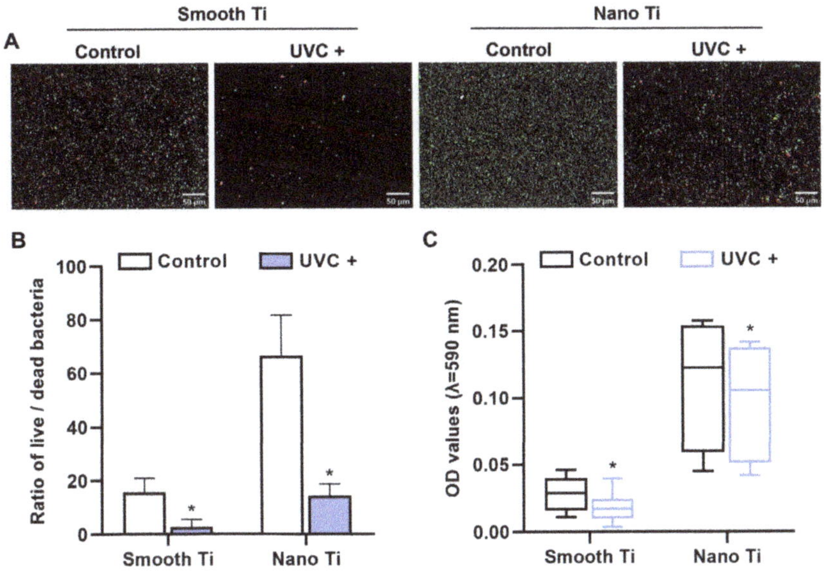

Figure 8. Response of *P. g.* on the sample surfaces. (**A**) Representative images of *P. g.* viability by fluorescent staining for live bacteria (green) and dead bacteria (red). (**B**) Quantitative comparison of live/dead ratio based on (**A**). (**C**) Quantitative comparison of biofilm formation by OD values of crystal violet with the subtraction of a blank. * represents $p < 0.05$ when compared to the control.

3.4. Co-Culture of Human Gingival Fibroblasts and *Porphyromonas gingivalis*

After 2 h of co-culture, HGFs (cytoskeleton in pink and nuclei in light blue) adhered to both smooth-based and nano-based Ti surfaces, which were previously covered with the *P. g.* colonies (diameter < 5 µm in light blue). The S + UVC group (Figure 9B,F) had more extensive cell spreading and less bacterial accumulation compared with the S group (Figure 9A,E), while the N + UVC group (Figure 9D,H) failed to show a significant change in cell spreading compared with the N group (Figure 9C,G), although the *P. g.* pre-accumulation declined.

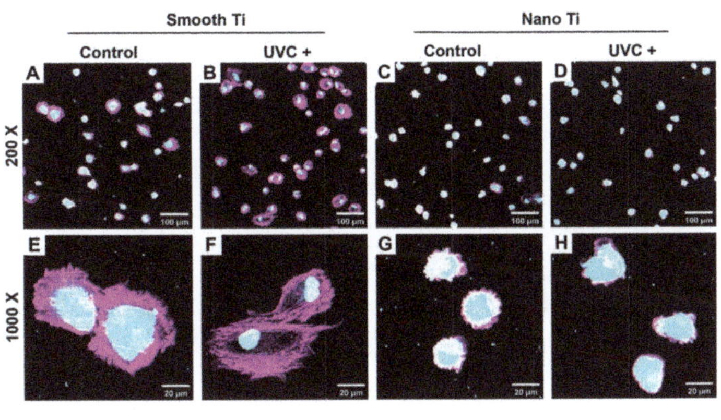

Figure 9. Representative fluorescent images of HGFs co-cultured with pre-accumulated *P. g.* at 2 h by fluorescent staining for cytoskeleton with phalloidin (pink) and dsDNA of cells and bacteria with DAPI (light blue): images in magnification 200× for (**A**) smooth Ti, (**B**) smooth + UVC Ti, (**C**) nano-Ti, (**D**) nano + UVC Ti and in magnification 1000× for (**E**) smooth Ti, (**F**) smooth + UVC Ti, (**G**) nano-Ti, and (**H**) nano + UVC Ti.

4. Discussion

UV photofunctionalization has been reported to improve cell adhesion and reduce biofilm formation on Ti-based surfaces, which may potentially be favorable for the soft tissue integration of Ti dental implants. In this study, smooth Ti and nano-Ti surfaces were chosen as two representative surface designs that are currently used in commercial abutments. Compared with the smooth-based Ti surfaces (Figure 1A,C) that exhibited scratch-like patterns on a micro scale, the nano-based Ti surfaces (Figure 1B,D) displayed a wide range of TiO_2 nanotubes with increased surface roughness (Figure 1E,F). It was previously evidenced that the introduction of nano-texture augmented the contact area and increased the surface wettability, which had a higher affinity for cells [32]. However, such multi-structured topography could be less effective for inhibiting bacterial accumulation and removing bacterial debris than the smooth surface [13]. For this reason, UVC pretreatment was applied in this study to optimize the bioactivity and antimicrobial activity of different Ti surfaces. After UVC pretreatment, both smooth and nano-Ti surfaces achieved a static WCA at 0° (Table 2) as evidence of a super-hydrophilic transition (Figure 1G,H) without significant changes in surface topography or roughness, which was consistent with the previous research [33]. This phenomenon was explained by the removal of carbohydrate contaminants and the production of hydrophilic chemicals [25]. It was also confirmed in our study by the results of EDS (Figure 2) and XPS (Figure 3), which exhibited increased hydroxyl groups (Figure 3A,B) and decreased carbon elements on the UVC-pretreated surfaces compared with the untreated ones. The related mechanism of the alteration in chemical properties may differ in the wavelengths of UV light. It was reported that UVA induced TiO_2 photocatalytic reactions while UVC generated hydrophilic groups by directly breaking the chemical bonds rather than photocatalysis [34].

HGFs belong to the major cellular constituents of the fibrous connective tissue and are responsible for collagen synthesis, which is essential for soft tissue integration [35]. In this study, UVC pretreatment encouraged the adhesion (Figure 4B), spreading (Figure 6), and proliferation (Figure 5A) of HGFs on smooth Ti surfaces, which broadly supported the work of other studies [20,36,37]. UV-induced superhydrophilicity accelerated the adsorption of extracellular molecules onto the substrates ahead of cell adherence, which contributed to an increased expression of focal adhesions in gingival fibroblasts [38–40]. Focal adhesion was considered as the mode cells contacting the implant surfaces [41]. It was initiated by the specific recognition of integrins with the pre-adsorbed extracellular matrix. As a result, FAK is recruited to the sites of integrins for activating phosphorylation to regulate the adhesion, spreading, proliferation, and migration of cells [26]. Additionally, vinculin played an essential role in the stabilization of focal adhesion, which acted as a linker between integrins and the cytoskeleton and was related to mechanical signal transduction [42]. As expected, our results showed that UVC pretreatment upregulated the mRNA levels of FAK, integrin β_1, and vinculin at 2 h on the smooth Ti surface (Figure 7A), indicating that the UVC-pretreated smooth Ti improved the expression of focal adhesion. Nonetheless, the mRNA level of FAK was slightly lower on the UVC-pretreated smooth Ti compared to the untreated one at 24 h (Figure 7C). It was consistent with the previous findings that the focal adhesion turnover might relate to the activation of cell migration and proliferation [43]. Thus, these results suggested that the UVC pretreatment enhanced the adhesive and proliferated behaviors of HGFs on smooth Ti.

Notably, our results also showed that the UVC-pretreated nano-Ti surface, compared with the untreated one, exhibited a lower speed of initial cell adhesion before 2 h (Figure 4C). Although it showed a similar degree of cell adherence with an increased extent of cell spreading after 2 h, the proliferation activity within a seven-day period (Figure 5B) displayed no significant difference between these two groups. Moreover, the results of RT-qPCR showed insignificant mRNA levels of FAK and integrin β_1 at 2 h between the UVC-pretreated nano-Ti and untreated ones (Figure 7B), although the change in the vinculin level conformed to the results of cell spreading (Figure 6). These results suggested that UVC pretreatment discouraged the initial adhesion of HGFs on the nano-Ti surface but had no significant adverse impact on cell proliferation. Undoubtedly, it was contradictory to our study that carried on the smooth Ti, and it was also different from the previous research which found that UVC pretreatment improved the initial adhesion of bone marrow cells and endothelial cells on the nano-porous network Ti surfaces [44]. Despite the types of cells, another possible reason might relate to the influence on wettability by surface structures. It has been extensively reported in the literature that surface topographies could severely modify surface wettability [45]. For example, a superhydrophilic surface with hierarchical structures exhibited a lower degree of cell adhesion than a moderate hydrophilic surface [46]. Such a phenomenon was explained by the rapid fluid spreading and hydrated layer formation, which consequently hindered the protein adsorption and cell attachment [32]. Nevertheless, the present study failed to demonstrate a distinction between the surface wettability of smooth and nano-samples by the detection of static WCA (Figure 1G,H). For proving this hypothesis, a dynamic measurement of WCA may be required. Moreover, other explanations such as the switched protein conformation induced by structural modification or UV irradiation should also be noticed [47,48].

P. g. is reportedly correlated with the occurrence of peri-implantitis [49]. In this study, UVC pretreatment inactivated *P. g.* both on smooth and nano-Ti surfaces (Figure 8), which is consistent with previous studies [50,51]. Related mechanisms were previously explained by the following facts. One of these was related to the production of reactive oxygen species (ROS) or direct damage to the DNA [37]. Another theory was associated with the formation of the hydrated layer as mentioned above [52]. On the other hand, the present study also demonstrated that under a condition of *P. g.* pre-accumulation, UVC pretreatment still enhanced the initial adhesion and spreading of HGFs on the smooth Ti (Figure 9A,B,E,F). While a reduced number of *P. g.* was observed on the UVC-activated nano-Ti compared with the non-activated nano-Ti, the impact on the early response of HGFs was insignificant (Figure 9C,D,G,H). These results might imply that UVC pretreatment had the potential to enhance the initial adhesion and spreading of HGFs on smooth Ti under the accumulation of *P. g.* On the nano-Ti surfaces, UVC pretreatment might reduce the pre-accumulated *P. g.* as well as maintain a good biocompatibility for HGFs at an early stage. It was confirmed with the recent study showing that UVC photofunctionalization improved the biocompatibility of TiO_2 surfaces, which were covered by the bacterial remnants. Nevertheless, our study was only undergone by a short-term observation and failed to show further changes in cellular functions in this co-culture model. It remains unknown how long this antibacterial effect lasts in an oral environment. It is also unclear whether the production of ROS would have a positive or negative impact on the peri-implant tissue, even though the tissue was not directly exposed to UVC during the pretreatment. While UV irradiation was proven to be safe and effective in anti-microbial photodynamic therapies by generating ROS [53], it could be different from UVC pretreatment when regarding the timing and dosage of UV. Therefore, the profound effects of UVC pretreatment need to be figured out in future studies. Other issues including time and cost efficiency as well as ozone pollution should also be considered.

In summary, our findings demonstrated that UVC pretreatment had different effects on HGFs while it had a similar tendency on *P. g.* between smooth and nano-Ti. Nevertheless, the biological response to the implant materials was extremely complicated, and it is still poorly understood for the optimal design of dental implant abutments. Our findings can provide new insights into the potential application of UVC pretreatment on Ti surfaces for enhancing soft tissue integration and preventing peri-implant diseases. The major limitations of this study are as follows. Firstly, the mechanism for the inhibited adhesive behavior of HGFs on UVC-pretreated nano-Ti has been not confirmed yet. Secondly, the molecular mechanisms of changes in the function of HGFs should be further explored. Thirdly, it remains unknown whether the long-term effects of UVC pretreatment exhibit the pros and cons of soft tissue sealing because most in vitro models were observed in short periods. In addition, the effect exerted on a flat disc sample might be significantly different from that on the curved abutment surface. More importantly, the present study lacks in vivo verification. To find out more evidence for promoting the clinical translation of UV pretreatment, the questions raised in this study may need to be solved in the future.

5. Conclusions

Within the limitations of this study, UVC pretreatment enhanced the adhesion and proliferation of HGFs and reduced the accumulation of *P. g.* on the smooth Ti surface compared with the untreated smooth Ti surface. Although the UVC-activated nano-Ti surface discouraged the initial adhesion of HGFs, it had no adverse effects on the proliferation of HGFs and still displayed an inactivation to *P. g.* compared with the untreated condition. It is suggested that UVC pretreatment could potentially be more favorable for improving the adhesion of HGFs onto smooth Ti. Further studies may be required for confirming the long-term effectiveness and biosafety of UVC pretreatment under the consideration of clinical applications.

Author Contributions: Y.W.: conceptualization, methodology, formal analysis, writing—original draft; H.D.: investigation, software, formal analysis, writing—original draft; J.L.: investigation, validation; X.Z.: validation, data curation; R.X.: resources, visualization; P.L.: supervision, writing—reviewing and editing; S.L.: conceptualization, supervision, funding acquisition. All authors have read and agreed to the published version of the manuscript.

Funding: This study was financially supported by the Natural Science Foundation of Xinjiang Uygur Autonomous Region in China [2020D01C004], the National Natural Science Foundation of China [81801008], the Scientific Research Project of Traditional Chinese Medicine Bureau of Guangdong Province (20221267), and the Science Research Cultivation Program of Stomatological Hospital, Southern Medical University [PY2021019 and PY2021003].

Data Availability Statement: Data are available on request from the corresponding author.

Conflicts of Interest: The authors declare no conflict of interest.

References

1. Haugen, H.J.; Chen, H. Is There a Better Biomaterial for Dental Implants than Titanium?—A Review and Meta-Study Analysis. *J. Funct. Biomater.* **2022**, *13*, 46. [CrossRef]
2. Hanawa, T. Zirconia versus titanium in dentistry: A review. *Dent. Mater. J.* **2020**, *39*, 24–36. [CrossRef]
3. Chackartchi, T.; Romanos, G.E.; Sculean, A. Soft tissue-related complications and management around dental implants. *Periodontology* **2019**, *81*, 124–138. [CrossRef]
4. Narimatsu, I.; Atsuta, I.; Ayukawa, Y.; Oshiro, W.; Yasunami, N.; Furuhashi, A.; Koyano, K. Epithelial and Connective Tissue Sealing around Titanium Implants with Various Typical Surface Finishes. *ACS Biomater. Sci. Eng.* **2019**, *5*, 4976–4984. [CrossRef]
5. Guo, T.; Gulati, K.; Arora, H.; Han, P.; Fournier, B.; Ivanovski, S. Race to invade: Understanding soft tissue integration at the transmucosal region of titanium dental implants. *Dent. Mater.* **2021**, *37*, 816–831. [CrossRef]
6. Yu, J.; Zhou, M.; Zhang, L.; Wei, H. Antibacterial Adhesion Strategy for Dental Titanium Implant Surfaces: From Mechanisms to Application. *J. Funct. Biomater.* **2022**, *13*, 169. [CrossRef]
7. Canullo, L.; Menini, M.; Santori, G.; Rakic, M.; Sculean, A.; Pesce, P. Titanium abutment surface modifications and peri-implant tissue behavior: A systematic review and meta-analysis. *Clin. Oral Investig.* **2020**, *24*, 1113–1124. [CrossRef]
8. Orapiriyakul, W.; Young, P.S.; Damiati, L.; Tsimbouri, P.M. Antibacterial surface modification of titanium implants in orthopaedics. *J. Tissue Eng.* **2018**, *9*, 1–16. [CrossRef]
9. Canullo, L.; Annunziata, M.; Pesce, P.; Tommasato, G.; Nastri, L.; Guida, L. Influence of abutment material and modifications on peri-implant soft-tissue attachment: A systematic review and meta-analysis of histological animal studies. *J. Prosthet. Dent.* **2021**, *125*, 426–436. [CrossRef]
10. Li, T.; Gulati, K.; Wang, N.; Zhang, Z.; Ivanovski, S. Bridging the gap: Optimized fabrication of robust titania nanostructures on complex implant geometries towards clinical translation. *J. Colloid Interface Sci.* **2018**, *529*, 452–463. [CrossRef]
11. Xu, R.; Hu, X.; Yu, X.; Wan, S.; Wu, F.; Ouyang, J.; Deng, F. Micro-/nano-topography of selective laser melting titanium enhances adhesion and proliferation and regulates adhesion-related gene expressions of human gingival fibroblasts and human gingival epithelial cells. *Int. J. Nanomed.* **2018**, *13*, 5045–5057. [CrossRef] [PubMed]
12. Amoroso, P.F.; Adams, R.J.; Waters, M.G.; Williams, D.W. Titanium surface modification and its effect on the adherence of Porphyromonas gingivalis: An in vitro study. *Clin. Oral Implant. Res.* **2006**, *17*, 633–637. [CrossRef]
13. Sousa, V.; Mardas, N.; Spratt, D.; Hassan, I.A.; Walters, N.J.; Beltran, V.; Donos, N. The Effect of Microcosm Biofilm Decontamination on Surface Topography, Chemistry, and Biocompatibility Dynamics of Implant Titanium Surfaces. *Int. J. Mol. Sci.* **2022**, *23*, 10033. [CrossRef] [PubMed]

14. Ogawa, T. Ultraviolet photofunctionalization of titanium implants. *Int. J. Oral Maxillofac. Implant.* **2014**, *29*, 95–102. [CrossRef]
15. Okubo, T.; Ikeda, T.; Saruta, J.; Tsukimura, N.; Hirota, M.; Ogawa, T. Compromised Epithelial Cell Attachment after Polishing Titanium Surface and Its Restoration by UV Treatment. *Materials* **2020**, *13*, 3946. [CrossRef]
16. Dini, C.; Nagay, B.E.; Magno, M.B.; Maia, L.C.; Barao, V.A.R. Photofunctionalization as a suitable approach to improve the osseointegration of implants in animal models-A systematic review and meta-analysis. *Clin. Oral Implant. Res.* **2020**, *31*, 785–802. [CrossRef]
17. Marasini, S.; Leanse, L.G.; Dai, T. Can microorganisms develop resistance against light based anti-infective agents? *Adv. Drug Deliv. Rev.* **2021**, *175*, 113822. [CrossRef]
18. Yamamura, K.; Miura, T.; Kou, I.; Muramatsu, T.; Furusawa, M.; Yoshinari, M. Influence of various superhydrophilic treatments of titanium on the initial attachment, proliferation, and differentiation of osteoblast-like cells. *Dent. Mater. J.* **2015**, *34*, 120–127. [CrossRef]
19. Huang, Y.; Zhang, H.; Chen, Z.; Wang, Y.; Yang, X.; Yu, H. Improvement in Osseointegration of Titanium Dental Implants After Exposure to Ultraviolet-C Light for Varied Durations: An Experimental Study in Beagle Dogs. *J. Oral Maxillofac. Surg.* **2022**, *80*, 1389–1397. [CrossRef]
20. Aita, H.; Att, W.; Ueno, T.; Yamada, M.; Hori, N.; Iwasa, F.; Tsukimura, N.; Ogawa, T. Ultraviolet light-mediated photofunctionalization of titanium to promote human mesenchymal stem cell migration, attachment, proliferation and differentiation. *Acta Biomater.* **2009**, *5*, 3247–3257. [CrossRef]
21. Aita, H.; Hori, N.; Takeuchi, M.; Suzuki, T.; Yamada, M.; Anpo, M.; Ogawa, T. The effect of ultraviolet functionalization of titanium on integration with bone. *Biomaterials* **2009**, *30*, 1015–1025. [CrossRef]
22. Liang, L.C.; Krieg, P.; Rupp, F.; Kimmerle-Muller, E.; Spintzyk, S.; Richter, M.; Richter, G.; Killinger, A.; Geis-Gerstorfer, J.; Scheideler, L. Osteoblast Response to Different UVA-Activated Anatase Implant Coatings. *Adv. Mater. Interfaces* **2019**, *6*, 1801720. [CrossRef]
23. Wu, J.; Zhou, L.; Ding, X.; Gao, Y.; Liu, X. Biological Effect of Ultraviolet Photocatalysis on Nanoscale Titanium with a Focus on Physicochemical Mechanism. *Langmuir* **2015**, *31*, 10037–10046. [CrossRef]
24. Li, S.; Ni, J.; Liu, X.; Zhang, X.; Yin, S.; Rong, M.; Guo, Z.; Zhou, L. Surface characteristics and biocompatibility of sandblasted and acid-etched titanium surface modified by ultraviolet irradiation: An in vitro study. *J. Biomed. Mater. Res. Part B Appl. Biomater.* **2012**, *100*, 1587–1598. [CrossRef]
25. Dini, C.; Nagay, B.E.; Cordeiro, J.M.; da Cruz, N.C.; Rangel, E.C.; Ricomini-Filho, A.P.; de Avila, E.D.; Barao, V.A.R. UV-photofunctionalization of a biomimetic coating for dental implants application. *Mater. Sci. Eng. C Mater. Biol. Appl.* **2020**, *110*, 110657. [CrossRef]
26. Palkowitz, A.L.; Tuna, T.; Bishti, S.; Boke, F.; Steinke, N.; Muller-Newen, G.; Wolfart, S.; Fischer, H. Biofunctionalization of Dental Abutment Surfaces by Crosslinked ECM Proteins Strongly Enhances Adhesion and Proliferation of Gingival Fibroblasts. *Adv. Healthc. Mater.* **2021**, *10*, 2100132. [CrossRef]
27. Bustin, S.A.; Benes, V.; Garson, J.A.; Hellemans, J.; Huggett, J.; Kubista, M.; Mueller, R.; Nolan, T.; Pfaffl, M.W.; Shipley, G.L.; et al. The MIQE guidelines: Minimum information for publication of quantitative real-time PCR experiments. *Clin. Chem.* **2009**, *55*, 611–622. [CrossRef]
28. Di Giulio, M.; Traini, T.; Sinjari, B.; Nostro, A.; Caputi, S.; Cellini, L. Porphyromonas gingivalis biofilm formation in different titanium surfaces, an in vitro study. *Clin. Oral Implants Res.* **2016**, *27*, 918–925. [CrossRef]
29. Kamble, E.; Pardesi, K. Antibiotic Tolerance in Biofilm and Stationary-Phase Planktonic Cells of Staphylococcus aureus. *Microb. Drug Resist.* **2021**, *27*, 3–12. [CrossRef]
30. Gao, Y.; Kang, K.; Luo, B.; Sun, X.; Lan, F.; He, J.; Wu, Y. Graphene oxide and mineralized collagen-functionalized dental implant abutment with effective soft tissue seal and romotely repeatable photodisinfection. *Regen. Biomater.* **2022**, *9*, 24. [CrossRef]
31. Rupp, F.; Gittens, R.A.; Scheideler, L.; Marmur, A.; Boyan, B.D.; Schwartz, Z.; Geis-Gerstorfer, J. A review on the wettability of dental implant surfaces I: Theoretical and experimental aspects. *Acta Biomater.* **2014**, *10*, 2894–2906. [CrossRef] [PubMed]
32. Cui, H.J.; Wang, W.S.; Shi, L.X.; Song, W.L.; Wang, S.T. Superwettable Surface Engineering in Controlling Cell Adhesion for Emerging Bioapplications. *Small Methods* **2020**, *4*, 2000573. [CrossRef]
33. Arroyo-Lamas, N.; Arteagoitia, I.; Ugalde, U. Surface Activation of Titanium Dental Implants by Using UVC-LED Irradiation. *Int. J. Mol. Sci.* **2021**, *22*, 2597. [CrossRef] [PubMed]
34. Johnson, H.A.; Williamson, R.S.; Marquart, M.; Bumgardner, J.D.; Janorkar, A.V.; Roach, M.D. Photocatalytic activity and antibacterial efficacy of UVA-treated titanium oxides. *J. Biomater. Appl.* **2020**, *35*, 500–514. [CrossRef] [PubMed]
35. Zhao, B.; van der Mei, H.C.; Subbiahdoss, G.; de Vries, J.; Rustema-Abbing, M.; Kuijer, R.; Busscher, H.J.; Ren, Y. Soft tissue integration versus early biofilm formation on different dental implant materials. *Dent. Mater.* **2014**, *30*, 716–727. [CrossRef] [PubMed]
36. Guo, L.; Smeets, R.; Kluwe, L.; Hartjen, P.; Barbeck, M.; Cacaci, C.; Gosau, M.; Henningsen, A. Cytocompatibility of Titanium, Zirconia and Modified PEEK after Surface Treatment Using UV Light or Non-Thermal Plasma. *Int. J. Mol. Sci.* **2019**, *20*, 5596. [CrossRef]
37. Yang, Y.; Zhang, H.; Komasa, S.; Morimoto, Y.; Sekino, T.; Kawazoe, T.; Okazaki, J. UV/ozone irradiation manipulates immune response for antibacterial activity and bone regeneration on titanium. *Mater. Sci. Eng. C Mater. Biol. Appl.* **2021**, *129*, 112377. [CrossRef]

38. Jeon, C.; Oh, K.C.; Park, K.H.; Moon, H.S. Effects of ultraviolet treatment and alendronate immersion on osteoblast-like cells and human gingival fibroblasts cultured on titanium surfaces. *Sci. Rep.* **2019**, *9*, 2581. [CrossRef]
39. Ikeda, T.; Ueno, T.; Saruta, J.; Hirota, M.; Park, W.; Ogawa, T. Ultraviolet Treatment of Titanium to Enhance Adhesion and Retention of Oral Mucosa Connective Tissue and Fibroblasts. *Int. J. Mol. Sci.* **2021**, *22*, 12396. [CrossRef]
40. Gittens, R.A.; Scheideler, L.; Rupp, F.; Hyzy, S.L.; Geis-Gerstorfer, J.; Schwartz, Z.; Boyan, B.D. A review on the wettability of dental implant surfaces II: Biological and clinical aspects. *Acta Biomater.* **2014**, *10*, 2907–2918. [CrossRef]
41. Katoh, K. FAK-Dependent Cell Motility and Cell Elongation. *Cells* **2020**, *9*, 192. [CrossRef]
42. Sero, J.E.; Bakal, C. Multiparametric Analysis of Cell Shape Demonstrates that beta-PIX Directly Couples YAP Activation to Extracellular Matrix Adhesion. *Cell Syst.* **2017**, *4*, 84–96.e6. [CrossRef] [PubMed]
43. Woods, A.J.; Kantidakis, T.; Sabe, H.; Critchley, D.R.; Norman, J.C. Interaction of paxillin with poly(A)-binding protein 1 and its role in focal adhesion turnover and cell migration. *Mol. Cell. Biol.* **2005**, *25*, 3763–3773. [CrossRef] [PubMed]
44. Hatoko, M.; Komasa, S.; Zhang, H.; Sekino, T.; Okazaki, J. UV Treatment Improves the Biocompatibility and Antibacterial Properties of Crystallized Nanostructured Titanium Surface. *Int. J. Mol. Sci.* **2019**, *20*, 5991. [CrossRef] [PubMed]
45. Rupp, F.; Liang, L.; Geis-Gerstorfer, J.; Scheideler, L.; Huttig, F. Surface characteristics of dental implants: A review. *Dent. Mater.* **2018**, *34*, 40–57. [CrossRef]
46. Stallard, C.P.; McDonnell, K.A.; Onayemi, O.D.; O'Gara, J.P.; Dowling, D.P. Evaluation of protein adsorption on atmospheric plasma deposited coatings exhibiting superhydrophilic to superhydrophobic properties. *Biointerphases* **2012**, *7*, 31. [CrossRef]
47. Hierro-Oliva, M.; Gallardo-Moreno, A.M.; Gonzalez-Martin, M.L. Surface Characterisation of Human Serum Albumin Layers on Activated Ti6Al4V. *Materials* **2021**, *14*, 7416. [CrossRef]
48. Firkowska-Boden, I.; Zhang, X.; Jandt, K.D. Controlling Protein Adsorption through Nanostructured Polymeric Surfaces. *Adv. Healthc. Mater.* **2018**, *7*, 1700995. [CrossRef]
49. Shibli, J.A.; Melo, L.; Ferrari, D.S.; Figueiredo, L.C.; Faveri, M.; Feres, M. Composition of supra- and subgingival biofilm of subjects with healthy and diseased implants. *Clin. Oral Implants Res.* **2008**, *19*, 975–982. [CrossRef]
50. de Avila, E.D.; Lima, B.P.; Sekiya, T.; Torii, Y.; Ogawa, T.; Shi, W.; Lux, R. Effect of UV-photofunctionalization on oral bacterial attachment and biofilm formation to titanium implant material. *Biomaterials* **2015**, *67*, 84–92. [CrossRef]
51. Ishijima, M.; de Avila, E.D.; Nakhaei, K.; Shi, W.; Lux, R.; Ogawa, T. Ultraviolet Light Treatment of Titanium Suppresses Human Oral Bacterial Attachment and Biofilm Formation: A Short-Term In Vitro Study. *Int. J. Oral Maxillofac. Implant.* **2019**, *34*, 1105–1113. [CrossRef] [PubMed]
52. Fan, H.; Guo, Z. Bioinspired surfaces with wettability: Biomolecule adhesion behaviors. *Biomater. Sci.* **2020**, *8*, 1502–1535. [CrossRef] [PubMed]
53. Komine, C.; Uchibori, S.; Tsudukibashi, O.; Tsujimoto, Y. Application of Reactive Oxygen Species in Dental Treatment. *J. Pers. Med.* **2022**, *12*, 1531. [CrossRef] [PubMed]

Journal of
Functional
Biomaterials

MDPI

Review

Research Advances on Stem Cell-Derived Extracellular Vesicles Promoting the Reconstruction of Alveolar Bone through RANKL/RANK/OPG Pathway

Xia Huang [1,2,3,†], Yuxiao Li [1,2,†], Hui Liao [1], Xin Luo [1], Yueping Zhao [2], Yadong Huang [1], Zhiying Zhou [2,3] and Qi Xiang [1,*]

1 Institute of Biomedicine and Guangdong Provincial Key Laboratory of Bioengineering Medicine, Jinan University, Guangzhou 510632, China
2 School of Stomatology, Jinan University, Guangzhou 510632, China
3 Department of Orthodontics, The First Affiliated Hospital of Jinan University, Guangzhou 510632, China
* Correspondence: txiangqi@jnu.edu.cn
† These authors contributed equally to this work.

Abstract: Periodontal bone tissue defects and bone shortages are the most familiar and troublesome clinical problems in the oral cavity. Stem cell-derived extracellular vesicles (SC-EVs) have biological properties similar to their sources, and they could be a promising acellular therapy to assist with periodontal osteogenesis. In the course of alveolar bone remodeling, the RANKL/RANK/OPG signaling pathway is an important pathway involved in bone metabolism. This article summarizes the experimental studies of SC-EVs applied for the therapy of periodontal osteogenesis recently and explores the role of the RANKL/RANK/OPG pathway in their mechanism of action. Their unique patterns will open a new field of vision for people, and they will help to advance a possible future clinical treatment.

Keywords: stem cell-derived extracellular vesicles; reconstruction of alveolar bone; osteogenesis; RANKL/RANK/OPG pathway

Citation: Huang, X.; Li, Y.; Liao, H.; Luo, X.; Zhao, Y.; Huang, Y.; Zhou, Z.; Xiang, Q. Research Advances on Stem Cell-Derived Extracellular Vesicles Promoting the Reconstruction of Alveolar Bone through RANKL/RANK/OPG Pathway. *J. Funct. Biomater.* **2023**, *14*, 193. https://doi.org/10.3390/jfb14040193

Academic Editors: Andreas K. Nüssler and Marco Tatullo

Received: 2 February 2023
Revised: 17 March 2023
Accepted: 27 March 2023
Published: 30 March 2023

1. Introduction

The periodontium plays an important role in people's daily lives. The entire periodontal structure consists of the gingiva exposed to the oral cavity, the alveolar bone covered by the gingiva, and the cementum covering the roots of the teeth. Periodontal ligaments are created when Sharpey's fibers enter the cementum-connected alveolar bone. One of the most infamous inflammatory diseases is periodontal disease, which starts as an external inflammatory reaction of the gingiva (gingivitis) and later proceeds to clinical attachment loss with the degeneration of the inflammatory in periodontal structure (periodontitis). The height of the alveolar ridge decreasing and bone loss are the disease's major defining characteristics [1]. Furthermore, external apical root resorption (EARR), a pathological side effect of orthodontic treatment, also causes bone loss and the reconstruction of the alveolar bone [2]. All of these conditions share the trait that the more advanced the progression, the more challenging they become to solve, as well as for regeneration to occur.

In the past few decades, people have explored bone regeneration treatments for alveolar bone defects. Currently, the main methods used to treat this problem are autologous or allogeneic bone transplantation and guided tissue regeneration (GTR). However, autologous bone grafts cannot be used for repairs in patients with large defects. Allogeneic bone fillings are accompanied by complications, such as bone defects and infections. GTR involves the use of absorbable collagen membranes to cover the defect area of alveolar bone, with the goal of achieving alveolar bone regeneration [3]. However, GTR treatments are not stable, owing to their multiple types of complications. Periodontal osteogenesis has received attention, as it relates to the treatment of common oral disorders, such as periodontal treatment, dental implants, orthodontic tooth movement, and disease recurrence. Therefore, we urgently need to find a more feasible way to promote treatments for periodontal osteogenesis.

Once, stem cell therapy was considered to be the most promising strategy for treating periodontal bone loss. In stem cell treatments, autologous or allogeneic stem cells are isolated and amplified in vitro and then planted on a natural or synthetic cell scaffold with a good biocompatibility, where they undergo gradual degradation until being absorbed by human body. This biomaterial scaffold provides a three-dimensional (3D) space for the cells to survive in, and the stem cells grow on the three-dimensional scaffold prefabricated before. Then, the hybrid material is implanted into periodontal defects to form alveolar bone-like tissues [4]. However, there are many problems with this clinical technique that have yet to be resolved, such as the possibility of autoimmune cell rejection, the high cost and technical difficulty, and the issue of stem cells holding low survival rate in stem-cell therapy, and the risk of cancer or tissue disease should also be taken in consideration [5]. Later, a large number experiments confirmed that extracellular vesicles derived from various stem cells (SC-EVs) have the same therapeutic efficacy or even more advantages than the original stem cells for tissue repair [6,7]. SC-EVs are nanosized membrane particles secreted by stem cells, such as pulp stem cells, bone marrow stem cells, and human umbilical cord mesenchymal stem cells, which carry diverse biological information, including DNA, RNA, bioactive lipids, proteins, and other biologically related molecules [8]. The use of SC-EV-assisted osteogenesis to treat periodontal bone loss [9–11] vastly withstands the risk of self-rejection and improves the safety and stability of treatments. It is noteworthy that stem cell therapy presents the common challenge of inability to resist cytotoxicity, either biobanking of the useful secretome at the inflammatory region [12].

Amid the course of alveolar bone remodeling, the RANKL (RANK ligand)/RANK (receptor activator of NF-κB)/OPG (osteoprotegerin) signal pathway is involved in bone resorption through its main work in osteoclast separation and enactment, as well as within the provocative reaction [6]. Trabecular bone resorption by osteoclasts is a necessary regulated process that frees an area for osteoblasts to form new bone [13]. Especially in orthodontic practice, the success of tooth movement is reliant on the RANKL/RANK/OPG signaling pathway, which applies pro-osteogenic and anti-inflammatory impacts within the handle of remaking of the alveolar bone.

An experimental study on the use of SC-EVs in periodontal osteogenesis, which explored the relationship between SC-EVs and the RANKL/RANK/OPG pathway, found that this signaling balance is an important contributing factor to bone remodeling using SC-EVs. The status of these signaling pathway axis is the key for the success of orthodontic and implant oral therapy. Further research progress on the use of SC-EVs for the treatment of alveolar bone osteogenesis via the RANKL/RANK/OPG pathway will help with the reconstruction of alveolar bone and the fabrication of more suitable and biocompatible scaffold materials based on SC-EVs. There are fewer antigens on the surface of exocrine membranes, which reduces immune rejection while retaining the bioactive components of stem cells. Therefore, we think the application of SC-EVs will have the best osteo-inductive properties and become a compelling new form of "stem cell-free" therapy.

2. The Reconstruction of Alveolar Bone

Alveolar bone remaking may be a persistent and slow preparation. It depends on the energetic adjust between bone retention and bone arrangement. At first, osteo-clast-mediated bone retention is actuated, and after that, unused osteoblast-mediated bone arrangement starts [14].

Bone retention and bone formation are interconnected, and they depend on the regulation of each jolt figure so that modern bone can supplant ancient bone and stabilize the volume of alveolar bone. Nearby components incorporate microdamage and aggravation, which initiate osteocytes to deliver variables that invigorate bone resorption. Specifically, the stimulus factors include: (i) biochemical factors established by a variation in hormone secretion, including the parathyroid hormone (PTH) [15] and estrogen [16]; (ii) an increase in some chemokines [17] and cytokines [18], which can indirectly promote osteoclastogenesis; and (iii) mechanical components that are set up by the mechanical stacking connected to bone [19,20]. Normally, an extracellular signal passes through the cell membrane, enters into the cell, and is converted into a chemical signal. Then, the chemical signal is transferred into the cell to activate osteoclast differentiation, proliferation, migration, and processes. However, in periodontitis, this balance is destroyed. Periodontitis mainly refers to pathological alveolar bone resorption and partial new bone formation. In the mechanism of periodontitis-related bone loss, virulent periodontal pathogens overwhelm the defense barrier of the oral mucosa. In response, the resident cells in the tissues, such as fibroblasts, keratinocytes, and dendritic cells, discharge incendiary cytokines, advance different incendiary cells to deliver anti-bacterial specialists, and create re-active oxygen species to dispose of the pathogens, but this, moreover, crushes the typical activities of alveolar bone remodeling [21]. Osteoclasts are initiated and enacted, whereas osteoblasts are repressed, which can crush the bone adjust and lead to a diminishment in bone volume.

Orthodontic tooth movement (OTM), moreover, involves alveolar bone reconstruction. OTM basically alludes to physical alveolar bone resorption and reconstruction. Amid OTM, bone resorption happens on the compression side of the alveolar bone, and osteogenesis happens on the pressure side of the alveolar bone. The orthodontic constraint exasperates the steady environment of the periodontal tendon space by changing the blood stream and nearby environment, causing a disturbance within the support of the alveolar bone status [22]. Orthodontic forces increase IL-1β and TNF-α in periodontal ligaments and induce PGE2 to be released from periodontal ligament cells and osteoblasts, which is also one of the important chemical mediators of compression-induced alveolar bone resorption [23]. These factors produced by the periodontal ligament cells on the compression side help RANKL promote bone resorption and experimental tooth movement. In contrast, tensile forces increase the osteogenic factors (e.g., TGF-β and IL-10) in periodontal ligaments, which may increase OPG and reduce RANKL production to inhibit osteoclast bone resorption.

In clinical cases, many patients have their damaged teeth extracted, but they cannot receive implant treatment precisely because the reconstruction of their alveolar bone is unsuccessful, such that mechanical stress on the implant teeth would have a negative impact on the insufficient bone tissue [24]. It is important to design an appropriate implant surface and alveolar bone regeneration materials in order to remove the osteoclasts and induce an inflammatory reaction to the implant, which is helpful to the alveolar bone. Many studies have focused on evaluating the effect of proteins coated on the surface of titanium implants and the effect of changing the surface roughness of implants on reducing osteoclasts. It has been found that coating the surface of inserts with a moo dosage of bone morpho-genetic proteins (BMPs) or OPG can improve bone arrangement and move forward bone integration [25].

Thus, we aimed at reconstructing alveolar bone with appropriate morphological characteristics. As for periodontitis, we expected to increase the bone formation and slow down the bone resorption rate, thus regaining the balance between them, and even resulting in reverse pathological processes in order to restore the alveolar bone (Figure 1).

At present, dentists tend to cure periodontitis by using GTR, subgingival debridement, subgingival scaling, root planning, etc. [26]. These treatments can effectively stop the inflammation, but they can do nothing to restore alveolar bone formation. As for OTM, if the mechanical force is not properly balanced, it can lead to a lack of bone coverage at the root of the tooth, which can then result in the reconstruction period of alveolar bone being too long to allow for correct remodeling. Therefore, it is an urgent problem for orthodontic treatment to accelerate bone reconstruction under conditions that allow for better bone repair.

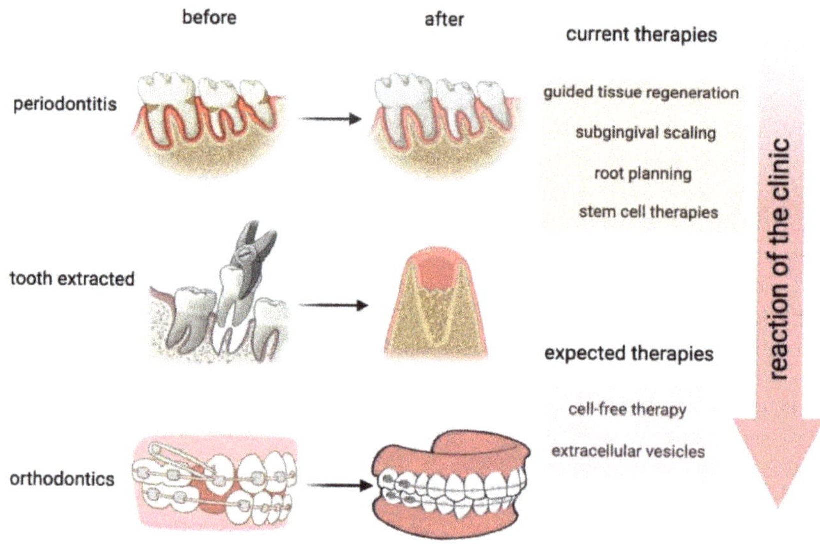

Figure 1. Development of methods for alveolar bone reconstruction. Above are typical scenes in clinical treatments, all of which are involved in alveolar bone reconstruction. We expect to achieve cell therapy by using extracellular vesicles to replace current therapies, such as guided tissue regeneration (GTR). (This figure was drawn using BioRender.com).

3. SC-EVs and Their Biological Characteristics

EVs are macromolecule lipid bilayer layer structure vesicles that are emitted by the larger part cells and are vital carriers of cell-to-cell communication. EVs carry a spread of biomolecules and speak to a substitution kind of data exchange, with subsequent proficiency and better specificity than flag transduction within the conventional sense. Moreover, EVs act as aggregates of information, extending the signal transduction to the intercellular level [27].

EVs are widely present in numerous body fluids, together with heterogeneous groups of membrane vesicles from various sources. They can carry a variety of biomolecules, including DNA fragments, RNA (including miRNA, mRNA, tRNA, and lncRNA), lipids, proteins, peptides, sugars, and metabolites [28]. Differences in the composition of the biomolecules determine the specificity of the EVs' function. Reckoning on the kind of cell from that it comes, an EV can reveal specific proteins for that cell [29].

Based on EVs biogenesis and size, EVs are typically divided into three classes: exosomes (Exos), microvesicles (MVs), and apoptotic bodies. Exosomes, that are 50–150 nm, are intraluminal vesicles fashioned by the endosomal membrane budding internally all through the development of multivesicular bodies (MVBs), and MVBs are secreted when fusion occurs with the cytomembrane [30]. MVs are 100–1000 nm in diameter and are formed by the plasma membrane budding outward. They are released directly into the extracellular matrix [31]. The diameter of apoptotic bodies is usually defined as 1–5 μm, and apoptotic bodies form following cytoplasmic membrane blebbing in cells undergo programmed cell death, or apoptosis [32].

All EVs are formed by a phospholipid bilayer and carry important biological information from related cells. After being released from cells, EVs can realize biomolecular transformation by recognizing specific proteins on the surface of the recipient cells or entering the recipient cells to perform functions, including repairing tissue damage and promoting angiogenesis. EVs are seen as a promising delivery system that can send signals into cells through different mechanisms. Instead of delivering cargo, EVs interact with the molecules on the surface of the cell membrane to trigger cascades of reactions within the cell [33].

The procedures regularly utilized for exosome partition basically contain ultra-centrifugation, ultrafiltration, immunoaffinity capture, charge neutralization-based polymer precipitation, size-exclusion chromatography, and microfluidic procedures [34]. Be that as it may, these ordinarily utilized separation strategies possess benefits and disadvantages. Ultra-centrifugation is the gold standard of exosome isolation approaches, and it promptly translates to exosome isolation [35]. In recent years, the wide application of microfluidic technology has provided a replacement plan for the isolation of exosomes [36], supporting variations in their biological and physical properties, which permit for a highly sensitive and high-speed separation; but, the yield is low, and this is often solely appropriate for diagnoses. Additionally, more and more industrial kits are developed for the isolation of exosomes, which are simple and convenient to operate without special equipment, and with the upgrading of products, the extraction potency and purification of exosomes has improved in a gradual way.

Stem cells are a category of undifferentiated pluripotent cells with broad prospects in regenerative medication thanks to their high self-renewal and multidirectional differentiation capabilities. EVs secreted by stem cells can show comparable capacities to stem cells, such as promoting tissue repair and regeneration [37]. Some examples of common mesenchymal stem cells (MSCs) include bone marrow MSCs (BMSCs), umbilical cord MSCs (UCMSCs), adipose-derived stem cells (ADSCs), and human periodontal ligament stem cells (hPDLSCs) (Figure 2).

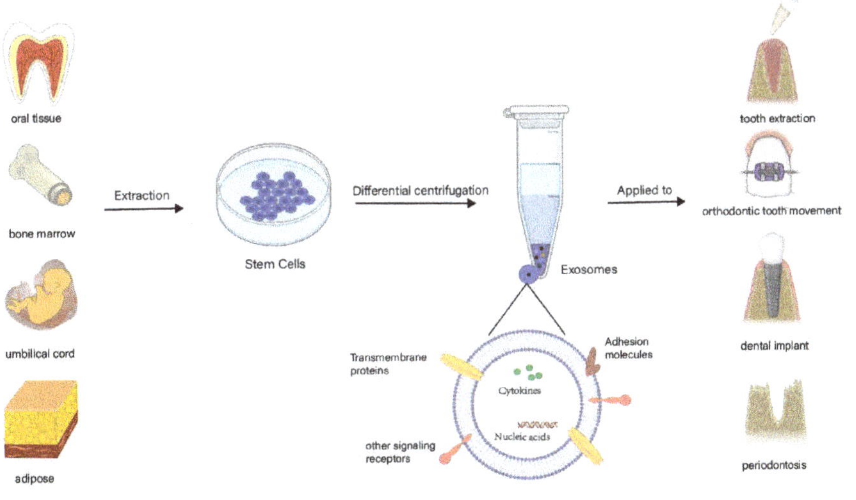

Figure 2. Use of SC-EVs within the field of oral medicine. Exos derived from some common stem cells act on teeth, as well as on the whole body. Of the stem cells derived from various tissues, oral stem cells are the most advantageous in terms of ethics and the level of difficulty in obtaining them. (This figure was drawn using BioRender.com and Adobe Illustrator 2021).

BMSC-EVs have been shown to have immunomodulating, proliferative, tissue repair, anti-inflammatory, and antioxidant effects; in addition, they can regulate osteoclast production and osteogenic differentiation [38]. ADSC-EVs are primarily associated with graft rejection, skin healing, and apoptosis [39]. UCMSC-EVs play a major role in promoting neovascularization, inflammatory responses, and nerve regeneration [40].

With the gradual deepening of EV research, it has been found that SC-EVs can not only effectively transmit the signals of maternal stem cell-induced regeneration, showing the same therapeutic effect, but they can also lower the risk of self-rejection and enhance the safety of treatments [6]. When attacked by inflammation, the body will produce an immune response. A damaging immune response is not only detrimental to the damage repair of defects, but it also aggravates the occurrence of damage. SC-EVs can inhibit the immune response's effect on damaged tissues [41]. In stem cell therapy, we use the pluridirectional differentiation potential of stem cells to realize the repair of tissue defects. Therefore, stemness is the most important aspect of stem cell-related therapy. Studies have shown that EVs extracted from embryonic stem cells (ESCs) can maintain the stemness of ESCs by activating the FAK pathway [42]. Thus, we suspect that SC-EVs can maintain the stemness of their original stem cells, such as human periodontal stem cells, which are often used to repair bone defects caused by periodontitis.

SC-EVs also have significant advantages in long-term stability. EVs or SC-EVs are usually frozen in PBS or normal saline, or in a specific cell-freezing medium, and they can remain stable during cryopreservation [43].

4. SC-EVs Promote Bone Tissue Repair and Regeneration

4.1. SC-EVs in Bone Tissue Repair and Regeneration

Due to SC-EVs having a wide range of sources and powerful biological functions, they are also widely used in tissue repair and regeneration, as well as biomarkers for disease diagnoses and other fields, and they have been verified in bone, heart, and other tissues. More importantly, SC-EVs are hydrophilic, which allows them to be loaded directly into soluble biomaterials, such as hydrogels. Electrostatic interactions also contribute to the ability of biomaterials to carry SC-EVs; for example, a positively charged biological scaffold can absorb negatively charged SC-Evs [44]. The bioactive adhesion of SC-Evs is beneficial because it allows the SC-Evs to adhere to the surface of different kinds of biological scaffolds, such as bioactive glass, bone matrix scaffolds, etc. [45].

Swanson et al. [10] extracted Exos from human dental pulp stem cells (hDPSCs), combined the Exos into nanofiber tissue-engineering scaffolds, and they implanted them into the skull defects of mice to examine the level of bone repair. This platform has great characteristics, such as huge surface range and huge porosity, permitting it to be embedded into Exos from three measurements, making it a perfect choice for functionalizing its surface by conveyance [46]. Then, a post-seeding technique was used to immobilize hDPSC-Exos on nanofibrous. The results of micro-CT scanning of the bone defect site showed that the bone density of the defect site increased significantly in the Exo stent group. Chitosan is a common medical biogel and a reliable controlled-drug-release material. Combined with EVs, chitosan based hydrogels promoted cell migration, angiogenesis, and re-epithelization [47]. When hDPSC-Exos are loaded into a chitosan hydrogel, DPSC-Exos can also play the role of osteosynthesis [48]. Diomede et al. [11] combined hPDLSC-derived EVs into a collagen membrane for skull defects in mice and found that, both in vitro and in vivo, EVs exhibited good osteogenic properties. hPDLSC-EVs effectively inherited the biological functions of hPDLSCs for osteogenic differentiation and the secretion of stromal cell-derived factor 1 (SDF1). As a result, the ponder effectively created an unused biocompatible osteogenic development, comprising a commercially accessible collagen film and hPDLSC-EVs or poly-ethylenimine (PEI)-engineered EVs (PEI-EVs). SC-EVs inferred from human umbilical line mesenchymal stem cells (hUCMSCs) and bone marrow-derived stem cells (BMSCs) can moreover proficiently express osteogenic markers [49]. Yang et al. [50] effectively repaired bone defects in mice by integrating Exos derived from hUCMSCs with hydrogels. Ideal vectors for EVs also include bone repair scaffolds with a stereoscopic structure, such as tricalcium β-phosphate (β-TCP) [51] and mesoporous bioactive glass (MBG) [52], which have more advantages in terms of adaptability to the EV size and the induction of osteogenesis [53]. Furthermore, these natural bone mimics are often biodegradable and exhibit a good biocompatibility. Recently, scientists developed a novel strategy to avoid the clearing of nanoparticle-fused EVs by the reticuloendothelial system (RES)/the mononuclear phagocytic system (MPS). The strategy involves saturating the MPS with macrophage-targeting EVs, and then engineering homing peptides to enhance the tissue-targeting features of EVs, which resulted in better therapeutic effects and the avoidance of side effects in the liver and spleen. This points out that safely and effectively generating bone-targeting EVs with low clearance rates is a future research direction [54,55].

4.2. SC-EVs in Alveolar Bone Tissue Repair and Regeneration

SC-EVs have been proven to have anti-inflammatory and immunosuppressive effects in different tissues. In addition, SC-EVs have strong bone tissue regeneration characteristics, and their safety level is high, which is necessary for a material for the regenerative treatment of alveolar bone loss.

Lei et al. [56] extracted small extracellular vesicles (sEVs) derived from PDLSCs (PDLSC-sEVs) from healthy periodontal subjects, and they used them in combination with a Matrigel hydrogel to repair bone defects in rat periodontitis models. They found that PDLSC-sEVs seem to save the osteogenic capacity of endogenous stem cells in inflammatory environments and contribute to alveolar bone regeneration. Watanabe et al. [57] studied the preventive and therapeutic effects of BMSC-EVs on bisphosphonate-associated jaw necrosis after a tooth extraction. The results stated that BMSC-EVs could reduce the number of senescent cells and inflammatory cytokines, increase the expression of the stem cell markers Bmi1 and Hmga2 and the vascular endothelial growth factor, prevent stem cells, osteoblasts, and fibroblasts from aging and the spread of chronic inflammation, promote the healing of the alveolar fossa, and effectively prevent the occurrence and development of bisphosphonate-related jaw necrosis. Nakao et al. [58] injected Exos derived from gingival mesenchymal stem cells (GMSC-Exos) and preconditioned with TNF-α (GMSC-Exo-TNF-α) into a ligation-induced periodontitis model in mice, and the effect of the treatment on inflammatory bone loss was observed. The results showed that a local injection of GMSC-Exos could lessen periodontal bone damage and decrease the amount of tartrate-resistant acid phosphatase(TRAP)-positive osteoclasts. GMSC-Exo-TNF-α further enhanced this effect, indicating that preconditioning with TNF-α was beneficial for regulating inflammation and osteoclast formation. A miRNA array analysis showed that TNF-α significantly stimulated the up-regulation of miR-1260b. miR-1260b can target Wnt5a and JNK1, which relate to the regulation of osteoclasts [59]. The above results show that SC-EVs are effective and practical non cellular therapies for periodontitis, and they can improve alveolar bone regeneration of periodontal support tissue.

5. RANKL/RANK/OPG Signaling Pathway and Alveolar Bone Osteogenesis

In the intricate signaling pathway network of bone metabolism, the RANKL/RANK/OPG signaling pathway is a key link. This signaling pathway participates in the whole process of alveolar bone metabolism through both the canonical pathways of NF-κB through TRAFs, as well as the non-canonical pathways of NF-κB, respectively, by TRAFs and NIK [60–62] (Figure 3). Some relative studies have observed that both RANK and RANKL expression was widely found in the gingival epithelium and the lamina propria, while OPG was mainly located in the connective tissue [63]. In addition, the RANKL/RANK/OPG pathway is vital in periodontitis, and one of the periodontal therapies is to reduce the level of RANKL in the gingival, or rather to decrease the RANKL/OPG ratio. Confirming whether the RANKL/RANK/OPG pathway is involved in the osteogenesis of EVs expression or not, as well as understanding its mechanism, are necessary requirements for SC-EVs to become a clinical therapy.

At the starting of the method of fortify osteoclast separation and actuation, bone–lining cells and other tranquil osteoblast cells obtain signals, and, at that point, they begin to discharge cytokines related to resorb bone, such as RANKL and the macrophage colony-stimulating calculate (M-CSF) [64]. M-CSF and RANKL are parts of extracellular signaling. During osteoclastogenesis, the combination of RANKL and RANK dynamically initiates intracellular signaling, including NF-κB signaling and MAPK signaling [65]. M-CSF stimulates the monocytes to differentiate into osteoclast precursors, and RANKL induces the formation of multinucleated osteoclasts [66]. Many inflammatory mediators are involved in the inflammatory bone resorption induced by periodontitis. When toll-like receptors on the surface of osteocytes recognize lipopolysaccharides on the surface of Gram-negative bacteria, the MAPK/ERK signaling pathway is stimulated, resulting in the upregulation of interleukin-6 (IL-6) expression [67], which triggers the activation of the Janus kinase (JAK). The JAK enters the nucleus by phosphorylating a transcriptional activator (STAT) to upregulate the RANKL expression in bone cells [68]. In addition to cytokines, there are many stimulating factors and signal pathways that regulate the activity of osteoblasts and osteoclasts [69]. Two mechanical forces, tensile stress and compressive stress, are involved in the OTM process. Osteoclasts and osteoblasts are activated on the compressive side and the tensile side, respectively, and they both are differentiated from stressed PDL stem cells [70].

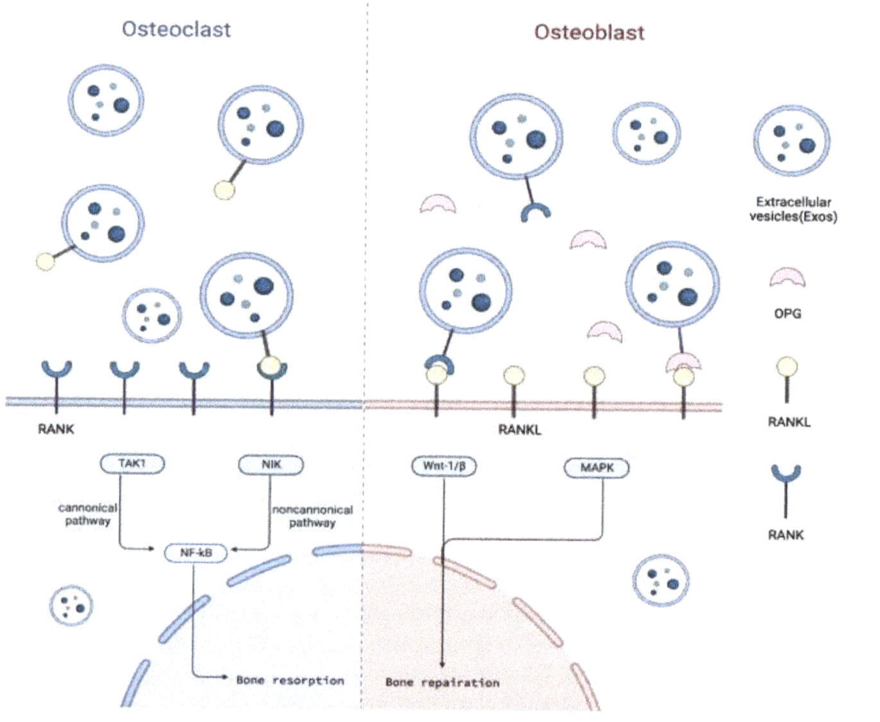

Figure 3. RANKL/RANK/OPG signaling pathway in alveolar bone. SC-EVs can act on osteoblasts and osteoclasts to adjust the balance and affect the reconstruction of alveolar bone through the RANKL/RANK/OPG pathway. (This figure was drawn using BioRender.com).

When bone resorption occurs to a certain extent, signal transduction reverses into osteoblasts. Osteoclasts are released from transforming growth factor-β, which could stimulate Wnt1 secretion and osteogenesis. Osteoclasts produce multiple factors to promote bone formation. Developed osteoclasts release exosomes, whose surfaces contain RANK, and then the exosomes combine with osteoblasts, logically activating osteoblasts. OPG released from osteoblasts blocks up excessive bone resorption by integrating self-expressed RANKL [71,72].

This cycle, among formation and resorption, is monitored precisely and builds the reconstruction of alveolar bone.

5.1. Outline of RANKL/RANK/OPG Signaling Pathway and Alveolar Bone Osteogenesis

RANK, also known as TNFRSF11, is a member of the tumor necrosis factor receptor family and is a type I trimerized transmembrane protein on osteoclasts. RANKL, also called TNFSF11, is a homomeric type II transmembrane protein on the surface of osteo-blasts.

During osteoclastogenesis, RANKL binds to RANK to activate NF-κB translocation into the nucleus, triggering osteoclast gene transcription [73]. As a soluble bait receptor of RANKL, OPG competes with RANK. Thus, it prevents the osteoclast process that is activated by the RANKL/RANK pathway and inhibits bone resorption [74].

The alveolar bone cells in the oral cavity are constantly undergoing cell metabolism, generating bone, or resorbing bone and causing the alveolar bone to undergo physiological and pathological remodeling. Physiological alveolar bone metabolism includes tooth eruptions and OTM.

Castaneda et al. [75] found that the overexpression of RANK in the alveolar bone can increase the amount of TRAP+ cells and improve osteoblast development regulator Runx2, as well as causing an acceleration of early tooth eruptions. Early in tooth and periodontal development, using a RANKL-blocking antibody could temporarily inhibit bone resorption and delay tooth eruptions [76]. It was deduced that RANKL/RANK are involved in the bone metabolism of tooth eruptions.

OTM is the process of the selective absorption and deposition of different alveolar bones under the action of force. Studies have found that, during OTM, the expression of RANK increases, and there is a certain link between RANK and osteoclastogenesis [77]. In another experiment, a low-energy laser was used to irradiate alveolar bone to stimulate osteogenesis and to speed up the movement of teeth. The research found that RANKL/RANK increased significantly in a targeted method, while OPG did not increase significantly [78].

In pathological alveolar bone metabolism, periodontitis is the main manifestation. Studies have found that WP9QY, a RANKL-binding peptide, can block RANK-induced osteoclasts, and WP9QY may be an effective drug for preventing bone loss in periodontitis. Researchers have used WP9QY to treat OPG-deficient mice with periodontitis, and micro-CT images confirmed that the interdental bone of the mandibular first molar had obvious osteogenesis and an increased bone density [79]. In an inflammatory environment, Th17 cells in periodontal connective tissue produce IL-17, which stimulates the combining of RANKL and RANK, leading to bone loss [80].

Through the above experimental results and research conclusions, it can be seen that the RANKL/RANK/OPG is involved in alveolar bone metabolism. The expression site, expression content, and inhibitors of RANKL/RANK/OPG can affect bone formation and resorption. These encourage us to still further figure out the osteogenic role of the RANKL/RANK/OPG pathway in SC-EVs.

5.2. SC-EVs in Alveolar Bone Osteogenesis through RANKL/RANK/OPG Pathway

The increased levels of RANKL in crevicular fluid or a high RANKL/OPG ratio often indicated osteopenia, a risk of alveolar bone resorption, or periodontitis [81–83]. Reducing the expression of RANKL or adjusting the propotion of RANKL/OPG is an effective strategy for preventing alveolar bone resorption and assisting osteogenesis. In the human body, osteocytes or osteoblasts tend to be spatially distant from osteoclasts, and RANKL tends to be distant from RANK, as well [84]. Tissue-engineering treatments implant stem cells or SC-EVs combined with active molecules/scaffolds into the bone defect area, which can directly promote the RANKL/RANK/OPG signaling pathway, locally, induce osteogenic differentiation, and enhance the formation of new bone [85].

Liu et al. [86] discovered periodontal osteogenic regeneration that was caused by BMSC-sEVs, which is favorable evidence for a cell-free therapy for periodontal regeneration and also affirms the effect of the RANKL/RANK/OPG pathway. They loaded the BMSC-sEVs into a hydrogel and then injected it into experimental periodontitis rats to verify the therapeutic effects. At four to eight weeks after administration, micro-CT illustrated that the alveolar bone misfortune within the BMSC-sEVs-hydrogel bunch were less than the control gather. Histological examination showed that aggravation and collagen destruction were, moreover, decreased. Encouragingly enough, the proportion of RANKL/OPG went down, and TGF-β1 and the extent of M2/M1 macrophage expanded when infused with BMSC-sEVs. From the examination, it was hypothesized that BMSC-sEVs advanced the recovery of periodontal tissue. BMSC-sEVs may intervene in the work of osteoclasts through the RANKL/RANK/OPG pathway to repress periodontal tissue harm, and they may influence macrophage polarization to control the incendiary safe reaction.

ADSCs are rich in the human body and easily extracted; they are a common source cells of SC-EVs [87,88]. Lee et al. [89] treated mice with osteoporosis in three groups: one that received an intravenous injection of ADSC-EVs (1 × 108 or 5 × 108 particles/100 µL PBS), one that received an intravenous injection of OPG-depleted ADSC-EVs (1 × 108 or 5 × 108 particles/100 µL PBS), and a control group. Bone loss was essentially diminished in ADSC-EV gathering, and BMSCs movement was advanced. In any case, the gather infused with OPG-depleted ADSC-EVs did not appear as an inclination towards a lessening in bone resorption, showing that OPG is imperative for the treatment impact of ADSC-EVs. As a non-cellular treatment for treating bone loss and assisting in osteogenesis, the restorative impact of SC-EVs depends to a certain degree on the substance of OPG.

Dental stem cells, such as GMSCs [90], apical papilla stem cells (SCAP) [91], and dental follicle stem cells (DFSCs) [92], are often used in oral stem cell therapy [93]. The advantages of using dental stem cells include the release of a large amount of EVs [93], immune regulation [94], and the ability to perform minimally invasive extractions to reduce patient discomfort [95].

Lipopolysaccharides (LPSs) are components of the outer membrane of Gram-negative bacteria, are components of the outer cell membrane of Gram-negative bacteria, and are one of the main inducements of periodontal injury. However, the damage to soft and hard tissues directly caused by LPSs is limited; they are more used as a bacterial virulence component to activate the host's immune inflammatory reaction. Studies have found that the inflammatory microenvironment induced by LPS can enhance the proliferation, differentiation, and adhesion abilities of dental follicle cell (DFC), which is conducive to periodontal regeneration, and this success could possibly be attributed to exosomes, as well [96]. The study showed that, in experimental periodontitis rats, LPS-preconditioned DFC-derived sEVs (L-DFC-sEVs) were loaded into hydrogel and applied to the treatment of periodontitis in vivo, which could partially reduce the expression of the RANKL/OPG. It is beneficial to repair the missing alveolar bone in the early stages of treatment, as well as preserve alveolar bone in the later period. Regrettably, the main components of L-DFC-sEVs and the underlying mechanism of osteogenesis were not explained by the experiment.

Based on this experimental plan, Huang et al. [97] performed a proteomic analysis on the components of L-DFC-sEVs and applied them to a canine periodontitis model. The results confirmed that, under inflammatory conditions, L-DFC-sEVs might inhibit the ROS/JNK signal pathway to decrease the proportion of PDLSCs RANKL/OPG and to promote the M2 polarization through ROS/ERK signal pathway, ultimately restraining alveolar bone loss and promoting osteogenesis. The above two experiments illustrate that the osteogenesis of SC-EVs involves the RANKL/RANK/OPG pathway, and at the same time, they emphasize that stem cells can still secrete EVs in an inflammatory environment to regulate the local inflammatory microenvironment. The collected SC-EVs had stronger antioxidant and anti-inflammatory effects than those collected from normal tissues.

6. Conclusions

In summary, many studies have shown that SC-EVs obtained from various tissues and combined with bioactive substances, such as hydrogels or scaffolds, can effectively promote osteogenesis and repair bone defects when used as bone tissue engineering technology. The possibility of using SC-EVs as a cell-free therapy is now a popular area of medical research. More and more studies have shown that the use of SC-EVs for alveolar bone osteogenesis is safe and effective, and it is expected to provide new ideas for oral clinical treatments. However, there are still some problems to be solved, such as the selection of the best source of stem cells, the selection of the appropriate diameter of EVs, the best transplantation method, the optimal number of EVs, the selection of new materials, and finally, which patients are suitable for treatment with SC-EVs, etc. Although there are still many problems to be solved, we already know that SC-EVs and the RANKL/RANK/OPG pathway have a significant connection in periodontal osteogenesis. They may represent an effective means to help clinicians solve problems, such as an insufficient periodontal bone mass, orthodontic tooth movement, or even inflammatory alveolar bone loss, such as that caused by periodontitis.

Author Contributions: X.H. and Y.L. contributed equally to this work and should be considered co-first authors. H.L., X.L., Y.Z., Y.H., Z.Z. and Q.X. made substantial, direct, and intellectual contributions to the work and approved it for publication. All authors have read and agreed to the published version of the manuscript.

Funding: This research was funded by the Natural Science Foundation of Guangdong Province, China (2021A1515012480), the Guangzhou Research and Development Plan in the Key Fields, China (202103030003), and the R&D Plan Project in the Key Fields of Guangdong Province, China (2022B1111080007). The study was also supported by the National Key Research and Development Program of China (2022YFC2403102).

Data Availability Statement: Data availability is not applicable to this article as no new data were created or analyzed in this study.

Acknowledgments: The authors thank Xu for helpful discussions on this work.

Conflicts of Interest: The authors declare that the research was conducted in the absence of any commercial or financial relationships that could be construed as potential conflict of interest.

References

1. Miron, R.J.; Moraschini, V.; Fujioka-Kobayashi, M.; Zhang, Y.; Kawase, T.; Cosgarea, R.; Jepsen, S.; Bishara, M.; Canullo, L.; Shirakata, Y.; et al. Use of platelet-rich fibrin for the treatment of periodontal intrabony defects: A systematic review and meta-analysis. *Clin. Oral Investig.* **2021**, *25*, 2461–2478. [CrossRef] [PubMed]
2. Kalra, S.; Gupta, P.; Tripathi, T.; Rai, P. External apical root resorption in orthodontic patients: Molecular and genetic basis. *J. Family Med. Prim. Care* **2020**, *9*, 3872–3882. [CrossRef] [PubMed]
3. Ul Hassan, S.; Bilal, B.; Nazir, M.S.; Naqvi, S.A.R.; Ali, Z.; Nadeem, S.; Muhammad, N.; Palvasha, B.A.; Mohyuddin, A. Recent progress in materials development and biological properties of GTR membranes for periodontal regeneration. *Chem. Biol. Drug Des.* **2021**, *98*, 1007–1024. [CrossRef] [PubMed]
4. Lu, J.; Hao, Y.; Zhao, W.; Lyu, C.; Zou, D. Molecular, Cellular and Pharmaceutical Aspects of Autologous Grafts for Peri-implant Hard and Soft Tissue Defects. *Curr. Pharm. Biotechnol.* **2017**, *18*, 85–94. [CrossRef] [PubMed]
5. Caplan, H.; Olson, S.D.; Kumar, A.; George, M.; Prabhakara, K.S.; Wenzel, P.; Bedi, S.; Toledano-Furman, N.E.; Triolo, F.; Kamhieh-Milz, J.; et al. Mesenchymal Stromal Cell Therapeutic Delivery: Translational Challenges to Clinical Application. *Front. Immunol.* **2019**, *10*, 1645. [CrossRef]
6. Bjørge, I.M.; Kim, S.Y.; Mano, J.F.; Kalionis, B.; Chrzanowski, W. Extracellular vesicles, exosomes and shedding vesicles in regenerative medicine—A new paradigm for tissue repair. *Biomater. Sci.* **2017**, *6*, 60–78. [CrossRef]
7. Russell, A.E.; Sneider, A.; Witwer, K.W.; Bergese, P.; Bhattacharyya, S.N.; Cocks, A.; Cocucci, E.; Erdbrügger, U.; Falcon-Perez, J.M.; Freeman, D.W.; et al. Biological membranes in EV biogenesis, stability, uptake, and cargo transfer: An ISEV position paper arising from the ISEV membranes and EVs workshop. *J. Extracell. Vesicles* **2019**, *8*, 1684862. [CrossRef]
8. Wen, S.W.; Lima, L.G.; Lobb, R.J.; Norris, E.L.; Hastie, M.L.; Krumeich, S.; Möller, A. Breast Cancer-Derived Exosomes Reflect the Cell-of-Origin Phenotype. *Proteomics* **2019**, *19*, e1800180. [CrossRef]
9. Yu, W.; Li, S.; Guan, X.; Zhang, N.; Xie, X.; Zhang, K.; Bai, Y. Higher yield and enhanced therapeutic effects of exosomes derived from MSCs in hydrogel-assisted 3D culture system for bone regeneration. *Biomater. Adv.* **2022**, *133*, 112646. [CrossRef]
10. Swanson, W.B.; Zhang, Z.; Xiu, K.; Gong, T.; Eberle, M.; Wang, Z.; Ma, P.X. Scaffolds with controlled release of pro-mineralization exosomes to promote craniofacial bone healing without cell transplantation. *Acta Biomater.* **2020**, *118*, 215–232. [CrossRef]
11. Diomede, F.; D'Aurora, M.; Gugliandolo, A.; Merciaro, I.; Ettorre, V.; Bramanti, A.; Piattelli, A.; Gatta, V.; Mazzon, E.; Fontana, A.; et al. A novel role in skeletal segment regeneration of extracellular vesicles released from periodontal-ligament stem cells. *Int. J. Nanomed.* **2018**, *13*, 3805–3825. [CrossRef]
12. Sharun, K.; Muthu, S.; Mankuzhy, P.D.; Pawde, A.M.; Chandra, V.; Lorenzo, J.M.; Dhama, K.; Amarpal; Sharma, G.T. Cell-free therapy for canine osteoarthritis: Current evidence and prospects. *Vet. Q.* **2022**, *42*, 224–230. [CrossRef] [PubMed]
13. Manzini, B.M.; Machado LM, R.; Noritomi, P.Y.; da Silva, J.V.L. Advances in Bone tissue engineering: A fundamental review. *J. Biosci.* **2021**, *46*, 17.
14. Kenkre, J.S.; Bassett, J. The bone remodelling cycle. *Ann. Clin. Biochem.* **2018**, *55*, 308–327. [CrossRef]
15. Hienz, S.A.; Paliwal, S.; Ivanovski, S. Mechanisms of Bone Resorption in Periodontitis. *J. Immunol. Res.* **2015**, *2015*, 615486. [CrossRef]

16. Streicher, C.; Heyny, A.; Andrukhova, O.; Haigl, B.; Slavic, S.; Schüler, C.; Kollmann, K.; Kantner, I.; Sexl, V.; Kleiter, M.; et al. Estrogen Regulates Bone Turnover by Targeting RANKL Expression in Bone Lining Cells. *Sci. Rep.* **2017**, *7*, 6460. [CrossRef] [PubMed]
17. Sucur, A.; Jajic, Z.; Artukovic, M.; Matijasevic, M.I.; Anic, B.; Flegar, D.; Markotic, A.; Kelava, T.; Ivcevic, S.; Kovacic, N.; et al. Chemokine signals are crucial for enhanced homing and differentiation of circulating osteoclast progenitor cells. *Arthritis Res. Ther.* **2017**, *19*, 142. [CrossRef]
18. Kitaura, H.; Marahleh, A.; Ohori, F.; Noguchi, T.; Shen, W.R.; Qi, J.; Nara, Y.; Pramusita, A.; Kinjo, R.; Mizoguchi, I. Osteocyte-Related Cytokines Regulate Osteoclast Formation and Bone Resorption. *Int. J. Mol. Sci.* **2020**, *21*, 5169. [CrossRef]
19. Buenzli, P.R.; Sims, N.A. Quantifying the osteocyte network in the human skeleton. *Bone* **2015**, *75*, 144–150. [CrossRef]
20. Ait Oumghar, I.; Barkaoui, A.; Chabrand, P. Toward a Mathematical Modeling of Diseases' Impact on Bone Remodeling: Technical Review. *Front. Bioeng Biotechnol.* **2020**, *8*, 584198. [CrossRef]
21. Huang, X.; Xie, M.; Xie, Y.; Mei, F.; Lu, X.; Li, X.; Chen, L. The roles of osteocytes in alveolar bone destruction in periodontitis. *J. Transl. Med.* **2020**, *18*, 479. [CrossRef] [PubMed]
22. Toms, S.R.; Lemons, J.E.; Bartolucci, A.A.; Eberhardt, A.W. Nonlinear stress-strain behavior of periodontal ligament under orthodontic loading. *Am. J. Orthod Dentofac. Orthop* **2002**, *122*, 174–179. [CrossRef] [PubMed]
23. Garlet, T.P.; Coelho, U.; Silva, J.S.; Garlet, G.P. Cytokine expression pattern in compression and tension sides of the periodontal ligament during orthodontic tooth movement in humans. *Eur. J. Oral Sci.* **2007**, *115*, 355–362. [CrossRef]
24. Urban, I.A.; Monje, A. Guided Bone Regeneration in Alveolar Bone Reconstruction. *Oral Maxillofac. Surg. Clin. N. Am.* **2019**, *31*, 331–338. [CrossRef] [PubMed]
25. Schwartz, Z.; Olivares-Navarrete, R.; Wieland, M.; Cochran, D.L.; Boyan, B.D. Mechanisms regulating increased production of osteoprotegerin by osteoblasts cultured on microstructured titanium surfaces. *Biomaterials* **2009**, *30*, 3390–3396. [CrossRef] [PubMed]
26. Sanz, M.; Herrera, D.; Kebschull, M.; Chapple, I.; Jepsen, S.; Beglundh, T.; Sculean, A.; Tonetti, M.S. Treatment of stage I-III periodontitis-The EFP S3 level clinical practice guideline. *J. Clin. Periodontol.* **2020**, *47* (Suppl. S22), 4–60. [CrossRef]
27. Jin, Y.; Ma, L.; Zhang, W.; Yang, W.; Feng, Q.; Wang, H. Extracellular signals regulate the biogenesis of extracellular vesicles. *Biol. Res.* **2022**, *55*, 35. [CrossRef]
28. Phillips, W.; Willms, E.; Hill, A.F. Understanding extracellular vesicle and nanoparticle heterogeneity: Novel methods and considerations. *Proteomics* **2021**, *21*, e2000118. [CrossRef]
29. van Niel, G.; D'Angelo, G.; Raposo, G. Shedding light on the cell biology of extracellular vesicles. *Nat. Rev. Mol. Cell Biol.* **2018**, *19*, 213–228. [CrossRef]
30. Yong, T.; Li, X.; Wei, Z.; Gan, L.; Yang, X. Extracellular vesicles-based drug delivery systems for cancer immunotherapy. *J. Control. Release* **2020**, *328*, 562–574. [CrossRef]
31. Stahl, P.D.; Raposo, G. Extracellular Vesicles: Exosomes and Microvesicles, Integrators of Homeostasis. *Physiology* **2019**, *34*, 169–177. [CrossRef] [PubMed]
32. Santavanond, J.P.; Rutter, S.F.; Atkin-Smith, G.K.; Poon, I.K.H. Apoptotic Bodies: Mechanism of Formation, Isolation and Functional Relevance. *Subcell. Biochem.* **2021**, *97*, 61–88. [CrossRef] [PubMed]
33. Mulcahy, L.A.; Pink, R.C.; Carter, D.R. Routes and mechanisms of extracellular vesicle uptake. *J. Extracell. Vesicles* **2014**, *3*, 24641. [CrossRef]
34. Théry, C.; Witwer, K.W.; Aikawa, E.; Alcaraz, M.J.; Anderson, J.D.; Andriantsitohaina, R.; Antoniou, A.; Arab, T.; Archer, F.; Atkin-Smith, G.K.; et al. Minimal information for studies of extracellular vesicles 2018 (MISEV2018): A position statement of the International Society for Extracellular Vesicles and update of the MISEV2014 guidelines. *J. Extracell. Vesicles* **2018**, *7*, 1535750. [CrossRef] [PubMed]
35. Yang, D.; Zhang, W.; Zhang, H.; Zhang, F.; Chen, L.; Ma, L.; Larcher, L.M.; Chen, S.; Liu, N.; Zhao, Q.; et al. Progress, opportunity, and perspective on exosome isolation—Efforts for efficient exosome-based theranostics. *Theranostics* **2020**, *10*, 3684–3707. [CrossRef]
36. Liga, A.; Vliegenthart, A.D.; Oosthuyzen, W.; Dear, J.W.; Kersaudy-Kerhoas, M. Exosome isolation: A microfluidic road-map. *Lab Chip* **2015**, *15*, 2388–2394. [CrossRef]
37. Tsiapalis, D.; O'Driscoll, L. Mesenchymal Stem Cell Derived Extracellular Vesicles for Tissue Engineering and Regenerative Medicine Applications. *Cells* **2020**, *9*, 991. [CrossRef]
38. Tang, Y.; Zhou, Y.; Li, H.J. Advances in mesenchymal stem cell exosomes: A review. *Stem. Cell Res. Ther.* **2021**, *12*, 71. [CrossRef]
39. Shukla, L.; Yuan, Y.; Shayan, R.; Greening, D.W.; Karnezis, T. Fat Therapeutics: The Clinical Capacity of Adipose-Derived Stem Cells and Exosomes for Human Disease and Tissue Regeneration. *Front. Pharmacol.* **2020**, *11*, 158. [CrossRef]
40. Zhang, Y.; Wang, W.T.; Gong, C.R.; Li, C.; Shi, M. Combination of olfactory ensheathing cells and human umbilical cord mesenchymal stem cell-derived exosomes promotes sciatic nerve regeneration. *Neural. Regen. Res.* **2020**, *15*, 1903–1911. [CrossRef]
41. Harrell, C.R.; Fellabaum, C.; Jovicic, N.; Djonov, V.; Arsenijevic, N.; Volarevic, V. Molecular Mechanisms Responsible for Therapeutic Potential of Mesenchymal Stem Cell-Derived Secretome. *Cells* **2019**, *8*, 467. [CrossRef] [PubMed]
42. Hur, Y.H.; Feng, S.; Wilson, K.F.; Cerione, R.A.; Antonyak, M.A. Embryonic Stem Cell-Derived Extracellular Vesicles Maintain ESC Stemness by Activating FAK. *Dev. Cell* **2021**, *56*, 277–291.e276. [CrossRef] [PubMed]

43. Trenkenschuh, E.; Richter, M.; Heinrich, E.; Koch, M.; Fuhrmann, G.; Friess, W. Enhancing the Stabilization Potential of Lyophilization for Extracellular Vesicles. *Adv. Healthc Mater.* **2022**, *11*, e2100538. [CrossRef]
44. Diomede, F.; Gugliandolo, A.; Cardelli, P.; Merciaro, I.; Ettorre, V.; Traini, T.; Bedini, R.; Scionti, D.; Bramanti, A.; Nanci, A.; et al. Three-dimensional printed PLA scaffold and human gingival stem cell-derived extracellular vesicles: A new tool for bone defect repair. *Stem. Cell Res. Ther.* **2018**, *9*, 104. [CrossRef] [PubMed]
45. Xie, H.; Wang, Z.; Zhang, L.; Lei, Q.; Zhao, A.; Wang, H.; Li, Q.; Cao, Y.; Jie Zhang, W.; Chen, Z. Extracellular Vesicle-functionalized Decalcified Bone Matrix Scaffolds with Enhanced Pro-angiogenic and Pro-bone Regeneration Activities. *Sci. Rep.* **2017**, *7*, 45622. [CrossRef] [PubMed]
46. Jin, Q.; Ma, P.X.; Giannobile, W.V. Platelet-Derived Growth Factor Delivery via Nanofibrous Scaffolds for Soft-Tissue Repair. *Adv. Skin Wound Care* **2010**, *1*, 375–381.
47. Qian, Z.; Bai, Y.; Zhou, J.; Li, L.; Na, J.; Fan, Y.; Guo, X.; Liu, H. A moisturizing chitosan-silk fibroin dressing with silver nanoparticles-adsorbed exosomes for repairing infected wounds. *J. Mater. Chem. B* **2020**, *8*, 7197–7212. [CrossRef]
48. Shen, Z.; Kuang, S.; Zhang, Y.; Yang, M.; Qin, W.; Shi, X.; Lin, Z. Chitosan hydrogel incorporated with dental pulp stem cell-derived exosomes alleviates periodontitis in mice via a macrophage-dependent mechanism. *Bioact. Mater.* **2020**, *5*, 1113–1126. [CrossRef]
49. Shang, F.; Liu, S.; Ming, L.; Tian, R.; Jin, F.; Ding, Y.; Zhang, Y.; Zhang, H.; Deng, Z.; Jin, Y. Human Umbilical Cord MSCs as New Cell Sources for Promoting Periodontal Regeneration in Inflammatory Periodontal Defect. *Theranostics* **2017**, *7*, 4370–4382. [CrossRef]
50. Yang, S.; Zhu, B.; Yin, P.; Zhao, L.; Wang, Y.; Fu, Z.; Dang, R.; Xu, J.; Zhang, J.; Wen, N. Integration of Human Umbilical Cord Mesenchymal Stem Cells-Derived Exosomes with Hydroxyapatite-Embedded Hyaluronic Acid-Alginate Hydrogel for Bone Regeneration. *ACS Biomater. Sci. Eng.* **2020**, *6*, 1590–1602. [CrossRef]
51. Wu, J.; Chen, L.; Wang, R.; Song, Z.; Shen, Z.; Zhao, Y.; Huang, S.; Lin, Z. Exosomes Secreted by Stem Cells from Human Exfoliated Deciduous Teeth Promote Alveolar Bone Defect Repair through the Regulation of Angiogenesis and Osteogenesis. *ACS Biomater. Sci. Eng.* **2019**, *5*, 3561–3571. [CrossRef] [PubMed]
52. Liu, A.; Lin, D.; Zhao, H.; Chen, L.; Cai, B.; Lin, K.; Shen, S.G. Optimized BMSC-derived osteoinductive exosomes immobilized in hierarchical scaffold via lyophilization for bone repair through Bmpr2/Acvr2b competitive receptor-activated Smad pathway. *Biomaterials* **2021**, *272*, 120718. [CrossRef]
53. Amengual-Tugores, A.M.; Ráez-Meseguer, C.; Forteza-Genestra, M.A.; Monjo, M.; Ramis, J.M. Extracellular Vesicle-Based Hydrogels for Wound Healing Applications. *Int. J. Mol. Sci.* **2023**, *24*, 4104. [CrossRef]
54. Song, X.; Xu, L.; Zhang, W. Biomimetic synthesis and optimization of extracellular vesicles for bone regeneration. *J. Control. Release* **2023**, *355*, 18–41. [CrossRef] [PubMed]
55. Belhadj, Z.; He, B.; Deng, H.; Song, S.; Zhang, H.; Wang, X.; Dai, W.; Zhang, Q. A combined "eat me/don't eat me" strategy based on extracellular vesicles for anticancer nanomedicine. *J. Extracell. Vesicles* **2020**, *9*, 1806444. [CrossRef]
56. Lei, F.; Li, M.; Lin, T.; Zhou, H.; Wang, F.; Su, X. Treatment of inflammatory bone loss in periodontitis by stem cell-derived exosomes. *Acta. Biomater.* **2022**, *141*, 333–343. [CrossRef] [PubMed]
57. Watanabe, J.; Sakai, K.; Urata, Y.; Toyama, N.; Nakamichi, E.; Hibi, H. Extracellular Vesicles of Stem Cells to Prevent BRONJ. *J. Dent. Res.* **2020**, *99*, 552–560. [CrossRef]
58. Nakao, Y.; Fukuda, T.; Zhang, Q.; Sanui, T.; Shinjo, T.; Kou, X.; Chen, C.; Liu, D.; Watanabe, Y.; Hayashi, C.; et al. Exosomes from TNF-α-treated human gingiva-derived MSCs enhance M2 macrophage polarization and inhibit periodontal bone loss. *Acta. Biomater.* **2021**, *122*, 306–324. [CrossRef]
59. David, J.P.; Sabapathy, K.; Hoffmann, O.; Idarraga, M.H.; Wagner, E.F. JNK1 modulates osteoclastogenesis through both c-Jun phosphorylation-dependent and -independent mechanisms. *J. Cell Sci.* **2002**, *115*, 4317–4325. [CrossRef]
60. Mori, K.; Mizokami, A.; Sano, T.; Mukai, S.; Hiura, F.; Ayukawa, Y.; Koyano, K.; Kanematsu, T.; Jimi, E. RANKL elevation activates the NIK/NF-κB pathway, inducing obesity in ovariectomized mice. *J. Endocrinol.* **2022**, *254*, 27–36. [CrossRef]
61. Sokos, D.; Everts, V.; de Vries, T.J. Role of periodontal ligament fibroblasts in osteoclastogenesis: A review. *J. Periodontal. Res.* **2015**, *50*, 152–159. [CrossRef] [PubMed]
62. Wang, J.; Wang, B.; Lv, X.; Wang, L. NIK inhibitor impairs chronic periodontitis via suppressing non-canonical NF-κB and osteoclastogenesis. *Pathog. Dis.* **2020**, *78*, ftaa045. [CrossRef] [PubMed]
63. Guo, Y.; Xu, C.; Wu, X.; Zhang, W.; Sun, Y.; Shrestha, A. Leptin regulates OPG and RANKL expression in Gingival Fibroblasts and Tissues of Chronic Periodontitis Patients. *Int. J. Med. Sci.* **2021**, *18*, 2431–2437. [CrossRef] [PubMed]
64. Robling, A.G.; Bonewald, L.F. The Osteocyte: New Insights. *Annu. Rev. Physiol.* **2020**, *82*, 485–506. [CrossRef] [PubMed]
65. Xu, H.; Liu, T.; Jia, Y.; Li, J.; Jiang, L.; Hu, C.; Wang, X.; Sheng, J. (-)-Epigallocatechin-3-gallate inhibits osteoclastogenesis by blocking RANKL-RANK interaction and suppressing NF-κB and MAPK signaling pathways. *Int. Immunopharmacol.* **2021**, *95*, 107464. [CrossRef] [PubMed]
66. Yao, Z.; Getting, S.J.; Locke, I.C. Regulation of TNF-Induced Osteoclast Differentiation. *Cells* **2021**, *11*, 132. [CrossRef]
67. Sakamoto, E.; Kido, J.I.; Takagi, R.; Inagaki, Y.; Naruishi, K.; Nagata, T.; Yumoto, H. Advanced glycation end-product 2 and Porphyromonas gingivalis lipopolysaccharide increase sclerostin expression in mouse osteocyte-like cells. *Bone* **2019**, *122*, 22–30. [CrossRef]

68. Mihara, M.; Hashizume, M.; Yoshida, H.; Suzuki, M.; Shiina, M. IL-6/IL-6 receptor system and its role in physiological and pathological conditions. *Clin. Sci.* **2012**, *122*, 143–159. [CrossRef]
69. Dallas, S.L.; Prideaux, M.; Bonewald, L.F. The osteocyte: An endocrine cell . . . and more. *Endocr. Rev.* **2013**, *34*, 658–690. [CrossRef]
70. Masella, R.S.; Meister, M. Current concepts in the biology of orthodontic tooth movement. *Am. J. Orthod. Dentofacial. Orthop.* **2006**, *129*, 458–468. [CrossRef]
71. Udagawa, N.; Koide, M.; Nakamura, M.; Nakamichi, Y.; Yamashita, T.; Uehara, S.; Kobayashi, Y.; Furuya, Y.; Yasuda, H.; Fukuda, C.; et al. Osteoclast differentiation by RANKL and OPG signaling pathways. *J. Bone Miner Metab* **2021**, *39*, 19–26. [CrossRef]
72. Weivoda, M.M.; Ruan, M.; Pederson, L.; Hachfeld, C.; Davey, R.A.; Zajac, J.D.; Westendorf, J.J.; Khosla, S.; Oursler, M.J. Osteoclast TGF-β Receptor Signaling Induces Wnt1 Secretion and Couples Bone Resorption to Bone Formation. *J. Bone Miner Res.* **2016**, *31*, 76–85. [CrossRef]
73. Liu, W.; Zhang, X. Receptor activator of nuclear factor-κB ligand (RANKL)/RANK/osteoprotegerin system in bone and other tissues (review). *Mol. Med. Rep.* **2015**, *11*, 3212–3218. [CrossRef]
74. Simonet, W.S.; Lacey, D.L.; Dunstan, C.R.; Kelley, M.; Chang, M.S.; Lüthy, R.; Nguyen, H.Q.; Wooden, S.; Bennett, L.; Boone, T.; et al. Osteoprotegerin: A novel secreted protein involved in the regulation of bone density. *Cell* **1997**, *89*, 309–319. [CrossRef]
75. Castaneda, B.; Simon, Y.; Jacques, J.; Hess, E.; Choi, Y.W.; Blin-Wakkach, C.; Mueller, C.; Berdal, A.; Lézot, F. Bone resorption control of tooth eruption and root morphogenesis: Involvement of the receptor activator of NF-κB (RANK). *J. Cell Physiol.* **2011**, *226*, 74–85. [CrossRef]
76. Lézot, F.; Chesneau, J.; Navet, B.; Gobin, B.; Amiaud, J.; Choi, Y.; Yagita, H.; Castaneda, B.; Berdal, A.; Mueller, C.G.; et al. Skeletal consequences of RANKL-blocking antibody (IK22-5) injections during growth: Mouse strain disparities and synergic effect with zoledronic acid. *Bone* **2015**, *73*, 51–59. [CrossRef]
77. Noguchi, T.; Kitaura, H.; Ogawa, S.; Qi, J.; Shen, W.R.; Ohori, F.; Marahleh, A.; Nara, Y.; Pramusita, A.; Mizoguchi, I. TNF-α stimulates the expression of RANK during orthodontic tooth movement. *Arch. Oral Biol.* **2020**, *117*, 104796. [CrossRef] [PubMed]
78. Fujita, S.; Yamaguchi, M.; Utsunomiya, T.; Yamamoto, H.; Kasai, K. Low-energy laser stimulates tooth movement velocity via expression of RANK and RANKL. *Orthod. Craniofac. Res.* **2008**, *11*, 143–155. [CrossRef] [PubMed]
79. Ozaki, Y.; Koide, M.; Furuya, Y.; Ninomiya, T.; Yasuda, H.; Nakamura, M.; Kobayashi, Y.; Takahashi, N.; Yoshinari, N.; Udagawa, N. Treatment of OPG-deficient mice with WP9QY, a RANKL-binding peptide, recovers alveolar bone loss by suppressing osteoclastogenesis and enhancing osteoblastogenesis. *PLoS ONE* **2017**, *12*, e0184904. [CrossRef] [PubMed]
80. Tsukasaki, M.; Takayanagi, H. Osteoimmunology: Evolving concepts in bone-immune interactions in health and disease. *Nat. Rev. Immunol.* **2019**, *19*, 626–642. [CrossRef] [PubMed]
81. Giannopoulou, C.; Martinelli-Klay, C.P.; Lombardi, T. Immunohistochemical expression of RANKL, RANK and OPG in gingival tissue of patients with periodontitis. *Acta. Odontol. Scand.* **2012**, *70*, 629–634. [CrossRef] [PubMed]
82. Paula-Silva, F.W.G.; Arnez, M.F.M.; Petean, I.B.F.; Almeida-Junior, L.A.; da Silva, R.A.B.; da Silva, L.A.B.; Faccioli, L.H. Effects of 5-lipoxygenase gene disruption on inflammation, osteoclastogenesis and bone resorption in polymicrobial apical periodontitis. *Arch. Oral Biol.* **2020**, *112*, 104670. [CrossRef] [PubMed]
83. Jiao, M.; Zhang, P.; Yu, X.; Sun, P.; Liu, M.; Qiao, Y.; Pan, K. Osteoprotegerin/receptor activator of nuclear factor-κB ligand are involved in periodontitis-promoted vascular calcification. *Exp. Ther. Med.* **2022**, *24*, 512. [CrossRef] [PubMed]
84. Holliday, L.S.; Patel, S.S.; Rody, W.J., Jr. RANKL and RANK in extracellular vesicles: Surprising new players in bone remodeling. *Extracell. Vesicles Circ. Nucl. Acids* **2021**, *2*, 18–28. [CrossRef]
85. Ho, M.L.; Hsu, C.J.; Wu, C.W.; Chang, L.H.; Chen, J.W.; Chen, C.H.; Huang, K.C.; Chang, J.K.; Wu, S.C.; Shao, P.L. Enhancement of Osteoblast Function through Extracellular Vesicles Derived from Adipose-Derived Stem Cells. *Biomedicines* **2022**, *10*, 1752. [CrossRef]
86. Liu, L.; Guo, S.; Shi, W.; Liu, Q.; Huo, F.; Wu, Y.; Tian, W. Bone Marrow Mesenchymal Stem Cell-Derived Small Extracellular Vesicles Promote Periodontal Regeneration. *Tissue Eng. Part A* **2021**, *27*, 962–976. [CrossRef] [PubMed]
87. Soltani, A.; Moradi, M.; Nejad, A.R.; Moradi, S.; Javandoost, E.; Nazari, H.; Jafarian, A. Adipose-derived stem cells: Potentials, availability and market size in regenerative medicine. *Curr. Stem Cell Res. Ther.* **2022**, *18*, 347–349. [CrossRef]
88. Si, Z.; Wang, X.; Sun, C.; Kang, Y.; Xu, J.; Wang, X.; Hui, Y. Adipose-derived stem cells: Sources, potency, and implications for regenerative therapies. *Biomed. Pharm.* **2019**, *114*, 108765. [CrossRef]
89. Lee, K.S.; Lee, J.; Kim, H.K.; Yeom, S.H.; Woo, C.H.; Jung, Y.J.; Yun, Y.E.; Park, S.Y.; Han, J.; Kim, E.; et al. Extracellular vesicles from adipose tissue-derived stem cells alleviate osteoporosis through osteoprotegerin and miR-21-5p. *J. Extracell. Vesicles* **2021**, *10*, e12152. [CrossRef]
90. Zarubova, J.; Hasani-Sadrabadi, M.M.; Dashtimoghadam, E.; Zhang, X.; Ansari, S.; Li, S.; Moshaverinia, A. Engineered Delivery of Dental Stem-Cell-Derived Extracellular Vesicles for Periodontal Tissue Regeneration. *Adv. Healthc Mater.* **2022**, *11*, e2102593. [CrossRef] [PubMed]
91. Su, X.; Yang, H.; Shi, R.; Zhang, C.; Liu, H.; Fan, Z.; Zhang, J. Depletion of SNRNP200 inhibits the osteo-/dentinogenic differentiation and cell proliferation potential of stem cells from the apical papilla. *BMC Dev. Biol.* **2020**, *20*, 22. [CrossRef]
92. Ma, L.; Rao, N.; Jiang, H.; Dai, Y.; Yang, S.; Yang, H.; Hu, J. Small extracellular vesicles from dental follicle stem cells provide biochemical cues for periodontal tissue regeneration. *Stem Cell Res. Ther.* **2022**, *13*, 92. [CrossRef] [PubMed]

93. Li, Y.; Duan, X.; Chen, Y.; Liu, B.; Chen, G. Dental stem cell-derived extracellular vesicles as promising therapeutic agents in the treatment of diseases. *Int. J. Oral Sci.* **2022**, *14*, 2. [CrossRef] [PubMed]
94. Bi, R.; Lyu, P.; Song, Y.; Li, P.; Song, D.; Cui, C.; Fan, Y. Function of Dental Follicle Progenitor/Stem Cells and Their Potential in Regenerative Medicine: From Mechanisms to Applications. *Biomolecules* **2021**, *11*, 997. [CrossRef] [PubMed]
95. Fawzy El-Sayed, K.M.; Dörfer, C.E. Gingival Mesenchymal Stem/Progenitor Cells: A Unique Tissue Engineering Gem. *Stem Cells Int.* **2016**, *2016*, 7154327. [CrossRef] [PubMed]
96. Shi, W.; Guo, S.; Liu, L.; Liu, Q.; Huo, F.; Ding, Y.; Tian, W. Small Extracellular Vesicles from Lipopolysaccharide-Preconditioned Dental Follicle Cells Promote Periodontal Regeneration in an Inflammatory Microenvironment. *ACS Biomater. Sci. Eng.* **2020**, *6*, 5797–5810. [CrossRef]
97. Huang, Y.; Liu, Q.; Liu, L.; Huo, F.; Guo, S.; Tian, W. Lipopolysaccharide-Preconditioned Dental Follicle Stem Cells Derived Small Extracellular Vesicles Treating Periodontitis via Reactive Oxygen Species/Mitogen-Activated Protein Kinase Signaling-Mediated Antioxidant Effect. *Int. J. Nanomed.* **2022**, *17*, 799–819. [CrossRef]

Journal of
Functional Biomaterials

Systematic Review

Cytotoxicity of Biodegradable Zinc and Its Alloys: A Systematic Review

Qian Liu [1,†], An Li [1,†], Shizhen Liu [2], Qingyun Fu [1], Yichen Xu [3], Jingtao Dai [1], Ping Li [1,*] and Shulan Xu [1,*]

1 Center of Oral Implantology, Stomatological Hospital, School of Stomatology, Southern Medical University, Guangzhou 510280, China
2 The School of Computing Science, University of Glasgow, Glasgow G12 8RZ, UK
3 State Key Laboratory of Oral Diseases & National Clinical Research Center for Oral Diseases, Department of Oral Prosthodontics, West China Hospital of Stomatology, Sichuan University, Chengdu 610041, China
* Correspondence: ping_li_88@smu.edu.cn (P.L.); xushulan_672588@smu.edu.cn (S.X.)
† These authors contributed equally to this work.

Abstract: Zinc-based biodegradable metals (BMs) have been developed for biomedical implant materials. However, the cytotoxicity of Zn and its alloys has caused controversy. This work aims to investigate whether Zn and its alloys possess cytotoxic effects and the corresponding influence factors. According to the guidelines of the PRISMA statement, an electronic combined hand search was conducted to retrieve articles published in PubMed, Web of Science, and Scopus (2013.1–2023.2) following the PICOS strategy. Eighty-six eligible articles were included. The quality of the included toxicity studies was assessed utilizing the ToxRTool. Among the included articles, extract tests were performed in 83 studies, and direct contact tests were conducted in 18 studies. According to the results of this review, the cytotoxicity of Zn-based BMs is mainly determined by three factors, namely, Zn-based materials, tested cells, and test system. Notably, Zn and its alloys did not exhibit cytotoxic effects under certain test conditions, but significant heterogeneity existed in the implementation of the cytotoxicity evaluation. Furthermore, there is currently a relatively lower quality of current cytotoxicity evaluation in Zn-based BMs owing to the adoption of nonuniform standards. Establishing a standardized in vitro toxicity assessment system for Zn-based BMs is required for future investigations.

Keywords: zinc; zinc alloy; biodegradable metals; absorbable metals; cytotoxicity

Citation: Liu, Q.; Li, A.; Liu, S.; Fu, Q.; Xu, Y.; Dai, J.; Li, P.; Xu, S. Cytotoxicity of Biodegradable Zinc and Its Alloys: A Systematic Review. *J. Funct. Biomater.* **2023**, *14*, 206. https://doi.org/10.3390/jfb14040206

Academic Editors: Masaaki Nakai and Elisa Boanini

Received: 16 February 2023
Revised: 18 March 2023
Accepted: 27 March 2023
Published: 7 April 2023

1. Introduction

Increasing attention has been dedicated to biodegradable metals (BMs) due to their potential to replace permanent implant materials for those of temporary function. BMs, mainly those based on magnesium (Mg), iron (Fe), zinc (Zn), and their alloys or composites, are expected to degrade gradually and leave no residues in vivo [1–3]. Zn-based BMs have been developed and investigated as potential implant materials since they have a moderate degradation rate and superior mechanical properties, which are suitable for clinical applications. Bowen et al. conducted a landmark study on the in vivo performance of pure Zn, demonstrating its excellent in vivo biocompatibility and suitable biodegradability for cardiac stent applications [4]. These findings have inspired researchers from various fields to investigate Zn-based BMs.

Currently, the main focus of the clinical application of novel Zn-based BMs lies in their use as vascular stents [5,6], surgical sutures [7,8], and craniomaxillofacial and orthopedic implants [9]. Excellent biocompatibility is a prerequisite for biomedical implant materials. Unlike Ti and its alloys, whose biocompatibility has been established for medical use [10,11], the toxicity of Zn and its alloys remains a subject of debate. Considering the 3Rs principles in animal research, i.e., to reduce, refine, or replace the use of animals in biomedical research, the in vitro assessment is indispensable in estimating the biocompatibility of the novel Zn-based BMs [12,13]. Although ISO 10993-5/12 standards provided a rapid and sensitive approach to assess the potential toxicity of substances [14], conflicting reports on the cytotoxicity of Zn-based BMs have thrown their biosafety into further confusion [15,16].

This systematic review aims to clarify the cytotoxicity of Zn and its alloys according to evidence-based biomaterials research retrieved through an extensive and comprehensive search strategy [17].

2. Materials and Methods

2.1. Quality Assurance and Criteria

This review was conducted in accordance with the Cochrane Handbook for Systematic Reviews of Interventions and is based on the handbook from the Office of Health Assessment and Translation (OHAT—NIH) for in vitro toxicological studies. We searched for the study articles according to the PICOS (patient, intervention, comparison, outcome, and study design) framework and reported according to PRISMA guidelines (Table S2). Two researchers strictly followed the PICOS strategy and extracted data independently after reviewing the titles, abstracts, and full-text articles. The quality assessment was performed using the Toxicological Data Reliability Assessment Tool (ToxRTool).

2.2. Search Strategy

The PICOS framework was followed as the basis of a search strategy, involving the following factors:

- Population (P): cells.
- Intervention (I): biodegradable Zn and its alloys.
- Comparison (C): nonbiodegradable metals, such as stainless steel, titanium, titanium alloy, and cobalt–chromium alloy; biodegradable polymers, such as polylactic acid; other biodegradable metals, such as Mg-based BMs.
- Outcome (O): cell viability.
- Study design (S): in vitro study.

An electronic search was performed utilizing the PubMed, Scopus, and Web of Science databases of articles published up to 28 September 2022. The search was implemented using a combination of medical subject heading (MeSH) terms and free words. The search strategies that were developed for each database are given in Table S1. Furthermore, a hand search of reference lists for potential eligibility of included articles was performed. A supplementary search was conducted on 1 February 2023 to update the references in a timely fashion.

2.3. Inclusion and Exclusion Criteria

The inclusion criteria were studies in the English language that conducted cytotoxicity assessment of biodegradable Zn or its alloys, in which experiments were implemented according to the ISO (International Organization for Standardization) 10993-5 or 10993-12 standards. The exclusion criteria were review articles, clinical studies, in vivo animal studies, case reports, retrospective studies, editorials, opinions, guidelines, conferences, and commentary articles.

2.4. Study Selection and Data Extraction

The database search outputs, generated using established search strategies, were imported into Endnote (Version X9.1) to remove the duplicate publications, and then the Rayyan website was used for blinded screening. Two researchers (Q.L. and P.L.) independently screened the literature and extracted the data in strict accordance with the inclusion/exclusion criteria, and then the preestablished data extraction checklist. Disagreements on the eligibility of studies were resolved through careful discussion, and any remaining disputes were resolved by a third researcher (A.L.).

For data extraction, the pre-established data extraction table, which contained basic and experimental information, was utilized for data extraction and analyzed by Microsoft

Office Excel 2013. The basic information comprised the year of publication and the author(s). The extracted experimental data were divided into three parts: the Zn-based material, tested cells, and test system. These items included test materials, material processing, types of cell lines, test methodologies, extract ratio and immersion time, dilution ratio, exposure time, use of controls, selected assays, other parameters, and study outcomes.

2.5. Assessment of Quality of Evidence

Two researchers (Q.L. and P.L.) independently conducted the reliability assessment using the ToxRTool, which was developed by the European Center for the Validation of Alternative Methods (ECVAM). ToxRTool provides comprehensive criteria and guidance for evaluating the inherent quality of toxicological data or reliability at the methodological level. The in vitro part of this tool consists of an 18-point rating checklist, which is grouped into the following five aspects: (1) test substance identification; (2) test substance characterization; (3) study design description; (4) study results documentation; (5) plausibility of study design and data. Each criterion can be graded as "1" (i.e., "criterion met") or "0" (i.e., "criterion not met" or not reported). Then, the results of 18 criteria were combined to determine the overall quality of the included articles. According to the reliability categorization, articles with 15–18 points were considered reliable without restrictions, studies with 11–14 points were reliable with possible restrictions, and studies with fewer than 11 points were considered unreliable. Furthermore, according to ToxRTool, items 1, 8–12, and 17 were highlighted in red, indicating particular importance. Regardless of the quality assessment score, studies that failed to meet all of the abovementioned red item criteria were classified as unreliable.

3. Results

3.1. Included Studies

The selection process for the included articles is presented in Figure 1. According to the inclusion and exclusion criteria, a total of 86 articles published between 2013 to 2023 were included in this review. The main characteristics of the included articles are summarized in Table 1.

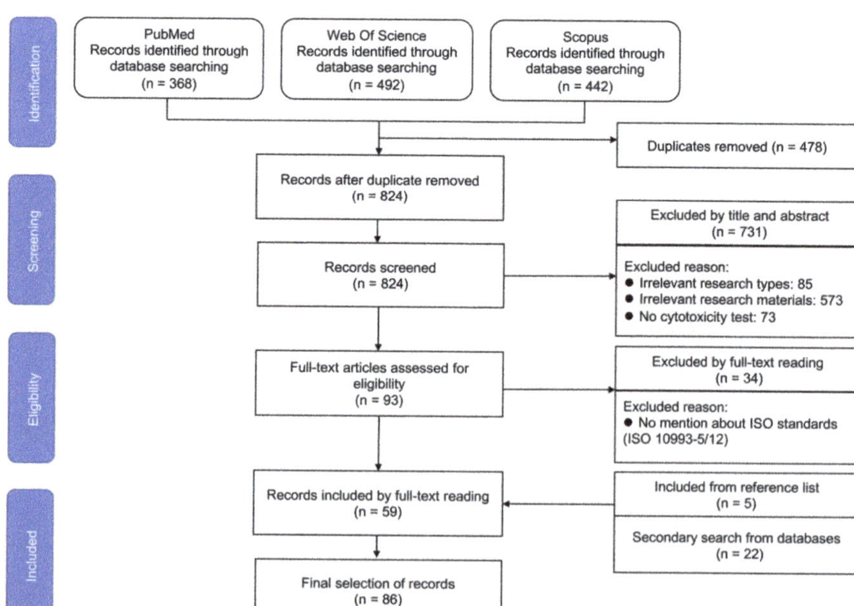

Figure 1. Flow diagram of the screening and selection process, according to the PRISMA statement.

Table 1. Main characteristics of the included articles.

Author /Year Ref.	Composition and Processing	Cell Line	Test Type (E, D)	Setup (SA: V, Time) and Extract Concentration (%)	Exposure Time	Negative Control	Positive Control	Assays	Others	Outcome
J. Cheng /2013 [18]	Zn (NA)	L929 ECV-304	E	1.25 cm^2/mL, 72 h NA	E: 1, 2, 4 days	CCM	CCM with 10% DMSO	MTT	SF	Zn showed no cytotoxicity toward ECV304 cells, but could significantly reduce the cell viability of L929 cells.
M.S. Dambatta /2015 [19]	Zn Zn–3Mg (AC)	NHOst (P7)	E	0.1, 0.5, 1.0, 2.0 mg/mL, 72 h NA	E: 1, 3, 7 days	CCM	NA	MTS	Filtered	The alloy's extract toward NHOst cells at low concentrations was cytocompatible (<0.5 mg/mL).
H. Gong /2015 [20]	Zn–1Mg (HE)	L929	E	Radio: NA, 72 h 6.25%	E: 24, 72 h	CCM	NA	MTS	SF, Filtered	Zn–1Mg alloy was biocompatible.
J. Kubásek /2015 [21]	Pure Zn Zn–0.8Mg Zn–1.6Mg (AC, HE)	U-2 OS L929 (P3)	E	NA 75%, 50%, 25%	E: 24 h	CCM with 5% FBS	CCM with 5% FBS and 0.64% phenol	WST-1		The maximum safe concentrations of Zn^{2+} for the U-2OS and L929 cells were 120 µM and 80 µM, respectively.
N.S. Murni /2015 [22]	Zn–3Mg Zn (AC)	NHOst primary cells	E	0.75 mg/mL, 72 h NA	E: 1, 3, 7 days	CCM	NA	MTS Annexin V/PI FITC–phalloidin		Zn–3Mg alloy extract exhibited adjustable cytotoxic effects on normal human osteoblast cells at the concentration of 0.75 mg/mL.
Z. Tang /2016 [23]	Zn Zn–3Cu–xMg (x = 0, 0.1, 0.5, 1.0 wt.%) (AC)	EA.hy926	E	1.25 cm^2/mL, 72 h 10%, 50%, 100%	E: 1, 3, 5 days	NA	NA	CCK-8		Zn–3Cu–xMg alloys were biocompatible.
J. Niu /2016 [24]	Zn–4wt.%Cu (AC, HE)	EA.hy926	E	1.25 cm^2/mL, 72 h 10%, 50%, 100%	E: 1, 3, 5 days	Ti	NA	CCK-8		Zn–4Cu presented acceptable toxicity toward human endothelial cells.
E. Jablonská /2016 [25]	Zn–1.5Mg (AC)	L929 U-2 OS	E & D	87.5 cm^2/mL, 24 h 100%, 50%	E: 1 day D: 24 h	E: CCM D: Untreated sample	NA	WST-1 DAPI	Pre-incubation	Pre-incubation significantly increased metabolic activity of L929 in indirect test, as well as number of U-2OS cells adhered to the surface of the alloy.

Table 1. *Cont.*

Author /Year Ref.	Composition and Processing	Cell Line	Test Type (E, D)	Setup (SA: V, Time) and Extract Concentration (%)	Exposure Time	Negative Control	Positive Control	Assays	Others	Outcome
C. Wang /2016 [26]	Zn ZA4-1 ZA4-3 ZA6-1 (HE)	HUVECs	E	1.25 mL/cm^2, 24 h 100%, 50%	E: 1, 2, 4 days	CCM	NA	CCK-8	SF	Cytotoxic effect was found in 100% extracts of both pure Zn and Zn alloys, while no cytotoxicity was observed after dilution.
C. Shen /2016 [27]	Zn–1.22%Mg (AC, HE)	HOS MG-63	E	1.25 cm^2/mL, 48 h 100%, 75%, 50%, 25%, 12.5%	E: 3 days	CCM	CCM with 5% DMSO	MTT		The as-extruded alloy had no potential cytotoxicity and tolerance in cellular applications.
G. Levy /2017 [28]	Zn–1%Mg Zn–1%Mg–0.5%Ca (AC)	Saos-2	E	1.25 cm^2/mL, 24 h NA	E: 24, 48 h	Cells in CCM	CCM with 10% DMSO	CCK-8	Pre-incubation	The safety of all the tested zinc alloys was established in terms of their toxic effect on cells.
Z. Tang /2017 [29]	Zn Zn–xCu (x = 1, 2, 3, 4 wt.%) (AC, HE)	EA.hy926	E	1.25 cm^2/mL, 72 h 10%, 50%, 100%	E: 1, 3, 5 days	NA	NA	CCK-8		Zn–xCu alloys were cytocompatible with human endothelial cells.
D. Zhu /2017 [30]	Pure Zn (NA)	hMSCs	E & D	1.25 mL/cm^2, 7 days Zn ion (20–30 μM)	E: 1, 7, 14 days D: 14 days	Cells in CCM	No cells in CCM	MTT Calcein- AM		Cell motility was higher on Zn than on AZ31.
T. Ren /2018 [31]	Zn–xMg0.5Zr (x = 0.5, 1, 1.5 wt.%) Zn (AC)	L929	E	Ratio: NA, 24 h 100%, 50%, 25%	E: 1, 2, 3 days	NA	NA	MTT		The Zn–Mg–Zr alloys showed nontoxicity through in vitro cytotoxicity tests.
X. Tong /2018 [32]	Zn Zn–Ge (HE, HR)	MC3T3-E1	E	1.25 mL/cm^2, 72 h 100%, 50%, 25%, 12.5%	E: 3 days	NA	NA	CCK-8		The <12.5% extracts of both the as-cast Zn–5Ge alloy and pure Zn showed grade 0 cytotoxicity.
N. Annonay /2018 [33]	Zn ZnZr (RF magnetron co-sputtering)	HUVECs	D	NA	E: 72 h	NA	NA	MTT Resazurin		Human endothelial cells indicated good cytocompatibility of both amorphous and crystalline films with zinc content above 80% at such thin metallic glass layers.

Table 1. *Cont.*

Author /Year Ref.	Composition and Processing	Cell Line	Test Type (E, D)	Setup (SA: V, Time) and Extract Concentration (%)	Exposure Time	Negative Control	Positive Control	Assays	Others	Outcome
Y. Chen /2018 [34]	Pure Zn (AE)	PRMECs	E	1.25 cm^2/mL, 24 h 100%, 80%, 60%, 40%, 20%	NA	Ti	NA	CCK-8 Calcein-AM/PI		100% and 80% pure zinc extracts were Grade 1, while 60%, 40%, and 20% extracts were Grade 0.
C. Xiao /2018 [35]	Zn Zn–0.05Mg (AC, AE)	L929	E	1.25 mL/cm^2, 72 h 100%, 50%, 10%	E: 1, 3, 5 days	CCM	CCM with 0.64% phenol	MTT		Zn and Zn–0.05Mg alloy were safe for cellular applications with a cytotoxicity grade of 0–1 to L929 cells.
P. Li /2018 [36]	Zn–4.0Ag Pure Zn (AC)	L929 Saos-2	E	3 mL/cm^2, NA 10%, 16.7%, 33.3%, 100%	E: 24, 48 h	Ti	Cu	XTT BrdU		A cytotoxic effect that decreased the viability and proliferation of L929 and Saos-2 cells was only observed in the undiluted extracts of the Zn–4Ag alloy.
X. Tong /2019 [37]	Zn–Cu foam (ED)	MC3T3-E1	E	0.2 g/mL, 72 h 100%, 50%, 25%, 12.5%	E: 1, 3, 5 days	NA	NA	CCK-8		The 100% and 50% concentrations of the extract showed clear cytotoxicity.
Y. Zhang /2019 [38]	Zn 0.5%Li (AC, HE)	BMSCs	E	Ratio: NA, 72 h 100%, 50%, 10%	NA	CCM	NA	CCK-8		The alloy was not toxic to BMSCs.
Y. Li /2019 [39]	Porous Zn (AM)	MG-63	D & E	0.2 g/mL, 72 h 10%	E: 0, 24, 48, 72 h D: 24 h	Ti	20% DMSO	MTS	Filtered	The AM porous Zn exhibited good biocompatibility in vitro.
H. Guo /2019 [40]	Pure Zn (HE, CD)	HUVECs	E	1.25 cm^2/mL, 24 h 100%, 50%, 10%	E: 1, 3, 5 days	CCM	NA	CCK-8	SF	The ϕ 0.3 mm pure Zn wire presented benign cytocompatibility in 100% concentration extract, whereas the ϕ 3.0 mm pure Zn wire exhibited higher cytotoxicity in 100% concentration extract.
Z. Shi /2019 [41]	Zn–0.8Mn Zn–0.8Mn–0.4X (X = Ag, Cu, Ca) (AC)	L929	E	0.2 g/mL 100%, 80%, 60%, 40% 20%	E: 48 h	100% HDPE extract	The medium with 10% FBS and 10% DMSO	MTT		The addition of Cu or Ca obviously alleviated the cytotoxic potential of Zn-0.8Mn alloy.

Table 1. *Cont.*

Author /Year Ref.	Composition and Processing	Cell Line	Test Type (E, D)	Setup (SA: V, Time) and Extract Concentration (%)	Exposure Time	Negative Control	Positive Control	Assays	Others	Outcome
P. Li /2019 [42]	Zn–4Ag Zn (AC)	TAg	E	1.25 cm^2/mL, NA 100%, 50%, 25%, 10%, 5%, 2%	E: 2, 6, 12 days	CCM and in osteogenic media	NA	CCK-8		Compared with pure Zn, the Zn–4Ag alloy seemed to exhibit no adverse cytotoxic effects on TAg cells.
P. Li /2019 [16]	Zn Zn–4Ag Zn–2Ag–1.8Au–0.2V (NA)	L929 Saos-2	E	3 cm^2/mL, NA NA	E: 24 h	Ti	Cu	FDA/EB CCK-8		Decreased cytotoxicity was observed in the extract media without FBS.
S. Lin /2019 [43]	Pure Zn Zn–0.02Mg (AC)	HUVECs	E	1.25 cm^2/mL, 72 h NA	E: 1, 3 days	NA	NA	MTS		Zn–0.02Mg alloy extracts promoted HUVEC activity after 1 and 3 days of incubation.
P. Li /2019 [44]	Zn Zn–xCu (x = 1, 2, 4 wt.%) (AC)	L929 Tag Saos-2	E	1.25 cm^2/mL, 24 h NA	E: 24 h	Ti	Cu	FDA/EB CCK-8 BrdU		As-rolled Zn–4Cu alloy exhibited no apparent cytotoxic effect toward L929, TAg, or Saos-2 cells.
Y. Zhang /2019 [45]	Zn–0.8%Li Zn–0.8%Li–0.2%X (X = Li, Ag) (AC)	L929 BMSCs	E	Ratio: NA, 72 h 100%, 50%, 10%	E: 1, 3, 5 days	CCM	CCM containing 0.64% phenol	CCK-8		The cytotoxicity of these extracts of Zn–Li–Ag alloy was of Grade 0–1.
D. Zhu /2019 [46]	Pure Zn Zn−1.5%Sr Zn−1.5%Mg (AC, HR, AE)	HCAECs (P4-6 HOBs hMSCs	D & E	NA 10%, 25%, 50%	E: 5 days D: 5 days	CCM	NA	MTT CyQUANT		The measured cell viability and proliferation of three different human primary cells fared better for Zn biomaterials than AZ31.
C. Shuai /2020 [47]	Zn–Al Zn–Al–2Sn (SLM)	MG-63	D & E	1.25 cm^2/mL, 72 h NA	E: 1, 3, 5 days D: 24 h	NA	NA	CCK-8		Zn–Al–2Sn alloy had acceptable cytocompatibility.
C. Chen /2020 [48]	Zn–1.5Cu–1.5Ag Zn (AC, AE)	EA.hy926	E	1.25 cm^2/mL, 72 h 20%, 50%	E: 1, 2, 3 days	NA	NA	CCK-8		The as-extruded alloy exhibited good biocompatibility at cellular level.

Table 1. *Cont.*

Author /Year Ref.	Composition and Processing	Cell Line	Test Type (E, D)	Setup (SA: V, Time) and Extract Concentration (%)	Exposure Time	Negative Control	Positive Control	Assays	Others	Outcome
O. Avior /2020 [49]	Zn Zn–2%Fe Zn–2%Fe–xCa (x = 0.3, 0.6, 1, 1.6 wt.%) (AC)	4T1	E	1.25 cm^2/mL, 24 h NA	E: 24, 48 h	CCM	CCM with 90% DMEM and 10% DMSO	XTT	Filtered	All the tested alloys can be noncytotoxic substances regarding 4T1 cells.
Z. Zhang /2020 [50]	Zn–0.3Fe (AC, BCWC)	HUVECs	E	1/3 mL/cm^2, 24 h 25%–100%	E: 24 h	CCM	NA	CCK-8		Both the alloys exhibited no cytotoxicity.
B. Jia /2020 [51]	Pure Zn Zn–xMn (x = 0.1, 0.4, 0.8 wt.%) (AE)	MC3T3-E1	E	1.25 mL/cm^2, 24 h 25%, 50%	CCK-8: 1, 3, 5, 7 days Live/dead: 3 days	NA	NA	CCK-8 DAPI/FITC–phalloidin Live/dead	Filtered	The addition of Mn significantly improved the cytocompatibility properties of pure Zn.
X. Xu /2020 [52]	Zn–0.8Li–0.2Ag (HR)	BMSCs	E	20 mL/cm^2 100%, 50%,10%	E: 1, 3, 5 days	CCM	CCM with 0.64% phenol	CCK-8		Zn–0.8Li–0.2Ag alloy showed no toxicity toward BMSCs in cytotoxicity test.
Y. Li /2020 [53]	Porous Zn (AM)	MG-63	D & E	0.2 g/mL, 72 h 10%	E: 0, 24, 48, 72 h D: 24 h	Ti	20% DMSO	MTS Live/dead	Filtered	The AM porous Zn exhibited good biocompatibility in vitro.
K. Wang /2020 [54]	Zn–xTi (x = 0.05, 0.1, 0.2, 0.3 wt.%) (AC, HR)	MG-63	E	1.25 cm^2/mL, 3 days 100%, 25%, 12.5%	E: 1 day	NA	NA	CCK-8		The extracts of both AC and HR Zn–xTi alloys at concentrations of $\leq$25% showed no cytotoxicity toward MG-63 cells.
J. Lin /2020 [55]	Zn–3Ge Zn–3Ge–0.5X (X = Cu, Mg, Fe) (AC, HR)	MG-63	E	1.25 cm^2/mL, 3 days 100%, 25%, 12.5%	E: 5 days	NA	NA	CCK-8		The cell viability of MG-63 cells in the extracts of all the Zn alloys at a concentration of 12.5% exceeded 90%.
J. Lin /2020 [56]	Zn–1Cu–0.1Ti Pure Zn (AC)	MC3T3-E1 MG-63	D & E	1.25 cm^2/mL, 72 h 100%, 25%, 12.5%	D: 24, 48 h E: 1, 3, 5 days	NA	NA	CCK-8		The extract of AC Zn–1Cu–0.1Ti alloy at a concentration $\leq$25% showed no significant cytotoxicity toward MC3T3-E1 and MG-63 cells.

Table 1. *Cont.*

Author /Year Ref.	Composition and Processing	Cell Line	Test Type (E, D)	Setup (SA: V, Time) and Extract Concentration (%)	Exposure Time	Negative Control	Positive Control	Assays	Others	Outcome
X. Tong /2020 [57]	Zn–1Mg Zn–1Mg–0.1RE (RE = Er, Dy, Ho) (AC, HR)	MC3T3-E1 MG-63	D & E	1.25 cm^2/mL, 72 h 100%, 25%, 12.5%	E: 1, 3, 5 days D: 24, 48 h	CCM	NA	CCK-8		The 12.5% concentration extracts of the HR Zn–1Mg and Zn–1Mg–0.1RE alloys showed good cell proliferation and growth of MG-63 without cytotoxicity.
H. Yang /2020 [58]	Zn–xMg Zn–xCa Zn–xSr Zn–xLi Zn–xMn Zn–xFe Zn–xCu Zn–xAg Zn (HE)	MC3T3-E1 HUVEC	D & E	1.25 mL/cm^2, 24 h 100%, 50%	E: 1, 2, 4 days D: 12 h	CCM	CCM with 10% DMSO	CCK-8 DAPI/FITC–phalloidin	SF	E: Pure Zn and other binary Zn alloys exhibited severe cytotoxicity except for Zn–0.8Ca and Zn–0.1Sr. D: MC3T3-E1 cell displayed a round and unhealthy shape on materials with good cytocompatibility.
P. Li /2020 [59]	Zn–2Ag–1.8Au–0.2V (AC)	L929 Saos-2	E	3 cm^2/mL, 24 h 33.3%, 16.7%, 10%	E: 24 h	Ti	Cu	XTT FDA/EB BrdU		It showed acceptable toxicity in the results obtained with cells exposed to 10% and 16.7% extracts and notable toxic effects in undiluted extracts.
R. Yue /2020 [60]	Zn Zn–3Cu Zn–3Cu–0.2Fe Zn–3Cu–0.5Fe (AE)	EA.hy926 A7r5	D & E	1.25 cm^2/mL, 3 days 10%, 50%, 75%, 100%	E: 3 days D: 12 h	No cells in CCM	Cells in CCM	CCK-8 LDH Live/Dead	Filtered	EA.hy926 cells were more tolerant than A7r5 cells to the extracts of Zn–3Cu–xFe alloys.
Z. Li /2020 [61]	Zn Zn–xLi (x = 0.2–1.4 wt.%) (AC)	L929	E	0.2 g/mL, 24 h 10%, 20%, 40%, 60%, 80%, 100%	E: 1 day	CCM	DMEM with 15% DMSO	MTT		The 10% extracts of Zn–Li alloys exhibited no cytotoxicity.
H. Guo /2020 [62]	Pure Zn (LC)	MC3T3-E1	E	1.25 cm^2/mL, 24 h 10%, 50%, 100%	E: 1, 3, 5 days	CCM	CCM with 10% DMSO	Calcein-AM/PI CCK-8	SF	Pure zinc membrane with 300 µm pores displayed acceptable MC3T3-E1 cytocompatibility in vitro.
C. Xiao /2020 [63]	Zn–0.05Mg–xAg (x = 0.5, 1.0 wt.%) (AC)	L929	E	3 mL/cm^2, 72 h 100%, 50%, 10%	E: 1, 3, 5 days	CCM	CCM with 0.64% phenol	MTT		L929 cells grew normally after culturing for 1, 3, and 5 days in the extracts of the alloys.

Table 1. *Cont.*

Author /Year Ref.	Composition and Processing	Cell Line	Test Type (E, D)	Setup (SA: V, Time) and Extract Concentration (%)	Exposure Time	Negative Control	Positive Control	Assays	Others	Outcome
L. Deng /2021 [64]	Zn–0.45Li Zn–2Li (AC, AE, AD)	L929	E	NA 25%, 100%	E: 24, 48, 72 h	NA	NA	MTT DAPI/FITC–phalloidin		The MTT cytotoxicity assay suggested a low corrosion rate and good cytocompatibility of the Zn–0.45Li alloys.
B. Jia /2021 [65]	Pure Zn Zn–xSr (x = 0, 0.1, 0.4, 0.8 wt.%) (HE)	MC3T3-E1	E	1.25 mL/cm^2, 24 h 50%, 25%	E: 1, 3, 5, 7 days	CCM	NA	CCK-8 Live/dead DAPI/FITC–phalloidin	Filtered	Pure Zn was mildly cytotoxic to MC3T3-E1 cells but Zn–Sr alloys could significantly improve cytocompatibility.
H. Wu /2021 [66]	Pure Zn Zn–Ag Zn–Mg–Ag (AC)	MC3T3	E	Ratio: NA 12.5%	E: 24, 48, 72, 96 h	NA	NA	CCK-8 DAPI/FITC–phalloidin		The Zn–0.04Mg–2Ag porous scaffold had excellent mechanical properties and biocompatibility.
E. Farabi /2021 [67]	Zn–Al–Li (AC, AE)	HuMSCs L929	E	2 mL, 21 days 50%, 100%	E: 3 h	NA	NA	MTS		The developed Zn–4Al–0.6Li and Zn–6Al–0.4Li alloys appeared to be cytocompatible with HuMSCs and L929 cells.
Y. Yang /2021 [68]	Zn Zn–xCe (x = 1, 2, 3 wt.%) (LPBF)	MG-63	E	1.25 cm^2/mL, 72 h NA	E: 1, 3, 7 days	CCM	NA	Calcein-AM CCK-8		Zn–Ce exhibited no obvious cell cytotoxicity.
X. Qu /2021 [69]	Pure Zn Zn–xAg (x = 0.5, 1, 2 wt.%) (HE)	MC3T3-K BMMs	E	1.25 cm^2/mL, 24 h 50%, 33.3%, 25%, 20%	E: 24, 72 h	NA	NA	CCK-8	Filtered	Zn–2Ag alloy significantly inhibited osteoclastic differentiation of BMMs cells in vitro.
A. Milenin /2021 [70]	Zn Zn–Mg (Properzi method)	hDPSC Saos-2	E	0.2 g/mL; 0.04 g/mL, NA NA	E: 24 h	TCP	NA	MTS		Mg content of 0.0026 wt.% in the Zn-based wire provided extracts that are toxic to cancer cells and nontoxic to healthy cells.

Table 1. *Cont.*

Author /Year Ref.	Composition and Processing	Cell Line	Test Type (E, D)	Setup (SA: V, Time) and Extract Concentration (%)	Exposure Time	Negative Control	Positive Control	Assays	Others	Outcome
J. Lin /2021 [71]	Zn–3Cu Zn–3Cu–0.2Ti (AC, HR, CR)	MG-63	E	1.25 cm^2/mL, 3 days 100%, 25%, 12.5%	E: 1 day	NA	NA	CCK-8		The extracts of both HR + CR Zn–3Cu and Zn–3Cu–0.2Ti alloys at a concentration of ≤25% showed no cytotoxicity toward MG-63 cells, and the Zn–3Cu–0.2Ti alloy exhibited higher cytocompatibility than Zn–3Cu.
J. Pinc /2021 [72]	Zn–0.8Mg–0.2Sr (HE)	NIH 3T3	E	1 mL/cm^2, 24 h 33.3%, 6.67%	E: 24, 48 h	Cells in CCM	NA	MTT Trypan blue	Pre-incubation	Poor cell viability in sample eluates was caused by the high Zn^{2+} ion release.
E. Jablonská /2021 [15]	Zn–0.8Mg (SPS)	U-2 OS L929 (P3-P20)	E	87.5 mm^2/mL, 24 h NA	E: 24 h	NA	NA	Resazurin	5%, 10%, or without FBS	The type of medium, the concentration of FBS, mode of exposition, and cell type all influenced the cytotoxicity of the extracts.
W. Zhang /2021 [73]	Zn Zn–0.5%Cu–xFe (x = 0, 0.1, 0.2, 0.4 wt.%) (AC)	L929 Saos-2 TAg	E & D	1.25 cm^2/mL, 24 h 100%	E: 24 h D: NA	Ti/CCM	Cu	CCK-8 FDA/EB		The extracts of Zn–0.5Cu–Fe (0.2 wt.%) alloys showed no cytotoxic effects toward tested cells.
P. Zhu /2021 [74]	Pure Zn (NA)	L929	D	-	D: 24 h	Ti	Cu	FDA/EB XTT BrdU	Pre-incubation	The direct cells cultured on Zn-based surfaces led to apparent misleading cytotoxicity with the CCK-8 assay.
P. Li /2021 [75]	Pure Zn Zn–3Cu (AC)	L929	E	1.25 cm^2/mL, 24 h NA	E: 24 h	Ti	Cu	FDA/EB CCK-8		The extract test indicated that gamma irradiation or H_2O_2 gas plasma sterilization did not induce cytotoxic effects toward L929 fibroblasts on Zn and Zn–Cu alloy.

Table 1. *Cont.*

Author /Year Ref.	Composition and Processing	Cell Line	Test Type (E, D)	Setup (SA: V, Time) and Extract Concentration (%)	Exposure Time	Negative Control	Positive Control	Assays	Others	Outcome
J. Capek /2021 [76]	Zn–0.8Mg–0.2Sr Zn (AC, HE)	L929 Saos-2 Tag	E	1.25 cm^2/mL, 24 h 100%, 50%, 25%	E: 24 h	Ti	Cu	FDA/EB CCK-8 BrdU		The 25% extracts of the Zn–0.8Mg–0.2Sr alloys had no apparent adverse effects on the cell viability and proliferation of L929, Tag, and Saos-2 cells.
O. Avior /2022 [77]	Zn–2%Fe–0.6%Ca (AC)	4T1	D	1.25 mL/cm^2, 24 h NA	D: 24 h, 48 h	Ti	NA	Live/dead	Pre-incubation	The tested alloy was suitable for cell growth under in vitro conditions, as seeded cells were adherent and viable on the alloy surface.
X. Tong /2022 [78]	Zn–1Mg–xGd (x = 0.1, 0.2, 0.3 wt.%) (AC, HR)	MG-63	E	1.25 cm^2/mL, 3 days 12.5%, 25%, 50%, 100%	E: 24 h	NA	NA	CCK-8		High-concentration (≥50%) extracts of Zn–1Mg–0.3Gd had clear inhibitory effects on MG-63 cells.
J. Jiang /2022 [79]	Pure Zn Zn–2.2wt.% Cu–xMn (x = 0, 0.4, 0.7, 1 wt.%) (AC)	EA.hy926 A7r5	E	1.25 cm^2/mL, 3 days 100%, 50%, 10%	E: 1, 3 days	No cells in CCM	Cells in CCM	CCK-8		Zn–2.2Cu–0.4Mn alloy exhibited acceptable in vitro cytocompatibility, comparable with pure Zn.
G. Bao /2022 [80]	Zn Zn–0.5Cu Zn–1Cu (NA)	HEECs HESCs	E	1.25 cm^2/mL, 24 h 100%, 50%, 10%	E: 1, 3, 5 days	CCM	CCM with 10% DMSO	CCK-8	SF	The Zn–0.5Cu exhibited slightly higher-level cell viability than Cu, however, it was much lower than pure Zn and Zu–1Cu.
H. Ren /2022 [81]	Zn Porous Zn–xCu (x = 0, 1, 2, 3) (APIM)	MC3T3-E1 L929	E	Ratio: NA, 24 h NA	E: 1, 2 days	NA	NA	MTT		The alloy exhibited good cytocompatibility at a low extract concentration.
Y. Qin /2022 [82]	Zn–xMg (x = 1, 2, 5 wt.%) (AC)	MC3T3-E1	E	1.25 cm^2/mL, 24 h 100%, 50%, 10%	E: 1, 3, 5 days	CCM	CCM with 10% DMSO	CCK-8 Calcein AM/PI		The cell viability increased with increasing Mg content.
Y. Xu /2022 [83]	Zn–0.5Cu–0.2Fe Zn (AC)	HUVEC RAW264.7 MC3T3-E1	E	1.25 cm^2/mL, 72 h 50, 25, 12.5%	E: 24 h	Ti	Cu	LDH FDA/EB		The hot extruded Zn–Cu–Fe alloy exhibited good performance in terms of cytocompatibility.

Table 1. *Cont.*

Author /Year Ref.	Composition and Processing	Cell Line	Test Type (E, D)	Setup (SA: V, Time) and Extract Concentration (%)	Exposure Time	Negative Control	Positive Control	Assays	Others	Outcome
Y. Zeng /2022 [84]	Zn–Fe–Si (AC)	HUVEC	E	1.25 cm^2/mL, 72 h 6.25%	E: 24, 72 h	CCM	NA	MTT		The biocompatibility of the test alloy was acceptable.
Y. Liu /2022 [85]	Zn–0.5Fe (AS)	MC3T3-E1	E	6 cm^2/mL, 24 h 12.5%, 25%, 50%	E: 1, 3, 5 days	CCM	NA	CCK-8		The Zn–0.5Fe alloy membrane had adequate biocompatibility.
D. Palai /2022 [86]	Zn Zn–xCu (x = 1, 2, 3 wt.%) (AC)	3T3 fibroblasts	E	Ratio: NA, 72 h 50%	E: 1,3, 5 days	NA	NA	MTT		The Zn–2Cu and Zn–3Cu alloys exhibited better cytocompatibility compared to pure Zn.
N.A. Gopal /2022 [87]	Zn–Ti–Cu–Ca–P (AS)	Vero cell	E	NA	E: 24, 48, 72 h	NA	NA	MTT EB/AO		The presented material can be used as a bio-implant.
N. Yang /2022 [88]	Zn–Cu–Ca Zn (AC, HR)	HUVEC L929	E	1.25 cm^2/mL, 24 h 100%, 50%, 25%, 12.5%	E: 1, 3 days	Cell in CCM	NA	CCK-8		The alloys had good cytocompatibility for the tested cell lines.
J. Duan /2022 [89]	Zn–2Cu–0.2Mn–xLi (x = 0, 0.1, 0.38 wt.%) (AC, HE)	MC3T3-E1		1.25 mL/cm^2, 24 h 100%, 50%, 25%	E: 1, 2, 3 days	CCM	NA	CCK-8 Calcein-AM/EthD-1 DAPI/FITC–phalloidin		MC3T3-E1 cells exhibited over 95% viability in the 25% extracts of all as-extruded alloys.
X. Zhu /2022 [90]	Zn–Mn Pure Zn (AC, HE)	L929	E	1.5 cm^2/mL, 24 h NA	E: 24, 48, 72 h	NA	NA	MTT		The concentration of Zn^{2+} in the 100% concentration extract exceeded the safety threshold, causing the relative growth rate of cells to be lower than 100%.
G.K. Levy /2019 [91]	Zn–1Mg Zn–1Mg–0.5Ca (DC)	MSCs	D&E	Ratio: NA, 24 h NA	E: 24, 48, 72 h D: 24 h	CCM	10% DMSO	CCK-8 Live/Dead		A short and simple 1 day surface stabilization treatment in cell growth medium significantly improved cell adhesion and viability.

Table 1. *Cont.*

Author /Year Ref.	Composition and Processing	Cell Line	Test Type (E, D)	Setup (SA: V, Time) and Extract Concentration (%)	Exposure Time	Negative Control	Positive Control	Assays	Others	Outcome
I. Cockerill /2019 [92]	Zn (AC)	MC3T3-E1	D&E	1.25 cm^2/mL, 72 h 10%	D: 24 h E: 1, 3, 5 days	CCM	NA	D: SEM E: MTT		The textured Zn samples supported the adhesion of pre-osteoblasts that exhibited flat morphologies with numerous cytoplasmic extensions, and cytocompatibility tests showed >75% cell viability in 10% extracts.
P. Li/2020 [93]	Pure Zn Zn–4Ag Zn–2Ag–1.8Au–0.2V (AC)	Saos-2	E	1.25 cm^2/mL, 24 h NA	E: 24 h	Ti	Cu	FDA/EB CCK-8		Samples treated with 250 μm sandblasting particles caused a mean decrease in viability below 70% of the control, i.e., classified as an apparent cytotoxic effect.
X. Tong /2022 [94]	Zn–xDy (x = 1, 3, 5 wt.%) Zn HR	MC3T3-E1	E	1.25 cm^2/mL, 48 h 100%, 25%, 12.5%	E: 3 days	CCM	NA	CCK-8 Calcein-AM/PI		The HR Zn–3Dy extract with 12.5% concentration showed the highest cell viability of ~102.1% toward MC3T3-E1 cells among all samples tested.
M. Wątroba /2022 [95]	Zn Zn–3Ag Zn–3Ag–0.5Mg (AC)	MG-63	E&D	1.25 cm^2/mL, 24 h 100%, 50%, 25%, 12.5%, 5%	E: 24 h D: 24 h	NA	CCM	WST-8 LDH Calcein-AM/DAPI		Cytotoxicity tests showed almost no significant differences between pure Zn and Zn alloys.
T. Di/2022 [96]	Zn–1Cu–xAg (x = 0.5, 1 wt.%) (HE)	MC3T3-E1	E&D	Ratio: NA, 24 h 100%, 50%, 25%, 12.5%, 6.25%	E: 1, 2, 3 days	CCM	NA	Hoechst 33342/PI MTT		The cytotoxicity grade of the twofold diluted extracts of Zn–1Cu–xAg alloy was 0–1, and the cytocompatibility met the requirements for orthopedic application.
Z. Wang /2022 [97]	Pure Zn Zn–Mg (HE)	MC3T3-E1 VEC	E	1.25 cm^2/mL, 24 h NA	E: 1, 3, 5 days	NA	NA	CCK-8		Zn–Mg alloys examined in this study exhibited good cytocompatibility in vitro with osteoblasts and endothelial cells.

Table 1. *Cont.*

Author /Year Ref.	Composition and Processing	Cell Line	Test Type (E, D)	Setup (SA: V, Time) and Extract Concentration (%)	Exposure Time	Negative Control	Positive Control	Assays	Others	Outcome
L.B. Tzion-Mottye /2022 [98]	Zn–2%Fe (AC)	Mus musculus 4T1	E	1.25 cm^2/mL, 24 h 10%	E: 24 h, 48 h	Ti	NA	XTT		Indirect cell viability assessment showed that the addition of Mn tended to increase cell viability in vitro.
Z. Zhang /2022 [99]	Zn–0.60Mn–0.064Mg Zn–0.81Mn–0.049Mg (HE)	MC3T3-E1	E	1.25 cm^2/mL, NA 100%, 25%–75%	E: 1, 3 days	CCM	Cells in CCM	CCK-8 Live/dead		Both alloys had biocompatibility.
L. Sheng /2022 [100]	Zn–1.5Fe (SPS)	MG-63	E	1.25 cm^2/mL, 24 h NA	E: 3, 5, 7 days	NA	NA	CCK-8		The viability of MG-63 on Zn–1.5Fe alloys was over 85%.
L. Jin /2022 [101]	Zn Zn–0.5Li (HR)	MC3T3-E1	E	1.25 cm^2/mL, 24 h 100%, 50%, 25%	E: 1, 3, 5 days	CCM	NA	CCK-8 Calcein-AM/PI	SF	The biocompatibility of Zn–0.5Li was higher than that of pure Zn.

Abbreviations: not available (NA); reference (Ref.); extract test (E); direct contact test (D); as-casting (AC); as-extruded (AE); hot-extruded (HE); hot-rolling (HR); bottom circulating water-cooled casting (BCWC); high-pressure solidification (HPS); laser cutting technology (LC); die-casting (DC); air pressure infiltration method (APIM); cold-drawing (CD); electro-deposition (ED); additively manufactured (AM); selective laser melting (SLM); hot-treatment (HT); laser powder bed fusion (LPBF); cells cultured in tissue culture plates (TCP); high-pressure solidification (HPS); spark plasma sintering (SPS); as-sintered (AS); cell culture medium (CCM); withdrawn supernatant fluid (SF); Ti–6Al–4V (Ti).

3.2. Quality Assessment According to the ToxRTool

Detailed total scores of each ToxRTool item in vitro criteria are presented in Figure 2a. Twenty-nine studies described the source of test substances. A total of 34 and 35 studies were graded as "1" in items 12 and 13, respectively. A total of 40 studies provided the source information of the test system. In addition, 55 studies set negative controls in cytotoxicity evaluation, and 73 articles met the 16th criterion detailing the statistical method for data analysis. The number of articles that met the remaining criteria ranged from 80 to 86 (over 90% of the total articles). As depicted in Figure 2b, it is clear that the quality assessment of included toxicity data was found to be relatively low. Furthermore, 31 articles were reliable without restrictions, three were deemed reliable with restrictions, and 52 articles were not reliable. All studies classified as not reliable failed to meet all of the essential criteria marked in red in the assessment tool.

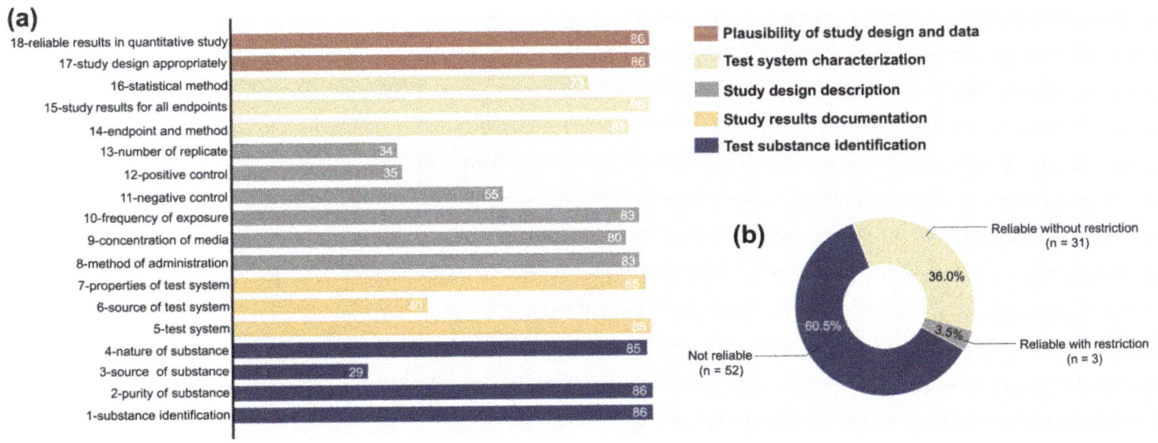

Figure 2. Results of quality assessment of included articles using the ToxRTool: (**a**) total score for each item of the ToxRTool in vitro criteria; (**b**) articles classified into three categories on the basis of quality evaluation results (reliable without restriction, reliable with restriction, and not reliable).

3.3. Main Characteristics of the Included Articles

3.3.1. Materials and Processing

As shown in Table 1, 51 articles involved pure Zn, and 77 investigated Zn-based alloys. The latter mainly consisted of binary and ternary alloys such as Zn–Mg [15,25–28,35,43,53,57,70,82], Zn–Cu [23,24,29,37,44,56,60,71,75,80,81,86], Zn–Ag [36,42,66,69], Zn–Mn [41,51,90], and Zn–Li alloys [38,61,64], with Zn–Mg and Zn–Cu being the predominant types. A small portion of Zn alloys comprised additional Al [47], Fe [49,50,84,85], Ca [58,77,88], Ge [55], Ti [54,87], Sr [65,72,76], Si [84], Zr [31,33], Sn [47], and V [59]. In addition, pure Zn was alloyed with rare-earth elements (REEs) such as Er, Dy, Ho, Ce, and Gd [57,68,78].

The morphological geometry of the test samples was varied, including discs, porous scaffolds [37,39,53], various thread types [60,70], foil [60], wires [40], membranes [85,87], and cylinders [25]. In terms of the preparation and processing of test samples, casting, hot-rolling, and hot-extruding were the main manufacturing processes for biodegradable Zn and its alloys. Other processing technologies included sintering [15], electrodeposited heat treatment [37], and additive manufacturing [39,47,53,68].

3.3.2. Tested Cell Types

Figure 3a displays the division of cell types tested into three main categories: cells pertaining to osteoblasts, cardiovascular-related cells, and others. A large proportion of studies used immortalized cell lines in evaluating toxicity, whereas only five used primary cells (with passage number ≤7) [21,22,34,46,69]. Regarding orthopedic investigations, 21 selected articles used the MC3T3 (pre-osteoblast cell line), which is usually used in osteogenic evaluation. Although the vast majority used MC3T3-E1, one report used another clone number [69] and one did not describe the specific clone number [66]. Fourteen studies adopted MG-63 in cytotoxicity evaluation, a cell line from human osteosarcoma, and nine studies investigated Saos-2 osteoblasts (human primary osteosarcoma cell line). A few studies used other cell types, such as bone marrow mesenchymal stem cells (MSCs) [30,46,67,91]. Human umbilical vein endothelial cells (HUVECs) and human endothelium-derived cell lines (EA.hy926) were the most commonly employed cell lines in vascular research. The mouse fibroblast cell line (L929) was investigated in 24 articles, accounting for a considerable fraction of the third category. Overall, more than 70% of the articles evaluated only one cell line, 19 articles (22.9%) compared the toxicity performance of two cell lines, and five studies (6.0%) selected three different cell lines in carrying out cytotoxicity tests.

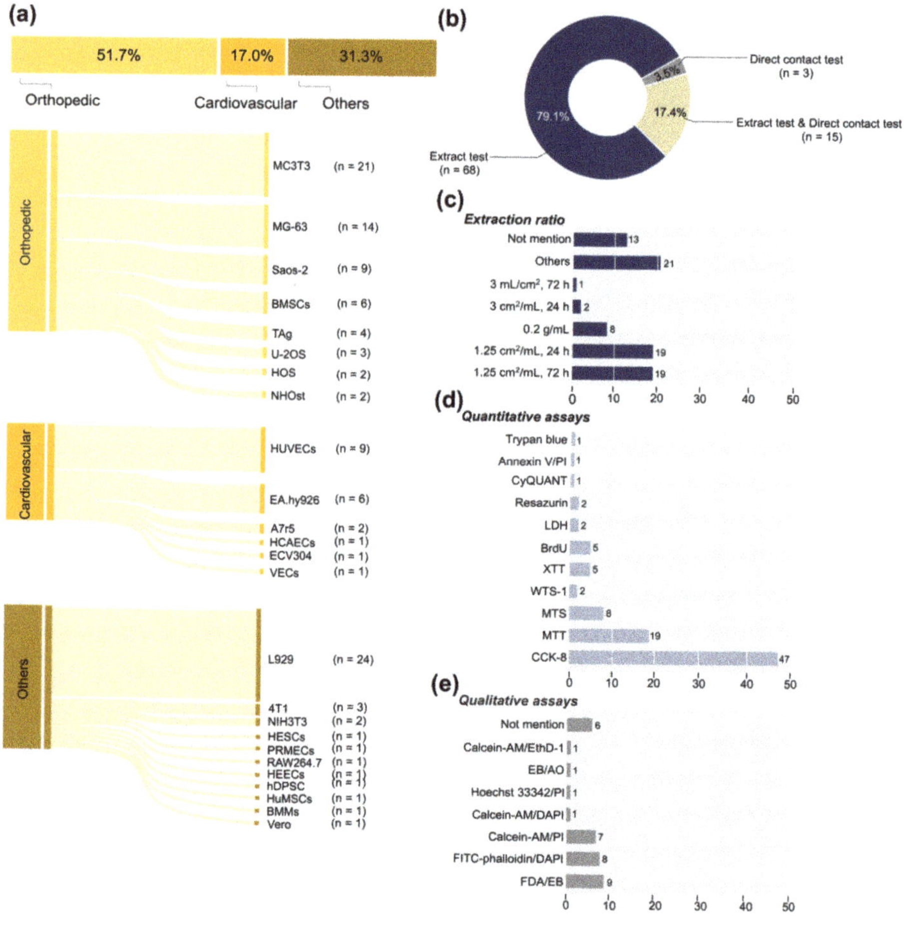

Figure 3. The main characteristics of the extracted experimental data: (**a**) selected cell types in studies primarily divided into three categories (cells related to cardiovascular, orthopedics, and other types); (**b**) cytotoxicity evaluation test methods; (**c**) extraction ratio of prepared extracts; (**d**,**e**) quantitative and qualitative assays used in cytotoxicity evaluation.

3.3.3. Test System

Concerning the extraction parameter, for evaluating the cytotoxicity of Zn-based BMs, 68 publications employed extract tests, while three studies exclusively conducted direct contact tests [33,74,77]. The remaining 15 studies combined the two approaches (Figure 3b). As shown in Figure 3c, various sample surface-to-volume extraction ratios and immersion times were used to prepare the sample extracts. Nineteen extract tests applied the ratio of 1.25 cm^2/mL with sample immersion times of 72 h and 24 h, respectively. Eight studies were conducted using the sample powder to evaluate cytotoxicity and adopted the extraction ratio of sample weight to medium volume [19,22,37,39,41,53,61,70]. Notably, 13 studies did not specify a certain ratio. Additionally, some researchers adopted the extraction ratios such as 3 mL/cm^2, 1.25 mL/cm^2, and 20 mL/cm^2 without providing any justification [32,52]. Extracts were often filtered via a membrane [19,20,39,49,51,53,60,65,69,72] or centrifuged to withdraw the supernatant fluid in certain studies [18,20,26,40,58,62,80,101,102]. Some studies investigated Zn-based samples precultured in the medium before the cytotoxicity test [25,28,72,74,77,92,102]. The effect of BSA (bovine serum albumin) on the cytotoxicity in the medium of pre-exposure samples was also examined [72]. Jablonska et al. confirmed the effect of FBS (fetal bovine serum) in the cell medium on cytotoxicity tests [15,83]. The effect of pretreatments such as pre-cultivation, stabilization treatment [91], sterilization treatment [75], sandblasting [93], and acid etching treatment [102] was also investigated.

Regarding the concentration of the extracts, approximately 76% of studies (63/83) set concentration gradients by dilution with the cell medium, while 18.4% (14/83) used undiluted extracts solutions for testing. It was ambiguous whether the extracts were diluted in six articles [68,70,77,81,87,90].

As shown in Figure 3d, cell viability tests could be divided into qualitative and quantitative assays. Among the assays used for quantitative tests, tetrazolium salt-based assays such as CCK-8 (WTS-8), MTT, WTS-1, and MTS were used frequently. Forty-seven studies used CCK-8 assays in cytotoxicity evaluation, ranking first. Cell survival was determined using the MTT assay in 19 studies and the MTS assay in eight studies. Two studies were tested using the lactate dehydrogenase (LDH) release assay, while five were tested using the bromodeoxyuridine (BrdU) incorporation assay to measure cell proliferation. In a few studies, cell viability was determined on the basis of a fluorometric resazurin reduction method, such as CyQUANT [46] and resazurin assays [15,33]. The assays for qualitative analyses are summarized in Figure 3e, in which live/dead staining FDA/EB dye [16,44,59], Calcein-AM/PI dye [30,82,101], and FITC–phalloidin/DAPI dye [25,51,58,64,65] were the primary assays to realize the visualization of cells. Six studies did not indicate the specific assays used in qualitative assessment.

With regard to the control groups, 32 studies set both positive and negative control groups, while 28 studies set only negative control groups, and the remaining 26 studies set neither negative nor positive control groups. The cells in the cell culture medium alone and supplemented with 5–20% dimethyl sulfoxide (DMSO) were usually chosen as negative and positive control groups, respectively. Some researchers also used Cu and Ti–6Al–4V alloy as negative and positive controls, respectively [59,73–76,83,93]. Five studies used the culture medium supplemented with 0.64% phenol as a positive control [21,35,45,52,63].

It is worth noting that the assessment criteria were inconsistent. In most studies, a reduction in cell viability of more than 30% was considered a cytotoxic effect, but the threshold for cytotoxicity was at 75% in multiple studies [68,92]. In addition, many of the included studies claimed to have graded the toxicity from quantitative results in extract tests according to ISO standards [32,34,35,45,52–54,86,99].

3.3.4. Outcome

The relationship between Zn-based BMs and cytotoxicity ranged from excellent biocompatibility to apparent cytotoxicity. Notably, the majority of the selected studies suggested that Zn-based BMs were nontoxic or produced toxic effects only under specific conditions, such as with a highly concentrated extraction solution. However, only two studies reported that pure Zn or its alloys were toxic [26,58].

4. Discussion

Zn-based BMs have been proposed and developed for biomedical implant materials. Alloying is a common way to improve the material properties of Zn-based BMs. This systematic review assessed the potential cytotoxic effects of Zn and its alloys. On the basis of the results, the current quality assessment of toxicity studies was assessed to be highly heterogeneous, with different study designs and non-standardized procedures making it difficult for quantitative analysis. A qualitative analysis showed that the cytotoxicity of Zn-based BMs is mainly determined by three factors: the Zn-based materials, tested cells, and test system (Figure 4).

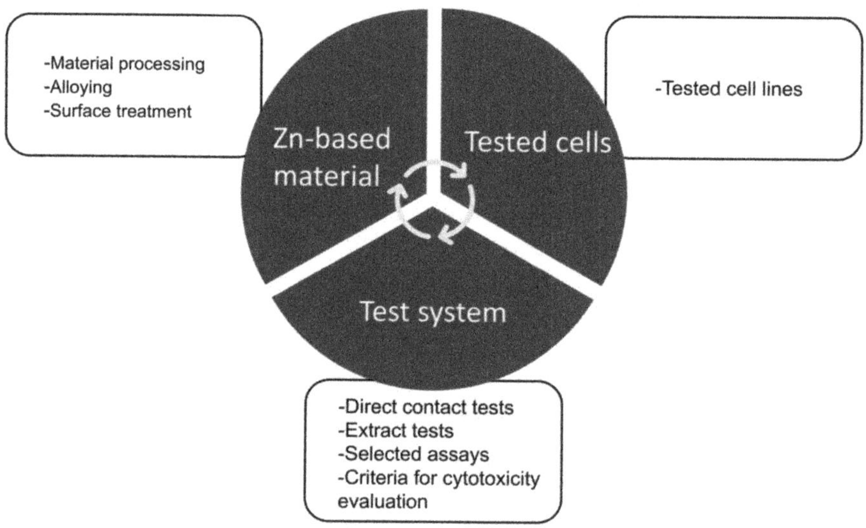

Figure 4. Schematic diagram of the assessment of the toxic effects of biodegradable Zn and its alloys according to three factors: Zn-based materials, tested cells, and test system.

4.1. *Effects of the Materials on Cytotoxicity*

4.1.1. Material Processing

Various processing techniques were used to improve the mechanical performance of Zn-based BMs. However, these techniques also changed the material microstructure, possibly influencing their corrosion behavior and biocompatibility. The as-cast alloys suffered from significant nonuniform micro-galvanic corrosion, although the biodegradation uniformity was improved by the hot extrusion or rolling processes [20,44]. After extrusion, primary dendritic phases were broken and distributed along the extrusion direction, thereby refining the grains distinctly due to dynamical recrystallization [24]. The more uniform corrosion and reduced corrosion rate brought about by the refinement of the second phase of the bottom circulating water-cooled casting method was also demonstrated, leading to higher cell viability than conventional casting in 100% extracts [50]. Nevertheless, the opposite was true for as-extruded Zn–1.2Mg alloy, which had lower cell survival than its as-cast alloys due to a higher concentration of Zn ions resulting from a higher corrosion rate [27]. This might be related to the Mg_2Zn_{11} phase, which is distributed relatively uniformly at grain boundaries and in the Zn matrix, involved in the formation of micro-galvanic cells [64,103]. Moreover, the corrosion resistance of the Zn matrix could be dramatically improved by appropriate heat treatment and plastic processing, resulting in a decrease in released metal ions and higher cytocompatibility [37,48]. Interestingly, one study confirmed that the crystallization process of the material affected its biocompatibility [33]. In brief, the effects of processing on the cytotoxicity of Zn-based BMs are mainly caused by changing a material's microstructure and corrosion behavior. When designing a novel biomaterial, the relationship between the improvement of mechanical properties and the consequent change in biocompatibility has to be evaluated.

4.1.2. Alloying and Its Micro-Galvanic Corrosion

The degradation behavior of Zn-based BMs in the body is intrinsically determined by their corrosion process [104]. The corrosion behavior of metals depends on metallurgical factors such as alloy composition, phase precipitation, and segregation of alloying elements and impurities [105,106]. Therefore, the elements added into Zn-based alloys can increase their cytotoxicity by promoting micro-galvanic corrosion.

Numerous studies demonstrated that adding Mg to pure Zn improved the biocompatibility of Zn-based alloys [23,43,82]. However, studies confirmed that the cell viability of the Zn–Mg alloys was not subject to monotonic variation with the Mg content [31,70], possibly due to the combined effect of grain refinement and passivation [82]. Copper is a component of numerous enzymes and plays a crucial role in the response to oxidative stress [107,108]. One study demonstrated that the cytocompatibility of the Zn–1Cu alloy was significantly higher than that of the pure Zn [80]. The beneficial effects of Ca, Sr, Fe, Ag, Mn, and Li as alloying elements on cytocompatibility have also been proven [55,58,69,90]. The positive effect on the biocompatibility of alloying elements was shown to be provided by Cu > Ca > Ag in decreasing order [41].

Alloying elements might interfere with the toxicity assessment by creating degradation reactions. As an example, the silver ions from Zn–Ag alloys could combine with chloride ions that existed in the medium to form a precipitate of silver chloride [66]. The concentration of Zn ions in the medium is decreased by alloying, but the toxic effect of the insoluble metal salts may be unknown. Excess intake of Al^{3+} is considered toxic. However, studies have reported that Zn–Al alloys have no harmful effects on HUVECs in diluted extracts, possibly because the Al^{3+} concentrations were negligible compared to the half-maximal inhibitory concentration (IC_{50}) of Al^{3+} [26]. Although a high concentration of rare-earth elements (REEs) inhibited ATPase activity and caused metabolic disorders in the body [109,110], the addition of REEs to the Zn matrix also led to favorable cytocompatibility [57,68,78,94]. Undoubtedly, the release rate and content of alloying elements are essential factors influencing the cytotoxicity of Zn-based alloys.

4.1.3. Surface Treatment

Pretreatment could alter the surface morphology, wettability, and roughness of the material and consequently impact its biocompatibility [111]. Among the studies included in the analysis, sample pre-cultivation was the most common pretreatment. Specifically, surface stabilization treatment resulted in a stable surface oxide film which inhibited the release of Zn^{2+}, decreasing the cytotoxic effect [44,48,91]. The components of the medium used for pre-cultivation are also crucial. The presence of BSA during pre-incubation resulted in the best wettability and the lowest ion release in the initial stages of the exposure. In this case, the biocompatibility was better than that of untreated groups [72]. Li et al. investigated the impact of sterilization treatments on the cytocompatibility of Zn-based BMs. Due to the excessive release of Zn ions and a local concentration over the cellular tolerance capacity, the autoclave-treated Zn matrix exhibited apparent cytotoxic effects on fibroblasts [75]. Furthermore, cell viability in extracts of the polished-textured samples was higher than those of the fine-textured and coarse-textured samples [92]. Likewise, sandblasting treatment of the surface of Zn-based alloy specimens decreased the cell viability due to localized corrosion of the samples [93]. Therefore, as a means of changing the microstructure of tested samples, pretreatment had prominent effects on cytotoxicity.

4.2. Effects of Tested Cells on Cytotoxicity

The various selected cell lines were mainly associated with the vascular and orthopedic research fields. Given that cytotoxicity is determined by the cellular tolerance of degradation products released from the Zn-based metals [108], the selection of tested cells was a vital factor in the toxicity assessment. Furthermore, metal ions usually affect cell growth in a concentration-dependent manner [60], and the tolerance of distinct cell types to metal ions is also inconsistent [76]. Hence, different cell lines could produce distinct outcomes in the toxicity assessment. For instance, Zn–Li alloy showed good biocompatibility with MC3T3-E1 cells and HUVECs [58], while its extracts showed significant cytotoxicity to L929 cells [61]. Milenin et al. determined that the viability of hDPSC was significantly higher than that of the Saos-2 under the same conditions [70]. Endothelial cells usually exhibited better cell viability compared to L929 cells, MC3T3-E1 cells, and vascular smooth muscle cells in toxicity evaluation [18,58,60], possibly because Zn, as an antioxidant and endothelial membrane stabilizer, could enhance endothelium integrity [112].

Several studies utilized stem cells with broad differentiation potential, and all these cell lines exhibited great vitality [30]. Even in the same research field, distinct cell lines have different tolerances. HOS cells showed a significant reduction in cell viability and induced cytotoxicity at a higher extract concentration, while the same conditions had a slight negative impact on the viability of MG-63 cells [27]. Also, MG-63 cells were more tolerant to Zn ions than MC3T3-E1 cells [56]. Moreover, MG-63 cells have been recommended for in vitro evaluation because they could closely simulate human cells [57]. However, this cell line may display heterogeneity among different cell populations due to donor factors. Considering the inhomogeneity of primary cells, established cell lines are more recommendable for use unless reproducibility and accuracy of the response can be demonstrated.

4.3. Effects of Test System on Cytotoxicity

4.3.1. Parameters of the Extract Tests

In addition to the corrosion properties of the alloy itself, the degradation rate mainly depends on the culture medium used. The ingredients of different media could be varied, and the selection of the cultivation medium is usually dictated by the cell type used. It was reported that the relatively low Zn ion release in McCoy's 5A medium than in DMEM or DMEM/F-12 could be attributed to increased passivation film formation by a high concentration of HPO_4^{2-} in the medium [44]. It is worth mentioning that extracts prepared by rinsing in α-MEM would help to stimulate the physiological environment [66]. Capek et al. confirmed that the $ZnCl_2$ was less toxic to L929 cells in DMEM than in α-MEM, possibly since DMEM contains more glucose, amino acids, and vitamins, consequently having a strong buffering effect [76]. To provide serum proteins in simulated body fluid, adding 10% FBS to the cell culture medium is a common practice. Notably, the presence of FBS in the extraction medium could accelerate the initial corrosion process of the Zn matrix, leading to the additional potential for cytotoxicity [16]. It could be due to rapid protein adsorption on the Zn surface inhibits initial surface passivation with a protective Zn phosphate layer [113].

Among the included studies, there was significant heterogeneity in the extraction ratio of prepared extracts, partly due to different versions of the ISO standards being referred to, but more often due to the adoption of unprecedented extraction ratios. Extraction for 24 h was thought not to be sufficient to obtain an extract that represents the tested material used in practice. Hence, an immersion time of 72 h is recommended in the latest ISO standard (10993-12: 2021). Few studies adopted this latest standard. It was observed that undiluted or high-concentration extracts could exhibit toxicity effects, which could be put down to high ion concentrations and osmotic pressure inhibiting cell adhesion and growth [37]. Wang et al. suggested using a minimum of 6–10 dilutions in evaluating Mg-based BMs. This range mimics the continual clearance of absorbable ions from the circulatory system [108]. Several studies followed this recommendation and reported no toxicity effect with 10% or

12.5% extracts of Zn matrix BMs. Notably, the concentration dependence of Zn ions is more pronounced than that of Mg [114,115]. Therefore, the most appropriate dilutions for toxicity testing of Zn-based BMs to mimic the in vivo environment are yet to be demonstrated.

4.3.2. Direct Contact Tests

Most included studies have the same study endpoints in both extraction and direct contact tests. However, one study reported that tested samples had good biocompatibility in extract experiments while exhibiting cytotoxicity in the direct contact experiments [56], possibly since in vitro direct cultivation of cells on the Zn matrix was hampered by rapid degradation and partial shedding of degradation products [73]. Even though the material is nontoxic, it might interfere with the proliferation of cells to some degree [67]. It is difficult to determine the primary factors leading to decreasing cell viability and adhesion. These factors include the increase in local pH, change in surface morphology, shedding of the corrosion layer, and surface composition. In addition, cell adhesion and proliferation largely depend on extracellular matrix deposition, which is controlled by the protein adsorption capacity of a matrix surface [86].

Although direct contact tests cannot capture all complexities in vitro, they are still necessary for rapid initial screening of cytocompatibility of medical devices.

Since tested cells can be influenced rapidly by released soluble corrosion products [110]. Since rapid protein adsorption occurs on the surface of the specimen after implantation, it is advisable to mimic this process by pre-culturing the sample in vitro when performing direct toxicity tests.

4.3.3. Selected Assays

Tetrazolium salt-based assays were used particularly frequently in toxicity assessment in the included studies. Furthermore, MTT and XTT are classical toxicity assessment methods recommended in ISO 10993 standards. However, existing studies have shown that tetrazolium-based assays can be confounded by the presence of metals, leading to false positive or negative effects [74,108]. Some authors used other parameters for toxicity assessment apart from cell viability and survival, such as the comet assay to detect the degree of DNA damage [21] and flow cytometry to evaluate the cell cycle [26]. The application of multiparametric assessment to support the observation of toxicity by a single endpoint requires more funding and research. With the development of in vitro assessments in toxicology, new paradigms of analysis, such as proteomics, genomics, and pathway analyses, contribute to the understanding of the mechanisms involved in the toxicity pathways beyond providing evidence for cell death [116]. Although these paradigms do not yet permeate the published toxicity assessments of Zn-based BMs, they might be a promising direction in the future.

4.3.4. The Criteria of Cytotoxicity Evaluation

The obtained data appears to be comparable only within the results of the same study or when stringently standardized. The criteria for toxicity evaluation in quantitative and qualitative tests were different. According to the ISO-10993 standard, the toxicity results were only graded in qualitative tests. By assessing the changes in tested cells, the change from normal morphology should be graded into five levels, and the numerical grade greater than two was considered a cytotoxic effect. Only one toxicity threshold was set in the quantitative evaluation, i.e., a reduction in cell viability of more than 30% was considered a cytotoxic effect. Non-standardized evaluation criteria make the quantitative analysis of toxicity data difficult and make a direct comparison of study results impossible.

4.4. Strengths and Limitations of This Work

To our knowledge, this is the first systematic review of the cytotoxicity of Zn and its alloys. First, this review used the ToxRTool tool to assess the quality of toxicity evidence from the included studies, revealing the relatively lower quality of in vitro toxicity assessment.

Subsequently, a detailed analysis was performed of the heterogeneity of toxicity testing in the included studies. However, this systematic review had three limitations, the first being the large volume of data from the included articles and the significant heterogeneity among the studies preventing a quantitative data analysis. Even though some trends regarding the relationship between cytotoxicity and physicochemical effects were identified, the specific influence and magnitude of each factor remain elusive. Secondly, our review was not registered in PROSPERO because RCT procedures do not apply to preclinical studies. Thirdly, this study only focused on the cytotoxicity of Zn-based BMs, while the biocompatibility is wider-ranging with many more elements than cytotoxicity. Therefore, systematic reviews regarding other aspects of biocompatibility (e.g., immunogenicity, inflammatory response, or tissue compatibility) should be performed in the future.

5. Conclusions

This systematic review aimed to provide insight into the available literature exploring the cytotoxicity of biodegradable Zn and its alloys. Within the limitations of this study, the following conclusions can be drawn:

1. High heterogeneity exists in the implementation of the included studies and the assessment results of the toxicity studies.
2. The qualitative analysis demonstrated that biodegradable Zn and its alloys had conditionally cytotoxic effects, mainly dependent on the Zn-based materials, tested cells, and test systems.
3. The material processing technologies and alloying elements had a potential effect on the toxicity of Zn-based BMs due to modifications in microstructure and corrosion characteristics.
4. Endothelial cells had better tolerance to the toxic effects of Zn-based BMs than other tested cells.
5. A standardized in vitro toxicity assessment system for biodegradable metals is still lacking, and further construction is required. In addition, researchers in this field need to comply with existing evaluation criteria and report test procedures in as much detail as possible to make the study data more informative and valuable to promote translational research and the long-term development of Zn-based BMs.

Supplementary Materials: The following supporting information can be downloaded at: https://www.mdpi.com/article/10.3390/jfb14040206/s1, Table S1: Electronic database and search strategy; Table S2: PRISMA_2020_checklist.

Author Contributions: Conceptualization, P.L. and S.X.; methodology, formal analysis, and investigation, P.L., Q.L., A.L. and Q.F.; software, Q.L. and A.L.; resources, S.X.; data curation, P.L.; writing—original draft preparation, Q.L., P.L. and A.L.; writing—review and editing, P.L. and S.X.; visualization, A.L., S.L., Y.X., Q.F. and J.D.; supervision, P.L. and S.X.; project administration, P.L. and A.L.; funding acquisition, P.L. and S.X. All authors have read and agreed to the published version of the manuscript.

Funding: This research was funded by the Guangdong Basic and Applied Basic Research Foundation (grant number 2021A1515111140, 2021B1515120059, and 2022A1515110379), the Science Research Cultivation Program of the Stomatological Hospital, Southern Medical University (grant number PY2021003 and PY2020007), the Scientific Research Project of Traditional Chinese Medicine Bureau of Guangdong Province (grant number 20221267), and the Science and Technology Projects in Guangzhou (grant number 202102080148).

Institutional Review Board Statement: Not applicable.

Informed Consent Statement: Not applicable.

Data Availability Statement: Experimental data from this study are available from the corresponding author upon reasonable request.

Conflicts of Interest: The authors declare no conflict of interest.

References

1. Han, H.S.; Loffredo, S.; Jun, I.; Edwards, J.; Kim, Y.C.; Seok, H.K.; Witte, F.; Mantovani, D.; Glyn-Jones, S. Current status and outlook on the clinical translation of biodegradable metals. *Mater. Today* **2019**, *23*, 57–71. [CrossRef]
2. Liu, Y.; Zheng, Y.F.; Chen, X.H.; Yang, J.A.; Pan, H.B.; Chen, D.F.; Wang, L.N.; Zhang, J.L.; Zhu, D.H.; Wu, S.L.; et al. Fundamental theory of biodegradable metals—definition, criteria, and design. *Adv. Funct. Mater.* **2019**, *29*, 1805402. [CrossRef]
3. Zhao, D.W.; Witte, F.; Lu, F.Q.; Wang, J.L.; Li, J.L.; Qin, L. Current status on clinical applications of magnesium-based orthopaedic implants: A review from clinical translational perspective. *Biomaterials* **2017**, *112*, 287–302. [CrossRef] [PubMed]
4. Bowen, P.K.; Drelich, J.; Goldman, J. Zinc exhibits ideal physiological corrosion behavior for bioabsorbable stents. *Adv. Mater.* **2013**, *25*, 2577–2582. [CrossRef] [PubMed]
5. Mostaed, E.; Sikora-Jasinska, M.; Drelich, J.W.; Vedani, M. Zinc-based alloys for degradable vascular stent applications. *Acta Biomater.* **2018**, *71*, 1–23. [CrossRef]
6. Bowen, P.K.; Shearier, E.R.; Zhao, S.; Guillory, R.J., 2nd; Zhao, F.; Goldman, J.; Drelich, J.W. Biodegradable Metals for Cardiovascular Stents: From Clinical Concerns to Recent Zn-Alloys. *Adv. Healthc. Mater.* **2016**, *5*, 1121–1140. [CrossRef]
7. Seitz, J.-M.; Durisin, M.; Goldman, J.; Drelich, J.W. Recent advances in biodegradable metals for medical sutures: A critical review. *Adv. Healthc. Mater.* **2015**, *4*, 1915–1936. [CrossRef]
8. Venezuela, J.J.D.; Johnston, S.; Dargusch, M.S. The Prospects for Biodegradable Zinc in Wound Closure Applications. *Adv. Healthc. Mater.* **2019**, *8*, e1900408. [CrossRef]
9. Xia, D.; Yang, F.; Zheng, Y.; Liu, Y.; Zhou, Y. Research status of biodegradable metals designed for oral and maxillofacial applications: A review. *Bioact. Mater.* **2021**, *6*, 4186–4208. [CrossRef]
10. Chen, J.; Yang, R.; Shi, B.; Xu, Y.; Huang, H. Obturator Manufacturing for Oronasal Fistula after Cleft Palate Repair: A Review from Handicraft to the Application of Digital Techniques. *J. Funct. Biomater.* **2022**, *13*, 251. [CrossRef]
11. Xu, Y.; Huang, H.; Wu, M.; Tian, Y.; Wan, Q.; Shi, B.; Hu, T.; Spintzyk, S. Rapid Additive Manufacturing of a Superlight Obturator for Large Oronasal Fistula in Pediatric Patient. *Laryngoscope* **2022**. [CrossRef]
12. Bedard, P.; Gauvin, S.; Ferland, K.; Caneparo, C.; Pellerin, E.; Chabaud, S.; Bolduc, S. Innovative Human Three-Dimensional Tissue-Engineered Models as an Alternative to Animal Testing. *Bioengineering* **2020**, *7*, 115. [CrossRef]
13. Przekora, A.; Kazimierczak, P.; Wojcik, M. Ex vivo determination of chitosan/curdlan/hydroxyapatite biomaterial osseointegration with the use of human trabecular bone explant: New method for biocompatibility testing of bone implants reducing animal tests. *Mater. Sci. Eng. C Mater. Biol. Appl.* **2021**, *119*, 111612. [CrossRef]
14. Jung, O.; Smeets, R.; Hartjen, P.; Schnettler, R.; Feyerabend, F.; Klein, M.; Wegner, N.; Walther, F.; Stangier, D.; Henningsen, A.; et al. Improved In Vitro Test Procedure for Full Assessment of the Cytocompatibility of Degradable Magnesium Based on ISO 10993-5/-12. *Int. J. Mol. Sci.* **2019**, *20*, 255. [CrossRef]
15. Jablonska, E.; Kubasek, J.; Vojtech, D.; Ruml, T.; Lipov, J. Test conditions can significantly affect the results of in vitro cytotoxicity testing of degradable metallic biomaterials. *Sci. Rep.* **2021**, *11*, 6628. [CrossRef]
16. Li, P.; Schille, C.; Schweizer, E.; Kimmerle-Muller, E.; Rupp, F.; Heiss, A.; Legner, C.; Klotz, U.E.; Geis-Gerstorfer, J.; Scheideler, L. Selection of extraction medium influences cytotoxicity of zinc and its alloys. *Acta Biomater.* **2019**, *98*, 235–245. [CrossRef]
17. Zhang, K.; Ma, B.; Hu, K.; Yuan, B.; Sun, X.; Song, X.; Tang, Z.; Lin, H.; Zhu, X.; Zheng, Y.; et al. Evidence-based biomaterials research. *Bioact. Mater.* **2022**, *15*, 495–503. [CrossRef]
18. Cheng, J.; Liu, B.; Wu, Y.H.; Zheng, Y.F. Comparative in vitro Study on Pure Metals (Fe, Mn, Mg, Zn and W) as Biodegradable Metals. *J. Mater. Sci. Technol.* **2013**, *29*, 619–627. [CrossRef]
19. Dambatta, M.S.; Murni, N.S.; Izman, S.; Kurniawan, D.; Froemming, G.R.; Hermawan, H. In vitro degradation and cell viability assessment of Zn-3Mg alloy for biodegradable bone implants. *Proc. Inst. Mech. Eng. H* **2015**, *229*, 335–342. [CrossRef]
20. Gong, H.; Wang, K.; Strich, R.; Zhou, J.G. In vitro biodegradation behavior, mechanical properties, and cytotoxicity of biodegradable Zn-Mg alloy. *J. Biomed. Mater. Res. B Appl. Biomater.* **2015**, *103*, 1632–1640. [CrossRef]
21. Kubasek, J.; Vojtech, D.; Jablonska, E.; Pospisilova, I.; Lipov, J.; Ruml, T. Structure, mechanical characteristics and in vitro degradation, cytotoxicity, genotoxicity and mutagenicity of novel biodegradable Zn-Mg alloys. *Mater. Sci. Eng. C Mater. Biol. Appl.* **2016**, *58*, 24–35. [CrossRef] [PubMed]
22. Murni, N.S.; Dambatta, M.S.; Yeap, S.K.; Froemming, G.R.A.; Hermawan, H. Cytotoxicity evaluation of biodegradable Zn-3Mg alloy toward normal human osteoblast cells. *Mater. Sci. Eng. C Mater. Biol. Appl.* **2015**, *49*, 560–566. [CrossRef] [PubMed]
23. Tang, Z.; Huang, H.; Niu, J.; Zhang, L.; Zhang, H.; Pei, J.; Tan, J.; Yuan, G. Design and characterizations of novel biodegradable Zn-Cu-Mg alloys for potential biodegradable implants. *Mater. Des.* **2017**, *117*, 84–94. [CrossRef]
24. Niu, J.; Tang, Z.; Huang, H.; Pei, J.; Zhang, H.; Yuan, G.; Ding, W. Research on a Zn-Cu alloy as a biodegradable material for potential vascular stents application. *Mater. Sci. Eng. C Mater. Biol. Appl.* **2016**, *69*, 407–413. [CrossRef]
25. Jablonska, E.; Vojtech, D.; Fousova, M.; Kubasek, J.; Lipov, J.; Fojt, J.; Ruml, T. Influence of surface pre-treatment on the cytocompatibility of a novel biodegradable ZnMg alloy. *Mater. Sci. Eng. C Mater. Biol. Appl.* **2016**, *68*, 198–204. [CrossRef]
26. Wang, C.; Yang, H.T.; Li, X.; Zheng, Y.F. In Vitro Evaluation of the Feasibility of Commercial Zn Alloys as Biodegradable Metals. *J. Mater. Sci. Technol.* **2016**, *32*, 909–918. [CrossRef]
27. Shen, C.; Liu, X.; Fan, B.; Lan, P.; Zhou, F.; Li, X.; Wang, H.; Xiao, X.; Li, L.; Zhao, S.; et al. Mechanical properties, in vitro degradation behavior, hemocompatibility and cytotoxicity evaluation of Zn–1.2Mg alloy for biodegradable implants. *RSC Adv.* **2016**, *6*, 86410–86419. [CrossRef]

28. Katarivas Levy, G.; Leon, A.; Kafri, A.; Ventura, Y.; Drelich, J.W.; Goldman, J.; Vago, R.; Aghion, E. Evaluation of biodegradable Zn-1%Mg and Zn-1%Mg-0.5%Ca alloys for biomedical applications. *J. Mater. Sci. Mater. Med.* **2017**, *28*, 174. [CrossRef]
29. Tang, Z.; Niu, J.; Huang, H.; Zhang, H.; Pei, J.; Ou, J.; Yuan, G. Potential biodegradable Zn-Cu binary alloys developed for cardiovascular implant applications. *J. Mech. Behav. Biomed. Mater.* **2017**, *72*, 182–191. [CrossRef]
30. Zhu, D.; Su, Y.; Young, M.L.; Ma, J.; Zheng, Y.; Tang, L. Biological Responses and Mechanisms of Human Bone Marrow Mesenchymal Stem Cells to Zn and Mg Biomaterials. *ACS Appl. Mater. Interfaces* **2017**, *9*, 27453–27461. [CrossRef]
31. Ren, T.; Gao, X.; Xu, C.; Yang, L.; Guo, P.; Liu, H.; Chen, Y.; Sun, W.; Song, Z. Evaluation of as-extruded ternary Zn–Mg–Zr alloys for biomedical implantation material: In vitro and in vivo behavior. *Mater. Corros.* **2019**, *70*, 1056–1070. [CrossRef]
32. Tong, X.; Zhang, D.; Zhang, X.; Su, Y.; Shi, Z.; Wang, K.; Lin, J.; Li, Y.; Lin, J.; Wen, C. Microstructure, mechanical properties, biocompatibility, and in vitro corrosion and degradation behavior of a new Zn-5Ge alloy for biodegradable implant materials. *Acta Biomater.* **2018**, *82*, 197–204. [CrossRef]
33. Annonay, N.; Challali, F.; Labour, M.N.; Bockelee, V.; Garcia-Sanchez, A.; Tetard, F.; Besland, M.P.; Djemia, P.; Chaubet, F. Thin films of binary amorphous Zn-Zr alloys developed by magnetron co-sputtering for the production of degradable coronary stents: A preliminary study. *Bioact. Mater.* **2018**, *3*, 385–388. [CrossRef]
34. Chen, Y.; Huang, P.; Chen, H.; Wang, S.; Wang, H.; Guo, J.; Zhang, X.; Zhang, S.; Yan, J.; Xia, J.; et al. Assessment of the Biocompatibility and Biological Effects of Biodegradable Pure Zinc Material in the Colorectum. *ACS Biomater. Sci. Eng.* **2018**, *4*, 4095–4103. [CrossRef]
35. Xiao, C.; Wang, L.; Ren, Y.; Sun, S.; Zhang, E.; Yan, C.; Liu, Q.; Sun, X.; Shou, F.; Duan, J.; et al. Indirectly extruded biodegradable Zn-0.05wt%Mg alloy with improved strength and ductility: In vitro and in vivo studies. *J. Mater. Sci. Technol.* **2018**, *34*, 1618–1627. [CrossRef]
36. Li, P.; Schille, C.; Schweizer, E.; Rupp, F.; Heiss, A.; Legner, C.; Klotz, U.E.; Geis-Gerstorfer, J.; Scheideler, L. Mechanical Characteristics, In Vitro Degradation, Cytotoxicity, and Antibacterial Evaluation of Zn-4.0Ag Alloy as a Biodegradable Material. *Int. J. Mol. Sci.* **2018**, *19*, 755. [CrossRef]
37. Tong, X.; Shi, Z.; Xu, L.; Lin, J.; Zhang, D.; Wang, K.; Li, Y.; Wen, C. Degradation behavior, cytotoxicity, hemolysis, and antibacterial properties of electro-deposited Zn-Cu metal foams as potential biodegradable bone implants. *Acta Biomater.* **2020**, *102*, 481–492. [CrossRef]
38. Zhang, Y.; Lu, Y.; Xu, X.; Chen, L.; Xiao, T.; Luo, X.; Yan, Y.; Li, D.; Dai, Y.; Yu, K. Microstructure, Corrosion Behaviors in Different Simulated Body Fluids and Cytotoxicity of Zn–Li Alloy as Biodegradable Material. *Mater. Trans.* **2019**, *60*, 583–586. [CrossRef]
39. Li, Y.; Pavanram, P.; Zhou, J.; Lietaert, K.; Taheri, P.; Li, W.; San, H.; Leeflang, M.A.; Mol, J.M.C.; Jahr, H.; et al. Additively manufactured biodegradable porous zinc. *Acta Biomater.* **2020**, *101*, 609–623. [CrossRef]
40. Guo, H.; Cao, R.H.; Zheng, Y.F.; Bai, J.; Xue, F.; Chu, C.L. Diameter-dependent in vitro performance of biodegradable pure zinc wires for suture application. *J. Mater. Sci. Technol.* **2019**, *35*, 1662–1670. [CrossRef]
41. Shi, Z.Z.; Yu, J.; Liu, X.F.; Zhang, H.J.; Zhang, D.W.; Yin, Y.X.; Wang, L.N. Effects of Ag, Cu or Ca addition on microstructure and comprehensive properties of biodegradable Zn-0.8Mn alloy. *Mater. Sci. Eng. C Mater. Biol. Appl.* **2019**, *99*, 969–978. [CrossRef] [PubMed]
42. Li, P.; Dai, J.; Schweizer, E.; Rupp, F.; Heiss, A.; Richter, A.; Klotz, U.E.; Geis-Gerstorfer, J.; Scheideler, L.; Alexander, D. Response of human periosteal cells to degradation products of zinc and its alloy. *Mater. Sci. Eng. C Mater. Biol. Appl.* **2020**, *108*, 110208. [CrossRef] [PubMed]
43. Lin, S.; Ran, X.; Yan, X.; Wang, Q.; Zhou, J.G.; Hu, T.; Wang, G. Systematical evolution on a Zn-Mg alloy potentially developed for biodegradable cardiovascular stents. *J. Mater. Sci. Mater. Med.* **2019**, *30*, 122. [CrossRef] [PubMed]
44. Li, P.; Zhang, W.; Dai, J.; Xepapadeas, A.B.; Schweizer, E.; Alexander, D.; Scheideler, L.; Zhou, C.; Zhang, H.; Wan, G.; et al. Investigation of zinccopper alloys as potential materials for craniomaxillofacial osteosynthesis implants. *Mater. Sci. Eng. C Mater. Biol. Appl.* **2019**, *103*, 109826. [CrossRef]
45. Zhang, Y.; Yan, Y.; Xu, X.; Lu, Y.; Chen, L.; Li, D.; Dai, Y.; Kang, Y.; Yu, K. Investigation on the microstructure, mechanical properties, in vitro degradation behavior and biocompatibility of newly developed Zn-0.8%Li-(Mg, Ag) alloys for guided bone regeneration. *Mater. Sci. Eng. C Mater. Biol. Appl.* **2019**, *99*, 1021–1034. [CrossRef]
46. Zhu, D.; Cockerill, I.; Su, Y.; Zhang, Z.; Fu, J.; Lee, K.W.; Ma, J.; Okpokwasili, C.; Tang, L.; Zheng, Y.; et al. Mechanical Strength, Biodegradation, and in Vitro and in Vivo Biocompatibility of Zn Biomaterials. *ACS Appl. Mater. Interfaces* **2019**, *11*, 6809–6819. [CrossRef]
47. Shuai, C.; Xue, L.; Gao, C.; Peng, S.; Zhao, Z. Rod-like Eutectic Structure in Biodegradable Zn-Al-Sn Alloy Exhibiting Enhanced Mechanical Strength. *ACS Biomater. Sci. Eng.* **2020**, *6*, 3821–3831. [CrossRef]
48. Chen, C.; Yue, R.; Zhang, J.; Huang, H.; Niu, J.; Yuan, G. Biodegradable Zn-1.5Cu-1.5Ag alloy with anti-aging ability and strain hardening behavior for cardiovascular stents. *Mater. Sci. Eng. C Mater. Biol. Appl.* **2020**, *116*, 111172. [CrossRef]
49. Avior, O.; Ben Ghedalia-Peled, N.; Ron, T.; Vago, R.; Aghion, E. The Effect of Ca on In Vitro Behavior of Biodegradable Zn-Fe Alloy in Simulated Physiological Environments. *Metals* **2020**, *10*, 1624. [CrossRef]
50. Shi, Z.Z.; Gao, X.X.; Chen, H.T.; Liu, X.F.; Li, A.; Zhang, H.J.; Wang, L.N. Enhancement in mechanical and corrosion resistance properties of a biodegradable Zn-Fe alloy through second phase refinement. *Mater. Sci. Eng. C Mater. Biol. Appl.* **2020**, *116*, 111197. [CrossRef]

51. Jia, B.; Yang, H.; Han, Y.; Zhang, Z.; Qu, X.; Zhuang, Y.; Wu, Q.; Zheng, Y.; Dai, K. In vitro and in vivo studies of Zn-Mn biodegradable metals designed for orthopedic applications. *Acta Biomater.* **2020**, *108*, 358–372. [CrossRef]
52. Xu, X.; Lu, Y.; Chu, X.; Yan, Y.; Liu, Y.; Xu, X.; Luo, X.; Chen, L.; Li, D.; Xiao, T.; et al. Microstructure, biodegradable behavior in different simulated body fluids, antibacterial effect on different bacteria and cytotoxicity of rolled Zn–Li–Ag alloy. *Mater. Res. Express.* **2020**, *7*, 055403. [CrossRef]
53. Li, Y.; Pavanram, P.; Zhou, J.; Lietaert, K.; Bobbert, F.S.L.; Kubo, Y.; Leeflang, M.A.; Jahr, H.; Zadpoor, A.A. Additively manufactured functionally graded biodegradable porous zinc. *Biomater. Sci.* **2020**, *8*, 2404–2419. [CrossRef]
54. Wang, K.; Tong, X.; Lin, J.; Wei, A.; Li, Y.; Dargusch, M.; Wen, C. Binary Zn–Ti alloys for orthopedic applications: Corrosion and degradation behaviors, friction and wear performance, and cytotoxicity. *J. Mater. Sci. Technol.* **2021**, *74*, 216–229. [CrossRef]
55. Lin, J.; Tong, X.; Sun, Q.; Luan, Y.; Zhang, D.; Shi, Z.; Wang, K.; Lin, J.; Li, Y.; Dargusch, M.; et al. Biodegradable ternary Zn-3Ge-0.5X (X=Cu, Mg, and Fe) alloys for orthopedic applications. *Acta Biomater.* **2020**, *115*, 432–446. [CrossRef]
56. Lin, J.; Tong, X.; Shi, Z.; Zhang, D.; Zhang, L.; Wang, K.; Wei, A.; Jin, L.; Lin, J.; Li, Y.; et al. A biodegradable Zn-1Cu-0.1Ti alloy with antibacterial properties for orthopedic applications. *Acta Biomater.* **2020**, *106*, 410–427. [CrossRef]
57. Tong, X.; Zhang, D.; Lin, J.; Dai, Y.; Luan, Y.; Sun, Q.; Shi, Z.; Wang, K.; Gao, Y.; Lin, J.; et al. Development of biodegradable Zn-1Mg-0.1RE (RE = Er, Dy, and Ho) alloys for biomedical applications. *Acta Biomater.* **2020**, *117*, 384–399. [CrossRef]
58. Yang, H.; Jia, B.; Zhang, Z.; Qu, X.; Li, G.; Lin, W.; Zhu, D.; Dai, K.; Zheng, Y. Alloying design of biodegradable zinc as promising bone implants for load-bearing applications. *Nat. Commun.* **2020**, *11*, 401. [CrossRef]
59. Li, P.; Schille, C.; Schweizer, E.; Kimmerle-Muller, E.; Rupp, F.; Han, X.; Heiss, A.; Richter, A.; Legner, C.; Klotz, U.E.; et al. Evaluation of a Zn-2Ag-1.8Au-0.2V Alloy for Absorbable Biocompatible Materials. *Materials* **2019**, *13*, 56. [CrossRef]
60. Yue, R.; Niu, J.; Li, Y.; Ke, G.; Huang, H.; Pei, J.; Ding, W.; Yuan, G. In vitro cytocompatibility, hemocompatibility and antibacterial properties of biodegradable Zn-Cu-Fe alloys for cardiovascular stents applications. *Mater. Sci. Eng. C Mater. Biol. Appl.* **2020**, *113*, 111007. [CrossRef]
61. Li, Z.; Shi, Z.Z.; Hao, Y.; Li, H.F.; Zhang, H.J.; Liu, X.F.; Wang, L.N. Insight into role and mechanism of Li on the key aspects of biodegradable ZnLi alloys: Microstructure evolution, mechanical properties, corrosion behavior and cytotoxicity. *Mater. Sci. Eng. C Mater. Biol. Appl.* **2020**, *114*, 111049. [CrossRef] [PubMed]
62. Guo, H.; Xia, D.; Zheng, Y.; Zhu, Y.; Liu, Y.; Zhou, Y. A pure zinc membrane with degradability and osteogenesis promotion for guided bone regeneration: In vitro and in vivo studies. *Acta Biomater.* **2020**, *106*, 396–409. [CrossRef] [PubMed]
63. Xiao, C.; Su, Y.; Zhu, X.; Yu, W.; Cui, D.; Wei, X.; Zhang, X.; Li, J.; Wang, F.; Ren, Y.; et al. Mechanical performance and biocompatibility assessment of Zn-0.05wt%Mg-(0.5, 1 wt%) Ag alloys. *J. Biomed. Mater. Res. B Appl. Biomater.* **2020**, *108*, 2925–2936. [CrossRef] [PubMed]
64. Deng, L.; Guo, P.; Li, F.; Yang, L.; Zhu, X.; Xu, C.; Zhang, Q.; Shi, Y.; Song, Z.; Sun, W.; et al. Ultrafine-grained Zn-0.45Li alloy with enhanced mechanical property, degradation behavior and cytocompatibility prepared by hot extrusion and multi-pass drawing. *Materialwiss. Werkst.* **2021**, *52*, 991–996. [CrossRef]
65. Jia, B.; Yang, H.; Zhang, Z.; Qu, X.; Jia, X.; Wu, Q.; Han, Y.; Zheng, Y.; Dai, K. Biodegradable Zn-Sr alloy for bone regeneration in rat femoral condyle defect model: In vitro and in vivo studies. *Bioact. Mater.* **2021**, *6*, 1588–1604. [CrossRef]
66. Wu, H.; Xie, X.; Wang, J.; Ke, G.; Huang, H.; Liao, Y.; Kong, Q. Biological properties of Zn–0.04Mg–2Ag: A new degradable zinc alloy scaffold for repairing large-scale bone defects. *J. Mater. Res. Technol.* **2021**, *13*, 1779–1789. [CrossRef]
67. Farabi, E.; Sharp, J.A.; Vahid, A.; Fabijanic, D.M.; Barnett, M.R.; Gallo, S.C. Development of high strength and ductile Zn-Al-Li alloys for potential use in bioresorbable medical devices. *Mater. Sci. Eng. C Mater. Biol. Appl.* **2021**, *122*, 111897. [CrossRef]
68. Yang, Y.; Yang, M.; He, C.; Qi, F.; Wang, D.; Peng, S.; Shuai, C. Rare earth improves strength and creep resistance of additively manufactured Zn implants. *Compos. B Eng.* **2021**, *216*, 108882. [CrossRef]
69. Qu, X.; Yang, H.; Jia, B.; Wang, M.; Yue, B.; Zheng, Y.; Dai, K. Zinc alloy-based bone internal fixation screw with antibacterial and anti-osteolytic properties. *Bioact. Mater.* **2021**, *6*, 4607–4624. [CrossRef]
70. Milenin, A.; Łukowicz, K.; Truchan, K.; Osyczka, A.M. In vitro cytotoxicity of biodegradable Zn-Mg surgical wires in tumor and healthy cells. *Acta Bioeng. Biomech.* **2021**, *23*. [CrossRef]
71. Lin, J.; Tong, X.; Wang, K.; Shi, Z.; Li, Y.; Dargusch, M.; Wen, C. Biodegradable Zn–3Cu and Zn–3Cu–0.2Ti alloys with ultrahigh ductility and antibacterial ability for orthopedic applications. *J. Mater. Sci. Technol.* **2021**, *68*, 76–90. [CrossRef]
72. Jan, P.; Španko, M.; Lacina, L.; Kubásek, J.; Ashcheulov, P.; Veřtát, P.; Školáková, A.; Kvítek, O.; Vojtěch, D.; Čapek, J. Influence of the pre-exposure of a Zn-0.8Mg-0.2Sr absorbable alloy in bovine serum albumin containing media on its surface changes and their impact on the cytocompatibility of the material. *Mater. Today Commun.* **2021**, *28*, 102556. [CrossRef]
73. Zhang, W.; Li, P.; Shen, G.; Mo, X.; Zhou, C.; Alexander, D.; Rupp, F.; Geis-Gerstorfer, J.; Zhang, H.; Wan, G. Appropriately adapted properties of hot-extruded Zn-0.5Cu-xFe alloys aimed for biodegradable guided bone regeneration membrane application. *Bioact. Mater.* **2021**, *6*, 975–989. [CrossRef]
74. Zhu, P.; Chen, J.; Li, P.; Xu, S. Limitation of Water-Soluble Tetrazolium Salt for the Cytocompatibility Evaluation of Zinc-Based Metals. *Materials* **2021**, *14*, 6247. [CrossRef]
75. Li, P.; Zhang, W.; Spintzyk, S.; Schweizer, E.; Krajewski, S.; Alexander, D.; Dai, J.; Xu, S.; Wan, G.; Rupp, F. Impact of sterilization treatments on biodegradability and cytocompatibility of zinc-based implant materials. *Mater. Sci. Eng. C Mater. Biol. Appl.* **2021**, *130*, 112430. [CrossRef]

76. Capek, J.; Kubasek, J.; Pinc, J.; Fojt, J.; Krajewski, S.; Rupp, F.; Li, P. Microstructural, mechanical, in vitro corrosion and biological characterization of an extruded Zn-0.8Mg-0.2Sr (wt%) as an absorbable material. *Mater. Sci. Eng. C Mater. Biol. Appl.* **2021**, *122*, 111924. [CrossRef]
77. Avior, O.; Ben Ghedalia-Peled, N.; Ron, T.; Goldman, J.; Vago, R.; Aghion, E. Stress Corrosion Analysis and Direct Cell Viability of Biodegradable Zn-Fe-Ca Alloy in In-Vitro Conditions. *Metals* **2022**, *12*, 76. [CrossRef]
78. Tong, X.; Zhu, L.; Wang, K.; Shi, Z.; Huang, S.; Li, Y.; Ma, J.; Wen, C.; Lin, J. Impact of gadolinium on mechanical properties, corrosion resistance, and biocompatibility of Zn-1Mg-xGd alloys for biodegradable bone-implant applications. *Acta Biomater.* **2022**, *142*, 361–373. [CrossRef]
79. Jiang, J.; Qian, Y.; Huang, H.; Niu, J.; Yuan, G. Biodegradable Zn-Cu-Mn alloy with suitable mechanical performance and in vitro degradation behavior as a promising candidate for vascular stents. *Biomater. Adv.* **2022**, *133*, 112652. [CrossRef]
80. Bao, G.; Wang, K.; Yang, L.; He, J.; He, B.; Xu, X.; Zheng, Y. Feasibility evaluation of a Zn-Cu alloy for intrauterine devices: In vitro and in vivo studies. *Acta Biomater.* **2022**, *142*, 374–387. [CrossRef]
81. Ren, H.; Pan, C.; Liu, Y.; Liu, D.; He, X.; Li, X.; Sun, X. Fabrication, in vitro and in vivo properties of porous Zn–Cu alloy scaffolds for bone tissue engineering. *Mater. Chem. Phys.* **2022**, *289*, 126458. [CrossRef]
82. Qin, Y.; Liu, A.; Guo, H.; Shen, Y.; Wen, P.; Lin, H.; Xia, D.; Voshage, M.; Tian, Y.; Zheng, Y. Additive manufacturing of Zn-Mg alloy porous scaffolds with enhanced osseointegration: In vitro and in vivo studies. *Acta Biomater.* **2022**, *145*, 403–415. [CrossRef] [PubMed]
83. Xu, Y.; Xu, Y.; Zhang, W.; Li, M.; Wendel, H.P.; Geis-Gerstorfer, J.; Li, P.; Wan, G.; Xu, S.; Hu, T. Biodegradable Zn-Cu-Fe Alloy as a Promising Material for Craniomaxillofacial Implants: An in vitro Investigation into Degradation Behavior, Cytotoxicity, and Hemocompatibility. *Front. Chem.* **2022**, *10*, 860040. [CrossRef]
84. Zeng, Y.; Guan, Z.; Linsley, C.S.; Pan, S.; Liu, J.; Wu, B.M.; Li, X. Experimental study on novel biodegradable Zn-Fe-Si alloys. *J. Biomed. Mater. Res. B Appl. Biomater.* **2022**, *110*, 2266–2275. [CrossRef] [PubMed]
85. Liu, Y.; Fu, Z.; Chu, X.; Lu, Y.; Zhang, J.; Huang, J.; Liu, Y.; Yan, Y.; Yu, K. Fabrication and characterization of A Zn-0.5Fe alloy membrane by powder metallurgy route for guided bone regeneration. *Mater. Res. Express.* **2022**, *9*. [CrossRef]
86. Palai, D.; Roy, T.; Prasad, P.S.; Hazra, C.; Dhara, S.; Sen, R.; Das, S.; Das, K. Influence of Copper on the Microstructural, Mechanical, and Biological Properties of Commercially Pure Zn-Based Alloys for a Potential Biodegradable Implant. *ACS Biomater. Sci. Eng.* **2022**, *8*, 1443–1463. [CrossRef]
87. Gopal, N.; Palaniyandi, P.; Ramasamy, P.; Panchal, H.; Ibrahim, A.M.M.; Alsoufi, M.S.; Elsheikh, A.H. In Vitro Degradability, Microstructural Evaluation, and Biocompatibility of Zn-Ti-Cu-Ca-P Alloy. *Nanomaterials* **2022**, *12*, 1357. [CrossRef]
88. Yang, N.; Balasubramani, N.; Venezuela, J.; Bielefeldt-Ohmann, H.; Allavena, R.; Almathami, S.; Dargusch, M. Microstructure refinement in biodegradable Zn-Cu-Ca alloy for enhanced mechanical properties, degradation homogeneity, and strength retention in simulated physiological condition. *J. Mater. Sci. Technol.* **2022**, *125*, 1–14. [CrossRef]
89. Duan, J.; Li, L.; Liu, C.; Suo, Y.; Wang, X.; Yang, Y. Novel Zn-2Cu-0.2Mn-xLi (x = 0, 0.1 and 0.38) alloys developed for potential biodegradable implant applications. *J. Alloys Compd.* **2022**, *916*, 165478. [CrossRef]
90. Zhu, X.; Ren, T.; Guo, P.; Yang, L.; Shi, Y.; Sun, W.; Song, Z. Strengthening mechanism and biocompatibility of degradable Zn-Mn alloy with different Mn content. *Mater. Today Commun.* **2022**, *31*, 103639. [CrossRef]
91. Katarivas Levy, G.; Kafri, A.; Ventura, Y.; Leon, A.; Vago, R.; Goldman, J.; Aghion, E. Surface stabilization treatment enhances initial cell viability and adhesion for biodegradable zinc alloys. *Mater. Lett.* **2019**, *248*, 130–133. [CrossRef]
92. Cockerill, I.; Su, Y.; Bitten, R.; Cloarec, B.; Aouadi, S.; Zhu, D.; Young, M.L. Salt Preform Texturing of Absorbable Zn Substrates for Bone-implant Applications. *JOM* **2020**, *72*, 1902–1909. [CrossRef]
93. Li, P.; Qian, J.; Zhang, W.; Schille, C.; Schweizer, E.; Heiss, A.; Klotz, U.E.; Scheideler, L.; Wan, G.; Geis-Gerstorfer, J. Improved biodegradability of zinc and its alloys by sandblasting treatment. *Surf. Coat. Technol.* **2021**, *405*, 126678. [CrossRef]
94. Tong, X.; Han, Y.; Zhou, R.; Jiang, W.; Zhu, L.; Li, Y.; Huang, S.; Ma, J.; Wen, C.; Lin, J. Biodegradable Zn-Dy binary alloys with high strength, ductility, cytocompatibility, and antibacterial ability for bone-implant applications. *Acta Biomater.* **2022**, *155*, 684–702. [CrossRef]
95. Watroba, M.; Bednarczyk, W.; Szewczyk, P.K.; Kawalko, J.; Mech, K.; Grunewald, A.; Unalan, I.; Taccardi, N.; Boelter, G.; Banzhaf, M.; et al. In vitro cytocompatibility and antibacterial studies on biodegradable Zn alloys supplemented by a critical assessment of direct contact cytotoxicity assay. *J. Biomed. Mater. Res. B Appl. Biomater.* **2023**, *111*, 241–260. [CrossRef]
96. Di, T.; Xu, Y.; Liu, D.; Sun, X. Microstructure, Mechanical Performance and Anti-Bacterial Activity of Degradable Zn-Cu-Ag Alloy. *Metals* **2022**, *12*, 1444. [CrossRef]
97. Wang, Z.; Wang, W.; Zhang, X.; Cao, F.; Zhang, T.; Bhakta Pokharel, D.; Chen, D.; Li, J.; Yang, J.; Xiao, C.; et al. Modulation of Osteogenesis and Angiogenesis Activities Based on Ionic Release from Zn-Mg Alloys. *Materials* **2022**, *15*, 7117. [CrossRef]
98. Ben Tzion-Mottye, L.; Ron, T.; Eliezer, D.; Aghion, E. The Effect of Mn on the Mechanical Properties and In Vitro Behavior of Biodegradable Zn-2%Fe Alloy. *Metals* **2022**, *12*, 1291. [CrossRef]
99. Shi, Z.-Z.; Li, X.-M.; Yao, S.-L.; Tang, Y.-Z.; Ji, X.-J.; Wang, Q.; Gao, X.-X.; Wang, L.-N. 300 MPa grade biodegradable high-strength ductile low-alloy (BHSDLA) Zn-Mn-Mg alloys: An in vitro study. *J. Mater. Sci. Technol.* **2023**, *138*, 233–244. [CrossRef]
100. Li, S.; Wang, X.; Ren, J.; Liu, C.; Hu, Y.; Yang, Y. Microstructure, mechanical property and corrosion behavior of biomedical Zn-Fe alloy prepared by low-temperature sintering. *J. Alloys Compd.* **2023**, *934*, 167812. [CrossRef]

101. Sun, J.-L.; Feng, Y.; Shi, Z.-Z.; Xue, Z.; Cao, M.; Yao, S.-L.; Li, Z.; Wang, L.-N. Biodegradable Zn-0.5Li alloy rib plate: Processing procedure development and in vitro performance evaluation. *J. Mater. Sci. Technol.* **2023**, *141*, 245–256. [CrossRef]
102. Xiang, E.; Gomez-Cerezo, M.N.; Ali, Y.; Ramachandra, S.S.; Yang, N.; Dargusch, M.; Moran, C.S.; Ivanovski, S.; Abdal-Hay, A. Surface Modification of Pure Zinc by Acid Etching: Accelerating the Corrosion Rate and Enhancing Biocompatibility and Antibacterial Characteristics. *ACS Appl. Mater. Interfaces* **2022**, *14*, 22554–22569. [CrossRef] [PubMed]
103. Vojtech, D.; Kubasek, J.; Serak, J.; Novak, P. Mechanical and corrosion properties of newly developed biodegradable Zn-based alloys for bone fixation. *Acta Biomater.* **2011**, *7*, 3515–3522. [CrossRef] [PubMed]
104. Zheng, Y.F.; Gu, X.N.; Witte, F. Biodegradable metals. *Mater. Sci. Eng. R Rep.* **2014**, *77*, 1–34. [CrossRef]
105. Liu, X.; Sun, J.; Qiu, K.; Yang, Y.; Pu, Z.; Li, L.; Zheng, Y. Effects of alloying elements (Ca and Sr) on microstructure, mechanical property and in vitro corrosion behavior of biodegradable Zn–1.5Mg alloy. *J. Alloys Compd.* **2016**, *664*, 444–452. [CrossRef]
106. Aghion, E.; Levy, G.; Ovadia, S. In vivo behavior of biodegradable Mg-Nd-Y-Zr-Ca alloy. *J. Mater. Sci. Mater. Med.* **2012**, *23*, 805–812. [CrossRef]
107. Berterame, N.M.; Martani, F.; Porro, D.; Branduardi, P. Copper homeostasis as a target to improve Saccharomyces cerevisiae tolerance to oxidative stress. *Metab. Eng.* **2018**, *46*, 43–50. [CrossRef]
108. Wang, J.; Witte, F.; Xi, T.; Zheng, Y.; Yang, K.; Yang, Y.; Zhao, D.; Meng, J.; Li, Y.; Li, W.; et al. Recommendation for modifying current cytotoxicity testing standards for biodegradable magnesium-based materials. *Acta Biomater.* **2015**, *21*, 237–249. [CrossRef]
109. Chen, Z.; Zhu, X. Accumulation of rare earth elements in bone and its toxicity and potential hazard to health. *J. Ecol. Rural Environ.* **2008**, *24*, 88–91.
110. Kirkpatrick, C.J.; Mittermayer, C. Theoretical and practical aspects of testing potential biomaterialsin vitro. *J. Mater. Sci. Mater. Med.* **1990**, *1*, 9–13. [CrossRef]
111. Wang, D.; Han, X.; Luo, F.; Thieringer, F.M.; Xu, Y.; Ou, G.; Spintzyk, S. Adhesive Property of 3D-Printed PEEK Abutments: Effects of Surface Treatment and Temporary Crown Material on Shear Bond Strength. *J. Funct. Biomater.* **2022**, *13*, 288. [CrossRef]
112. Hennig, B.; Toborek, M.; McClain, C.J. Antiatherogenic properties of zinc: Implications in endothelial cell metabolism. *Nutrition* **1996**, *12*, 711–717. [CrossRef]
113. Beaussant Törne, K.; Örnberg, A.; Weissenrieder, J. Characterization of the protective layer formed on zinc in whole blood. *Electrochim. Acta* **2017**, *258*, 1476–1483. [CrossRef]
114. Livingstone, C. Zinc: Physiology, deficiency, and parenteral nutrition. *Nutr. Clin. Pract.* **2015**, *30*, 371–382. [CrossRef]
115. Karunakaran, R.; Ortgies, S.; Tamayol, A.; Bobaru, F.; Sealy, M.P. Additive manufacturing of magnesium alloys. *Bioact. Mater.* **2020**, *5*, 44–54. [CrossRef]
116. Bushnell, P.J.; Kavlock, R.J.; Crofton, K.M.; Weiss, B.; Rice, D.C. Behavioral toxicology in the 21st century: Challenges and opportunities for behavioral scientists. Summary of a symposium presented at the annual meeting of the neurobehavioral teratology society, June, 2009. *Neurotoxicol. Teratol.* **2010**, *32*, 313–328. [CrossRef]

Journal of
Functional Biomaterials

MDPI

Article

Self-Assembling Peptide RADA16 Nanofiber Scaffold Hydrogel-Wrapped Concentrated Growth Factors in Osteogenesis of MC3T3

Renjie Yang [1], Jiali Chen [2], Dingjie Wang [3], Yichen Xu [4,*] and Guomin Ou [3,*]

1 State Key Laboratory of Oral Diseases & National Clinical Research Center for Oral Diseases, Eastern Clinic, West China Hospital of Stomatology, Sichuan University, Chengdu 610051, China

2 State Key Laboratory of Oral Diseases & National Clinical Research Center for Oral Diseases, Department of Oral and Maxillofacial Surgery, West China Hospital of Stomatology, Sichuan University, Chengdu 610041, China

3 State Key Laboratory of Oral Diseases & National Clinical Research Center for Oral Diseases, Department of Oral Implantology, West China Hospital of Stomatology, Sichuan University, Chengdu 610041, China

4 State Key Laboratory of Oral Diseases & National Clinical Research Center for Oral Diseases, Department of Oral Prosthodontics, West China Hospital of Stomatology, Sichuan University, Chengdu 610041, China

* Correspondence: yichen.xu@scu.edu.cn (Y.X.); ouguom@scu.edu.cn (G.O.)

Abstract: Concentrated growth factors (CGFs) are widely used in surgery with bone grafting, but the release of growth factors from CGFs is rapid. RADA16, a self-assembling peptide, can form a scaffold that is similar to the extracellular matrix. Based on the properties of RADA16 and CGF, we hypothesized that the RADA16 nanofiber scaffold hydrogel could enhance the function of CGFs and that the RADA16 nanofiber scaffold hydrogel-wrapped CGFs (RADA16-CGFs) would perform a good osteoinductive function. This study aimed to investigate the osteoinductive function of RADA16-CGFs. Scanning electron microscopy, rheometry, and ELISA were performed, and MC3T3-E1 cells were used to test cell adhesion, cytotoxicity, and mineralization after administration with RADA16-CGFs. We found that RADA16 endowed with the sustained release of growth factors from CGFs, which can help maximize the function of CGFs in osteoinduction. The application of the atoxic RADA16 nanofiber scaffold hydrogel with CGFs can be a new therapeutic strategy for the treatment of alveolar bone loss and other problems that require bone regeneration.

Keywords: alveolar bone cleft; alveolar bone grafting; RADA16; CGFs; osteogenesis

Citation: Yang, R.; Chen, J.; Wang, D.; Xu, Y.; Ou, G. Self-Assembling Peptide RADA16 Nanofiber Scaffold Hydrogel-Wrapped Concentrated Growth Factors in Osteogenesis of MC3T3. *J. Funct. Biomater.* **2023**, *14*, 260. https://doi.org/10.3390/jfb14050260

Academic Editors: Masami Okamoto and Ik-Hwan Kim

Received: 28 March 2023
Revised: 27 April 2023
Accepted: 5 May 2023
Published: 8 May 2023

1. Introduction

The alveolar cleft is one common congenital craniofacial defect in patients with cleft lip and cleft palate and is often accompanied by tooth agenesis. Alveolar bone grafting is commonly performed to reconstruct the alveolar ridge, stabilize the maxillary arch, and guarantee implant treatment [1]. Iliac cancellous bone grafting is the gold standard treatment for alveolar clefts [2], providing osteoinductive (recruitment of host mesenchymal stem cells from the surrounding tissue) and osteoconductive (scaffolding for cells to climb in and vascularization) [3]. This process is facilitated by the presence of growth factors within the autogenous bone material.

For some patients, due to the lack of donor site morbidity and large volume, allogenic bone and substitute bone could be potential alternative choices [4]. Allogenic bone and substitute bone grafts avoid second surgical sites to simplify the operation and relieve pain [5]. However, compared with autogenous bone grafts, these materials lack osteogenic and osteoinductive properties. Thus, we expect to design a complex bone substitute to provide both scaffold and bioactivity to cover this shortage [6].

Concentrated growth factors (CGFs) are the platelet-rich fibrin (PRF) derivatives that have been applied in the clinic [7], which can release Transforming Growth Factor-β (TGF-β), Vascular Endothelial Growth Factor (VEGF), Platelet-Derived Growth Factor (PDGF), Bone Morphogenetic Protein 2 (BMP2) and Insulin-like Growth Factor (IGF-1) [8,9]. These growth factors can increase proliferation and extracellular matrix mineralization [10]. Studies have shown the osteogenic effect of CGFs in vivo [11–14] and in vitro [12,15,16]. However, the release of growth factors from CGFs is rapid, while bone formation is slow and complex [16]. To match the process of bone formation, extending the releasing process of the growth factors of CGFs becomes important. Studies have shown that CGF-wrapped scaffolds can help extend the releasing process [17], and different hydrogels, such as chitosan and sodium alginate, have been tested [18].

Self-assembling peptides (SAPs) are oligopeptides that can spontaneously self-assemble into nanostructures under appropriate conditions [19,20]. A series of SAPs have been synthesized, and RADA16 became distinctive because its structure is exceedingly orderly, and its self-assembling can be controlled in a neutral pH solution or a physiological saline solution. RADA16 can form a stable scaffold structure similar to the extracellular matrix (ECM), with excellent biocompatibility and low immunogenicity [21,22]. RADA16 nanofiber scaffold hydrogel can help in cell adhesion and proliferation and provide a suitable microenvironment for cell differentiation [23]. Studies have shown that RADA16 nanofiber scaffold hydrogel could promote osteogenesis and control the release of functional factors when combined with the application of growth factors, such as Bone Morphogenetic Protein-2 (BMP-2) [22,24–26].

Based on the properties of RADA16 and CGF, we hypothesized that the RADA16 nanofiber scaffold hydrogel could enhance the function of CGFs and that the RADA16 nanofiber scaffold hydrogel-wrapped CGFs (RADA16-CGFs) would perform osteogenic and osteoinductive functions. Therefore, this study aimed to investigate the osteogenesis function of RADA16-CGFs in vitro to provide a new strategy for the clinic.

2. Materials and Methods

2.1. Blood Sample Centrifugation and CGF Preparation

Vein blood samples were obtained from 6 healthy volunteers who were nonsmokers from the West China College of Stomatology, Sichuan University. Sample collection was performed with the approval of the Ethics Committee of West China Hospital of Stomatology, Sichuan University (WCHSIRB-D-2021-534). All potential participants signed an informed consent form prior to sample collection and were informed of the potential benefits and risks of participating. The CGFs were produced as follows: 9 mL of blood was drawn from the arm vein in 10 mL blood collection tubes without an anticoagulant solution. The tubes were immediately centrifuged in a special machine (Medifuge MF200, Silfradent Srl, Santa Sofia, Italy) under the CGF preparation procedure. At the end of the process, there were three blood fractions. The middle fraction was extracted as the CGF composition and squashed to form the CGF membrane (Figure 1).

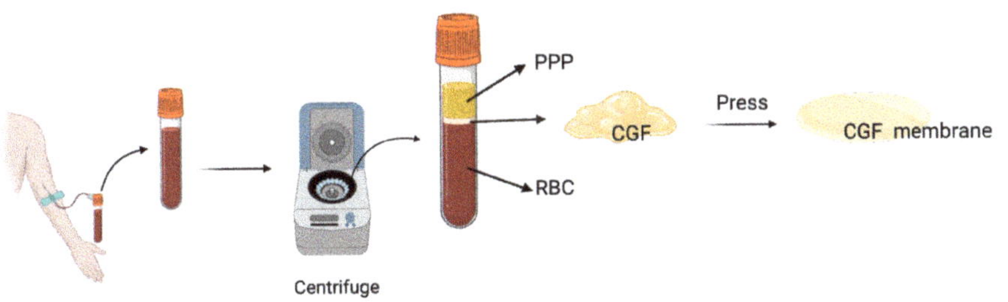

Figure 1. CGF preparation. Blood was drawn from the arm vein, and the tubes were centrifuged under the CGF preparation procedure. Three blood fractions were found at the end of the process. The middle fraction was extracted as the CGF composition and squashed to form the CGF membrane.

2.2. RADA16 Synthesis

The RADA16 applied in this study is L-RADA16, of which the peptide sequence is AcN-RADARADAR ADARADA-CONH2 (molecular weight: 1671.76). The RADA16 peptide was custom-synthesized commercially using solid-phase peptide synthesis (Hangzhou Jiatai Co., Ltd., Hangzhou, China). The C-terminus and N-terminus of this peptide were amidated and acetylated, respectively [20]. The peptide was purified using high-performance liquid chromatography and characterized using mass spectroscopy. Samples were dissolved in PBS (pH 7.4) at a concentration of 20.0 mg/mL (2% *w*/*v* RADA16) and stored at 4 °C in a refrigerator for further study (Figure 2a).

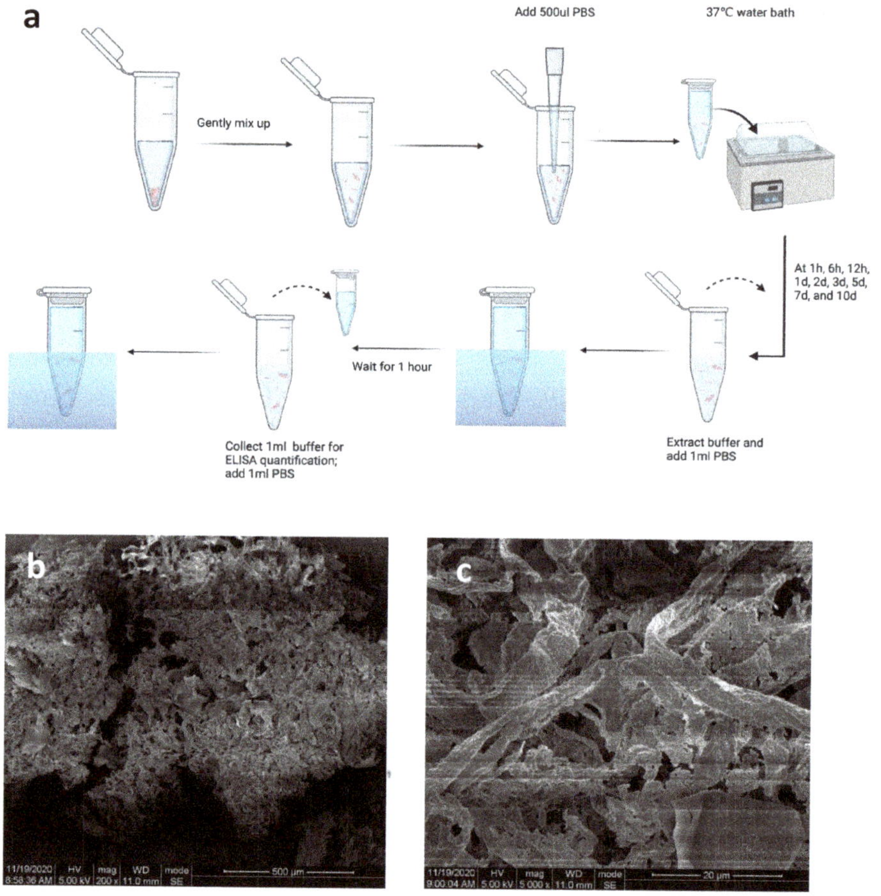

Figure 2. Properties of CGFs. (**a**) Sample preparation for ELISA quantification. Shredded CGF membrane and 2% *w*/*v* RADA16 solution (500 µL) were softly mixed in a 1.5 mL EP tub. Then, 500 µL PBS was added to the tube and put into a 37 °C water bath. The buffer was extracted at 1 h, 6 h, 12 h, 1 d, 2 d, 3 d, 5 d, 7 d, and 10 d, and another 1 mL PBS was added into the tubes for 1 h, and extracted for ELISA quantification. (**b**,**c**) SEM image of CGFs. Bars, 500 µm and 20 µm, respectively.

2.3. RADA16-CGF Fabrication

Shredded CGF membrane and 2% *w*/*v* RADA16 solution (500 µL) were softly mixed in a 1.5 mL EP tube and prepared for further study.

2.4. Scanning Electron Microscopy (SEM) Assay

To observe the CGFs, the CGF membrane was dried for 12 h in a lyophilizer before the SEM assay. To observe the RADA16 nanofiber scaffold hydrogel, the peptide solution (2% *w*/*v* RADA16) was mixed with PBS (pH 7.4) at a volume ratio of 1:1 and stored in a 1.5 mL EP tube (1% *w*/*v* RADA16) with sonication for 30 min to decrease viscidity. The peptide solution sample was added to 400 mL PBS and stored at 37 °C overnight. The hydrogel was then dried for 12 h in a lyophilizer. The samples were observed under a scanning electron microscope (Hitachi S-3400N; Hitachi, Ltd., Tokyo, Japan).

2.5. Rheometry

Rheology assays were carried out using a 1° stainless steel cone-controlled rheometer with a diameter of 20 mm (Thermo Fisher Scientific, Waltham, MA, USA). Samples were dissolved in PBS (pH 7.4) at a concentration of 10.0 mg/mL the peptide solution (2% *w*/*v* RADA16) was mixed with PBS (pH 7.4) at a volume ratio of 1:1 and stored at 4 °C in a refrigerator overnight, and 150–200 μL of samples were used for analysis at 25 °C.

2.6. ELISA Quantification for Growth Factors Released from RADA16-CGFs

One milliliter of PBS was added to the RADA16-CGF tube. The buffer was extracted at 1 h, 6 h, 12 h, 1 d, 2 d, 3 d, 5 d, 7 d, and 10 d, and another 1 mL PBS was added into the tubes for 1 h, and extracted for ELISA quantification. When all samples were collected, quantifications of platelet-derived growth factor-BB (PDGF-BB) and vascular endothelial growth factor (VEGF) were performed using commercially available ELISA kits (Beyotime, Biotech Co., Ltd., Nanjing, China).

2.7. Cells

The MC3T3-E1 cells (cat. # CRL-2593) were used in this study, which was purchased from American Type Culture Collection (ATCC) (Manassas, VA, USA). MC3T3-E1 cell cultures were maintained in minimum essential medium (MEM, Gibco; Thermo Fisher Scientific, Inc.) supplemented with 10% fetal bovine serum (FBS, Gibco; Thermo Fisher Scientific, Inc.), and 1% (*v*/*v*) penicillin/streptomycin (Gibco; Thermo Fisher Scientific, Inc.) at 37 °C in a humidified CO_2 incubator. Cells at approximately 80% confluence were passaged by trypsin digestion and expanded through two passages before being used for the study.

2.8. Cell Culture Medium Preparation

CGFe (CGF extracts) preparation: CGFs after lyophilization overnight were pulverized and immersed in 45 mL DMEM for incubation at 4 °C for 24 h. CGFe was collected after centrifugation and bacteriological filtration.

CGF-conditioned medium (CCM) preparation: CCM was completed with CGFe supplemented with 10% fetal bovine serum and 1% antibiotic.

Osteogenic medium (OM) preparation: 0.5 g vitamin C (Vc) was dissolved in 10 mL PBS (Solution I). A total of 3.0611 g $C_3H_7Na_2O_6P_5$ (H_2O) was dissolved in 9.1 mL PBS (Solution II). One milligram of hexadecadrol was dissolved in 1 mL ethanol, and 200 μL of this solution was added into 10 mL α-MEM (Solution III). Then, 50 mL of medium was mixed with 50 μL Solution I, 500 μL Solution II, and 500 μL Solution III.

CGF-containing Osteogenic medium (CCOM) preparation: CGFe was mixed with 10% fetal bovine serum, 1% antibiotic, 50 μL Solution I, 500 μL Solution II, and 500 μL Solution III.

2.9. Cell Adhesion Experiment

24 well plates were prepared. A total of 3×10^4 MC3T3-E1 cells were seeded in each well. Four groups were designed: (1) Control group: 1 mL α-MEM was added into the well; (2) CGFs group: 1 mL Solution IV was added into the well; (3) RADA16 group: 200 μL 1% *w*/*v* RADA16 was added to the well and stored under 37 °C, 1 mL α-MEM was added and replaced once 15 min until the pH of the medium was close to 7.4, and stored in the incubator overnight before

adding the cells; (4) RADA16-CGFs group, the fabrication process was similar to RADA16 group, but change the final time of the addition of 1 mL α-MEM to 1 mL Solution IV. After cell seeding, the cells of each well were observed at 10 min, 30 min, 60 min, and 90 min after incubation. The number of adhesive cells was measured.

2.10. Cytotoxicity Experiment

MC3T3-E1 cells were seeded in a 96-well tissue culture plate at a density of 3000 cells per well. The design of the groups was the same as in the cell adhesion experiment, but the total medium was changed to 200 μL per well with the same ratio of different solutions as before. On days 1, 3, 5, and 7, the Cell Counting Kit-8 (CCK-8) assay was applied to determine the overall proliferation cytotoxicity. The absorbance (optical density (OD) value at 450 nm) was read by a microplate reader (Thermo Fisher Scientific, Inc.). To avoid the interference of RADA16, the medium was transferred to a new 96-well plate for measurement.

2.11. Alkaline Phosphatase (ALP) Staining

MC3T3-E1 cells were seeded in a 24-well tissue culture plate at a density of 3×10^4 cells per well with the corresponding reagents (1 mL). Five groups were designed: (1) Negative control with α-MEM; (2) Positive control with OM; (3) CGF-OM group; (4) RADA16-OM group; and (5) RADA16-CGFs-OM group. ALP staining (Beyotime, Biotech Co., Ltd., Nanjing, China) was performed after 7 d of incubation.

2.12. Alizarin Red S Staining

To identify the mineralization nodules, Alizarin Red S (Solarbio, Beijing, China) staining was performed after the MC3T3-E1 cells were seeded at a density of 3×10^4 cells per well and grew for 21 days in five groups. After gently rinsing with ddH2o, the cells were stained in a solution of 2% ARS at pH 4.1 for 20 min and then washed with ddH_2O. The samples were air-dried, and images were captured under a light microscope (magnification, ×6). Additionally, the bound dye was dissolved with 10% cetylpyridinium chloride, and the ARS in the samples was quantified by measuring the absorbance at 562 nm.

2.13. Alp Gene Expression

For the detection of *Alp* genes (ALP, F: CAGTTCGTATTCCACATCAGTTC R: CAAG-GACATCGCATATCAGCT; GAPDH, F: AAGAAGGTGGTGAAGCAGG R: GAAGGTG-GAAGAGTGGGAGT), MC3T3-E1 cells were plated at a density of 1×10^4 cells per well in separate 6-well plates in different media of five groups. Total RNA from all groups was extracted using TRIzol reagent after 14 d of culture and analyzed by reverse transcription-quantitative (RT-q) PCR.

2.14. Statistical Analysis

All analyses were performed using SPSS 25 software (IBM Corp., Armonk, NY, USA). All experiments were performed in at least three independent repeats. All data are shown as the mean and standard deviation (SD) and were analyzed using one-way ANOVA or a nonparametric test followed by the least significant difference post hoc test. The levels of significance were set at * $p < 0.05$, ** $p < 0.01$, and *** $p < 0.001$ (as indicated in the figures and legends with the corresponding symbols).

3. Results

3.1. Characterization of CGFs

After blood sample centrifugation, three blood fractions, including the fractions of plate-poor plasma, CGFs, and erythrocytes, were found in the tubes. The middle CGF fraction was extracted. The SEM images demonstrated the architecture and morphology of the CGFs (Figure 2b,c).

3.2. Characterization of RADA16

RADA16 peptide sequences were confirmed by mass spectroscopy and purified by high-performance liquid chromatography (HPLC), and the final purity was 95.22% (Figure 3a,b). SEM images demonstrated the nanostructure of 1% w/v RADA16 the nanofiber scaffold hydrogel in PBS, which showed that RADA16 peptides were able to self-assemble into nanofibers at 25 °C. These nanofibers were about 20 nm in diameter and formed pores of about 200 nm (Figure 3c,d). The rheological test of 1% w/v RADA16 nanofiber scaffold hydrogel showed the storage modulus G′ was always greater than the loss modulus G″, and the frequency G′ was always over zero, in 0.1–10 rad/s, which confirmed that saline solution such as PBS can induce RADA16 to self-assemble into nanofiber hydrogel (Figure 3e).

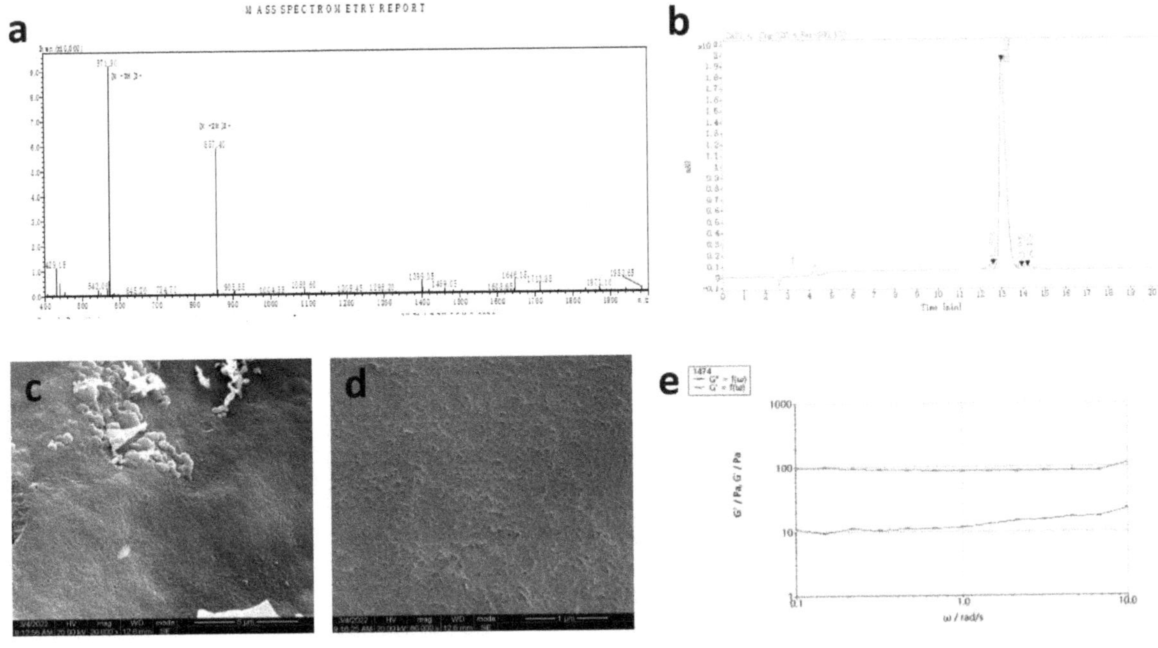

Figure 3. Properties of RADA16. (**a**) Mass spectroscopy of the synthetic peptide. (**b**) High-performance liquid chromatography of synthetic peptides. (**c**,**d**) SEM images of RADA16 (1%). Bars, 5 μm and 1 μm, respectively. (**e**) Rheometry performance of RADA16 (1%).

3.3. Growth Factor Releasing Process of RADA16-CGF

The growth factors releasing CGFs are rapid [27], and the application of RADA16 nanofiber scaffold hydrogel aims to decrease the release. The dynamics of PDGF-BB and VEGF release of RADA16-CGFs are shown in Figure 4a,b. The concentrations of PDGF-BB released at 5 d and 7 d were significantly different from the release at 1 h, while the releases at other time points were not different from the release at 1 h. No difference was found between any time points of the VEGF release. Thus, the releasing process of RADA16-CGFs could be stable and much slower than that of pure CGFs [27].

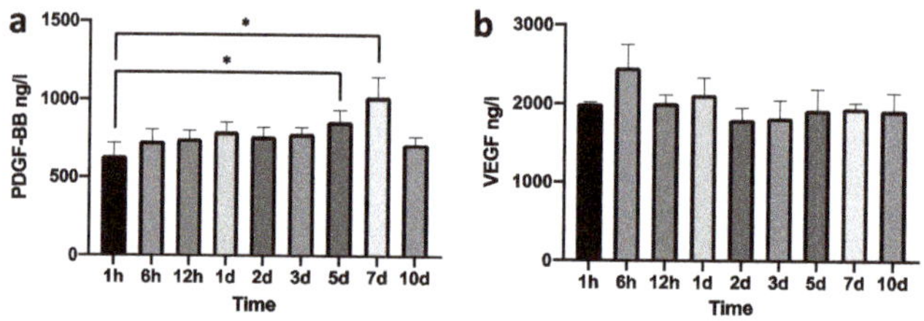

Figure 4. Growth factors released from RADA-CGFs. (**a**) Concentration of PDGF-BB at each time point. *, $p < 0.05$. (**b**) Concentration of VEGF at each time point.

3.4. Cell Proliferation of MC3T3 with RADA16-CGF

Cell culturing with CGFs, RADA16, and RADA16-CGFs did not affect the cell morphology, and the RADA16-CGF group had better cell proliferation than the other three groups after 72 h (Figure 5a). By measuring the cell adhesion, we found that in the early stage (10–30 min), CGFs could help the cell adhesion, while the RADA16 and RADA16-CGFs could decrease the cell adhesion; but later (90 min), all CGFs, RADA16, and RADA16-CGFs groups had better cell adhesion than the control group, and RADA16-CGFs had the best effect, which further indicated the function of the RADA16 nanofiber scaffold hydrogel (Figure 5b). By analyzing the results of CCK-8 assays, there was no statistical difference between each group at 1 day and 3 days, thus indicating similar proliferation capacity in each group at the early time. However, at 7 days, except for the control and RADA16 groups, every two groups showed significantly different proliferation capacities, and RADA16-CGFs had the best effect, while the control group had the worst (Figure 5c). Thus, RADA16 and CGFs could both accelerate cell proliferation, and when the two are used in combination, they synergistically promote cell proliferation.

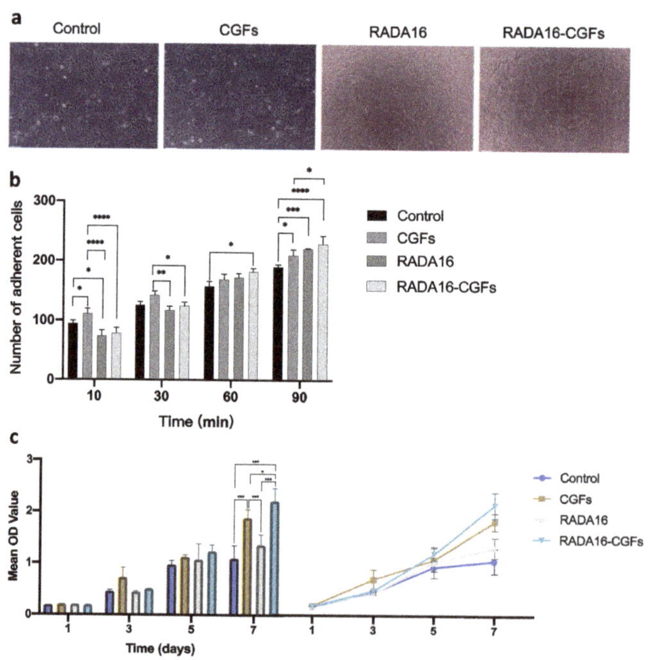

Figure 5. Cell proliferation of MC3T3 with RADA16-CGF. (**a**) MC3T3 cell morphology co-cultured with different composites after 3 days. Scale bars, 100 μm. (**b**) Comparison of the number of adherent cells between the control, CGFs, RADA16, and RADA16-CGF groups at different timepoints. *, $p < 0.05$; **, $p < 0.005$; ***, $p < 0.001$; ****, $p < 0.0001$. (**c**) Comparison and tendency chart of the CCK-8 assays for the proliferation of MC3T3 cells co-cultured with different composites. *, $p < 0.05$; ***, $p < 0.001$.

3.5. Effect of RADA16-CGFs on Mineralization Capability

Mineralization at 7 d in the different groups was confirmed by ALP staining. It demonstrated that the number of mineralized nodules in the RADA16-CGFs-OM group was the highest at 7 d, while the CGFs-OM group was more than the positive control, and the RADA16-OM group was similar to the positive control (Figure 6a). ARS staining was applied to further confirm the mineralization capability at 21d, which determined that CGFs alone could enhance the mineralization of MC3T3-E1 cells, but RADA16 alone might not have such enhancement (Figure 6b,c). RADA16-CGFs-OM, again, showed the best mineralization capability that the others had on MC3T3-E1 cells. The q-PCR of *Alp* again confirmed that CGFs-OM, RADA16-OM, and RADA16-CGFs-OM could promote mineralization capability (Figure 6d).

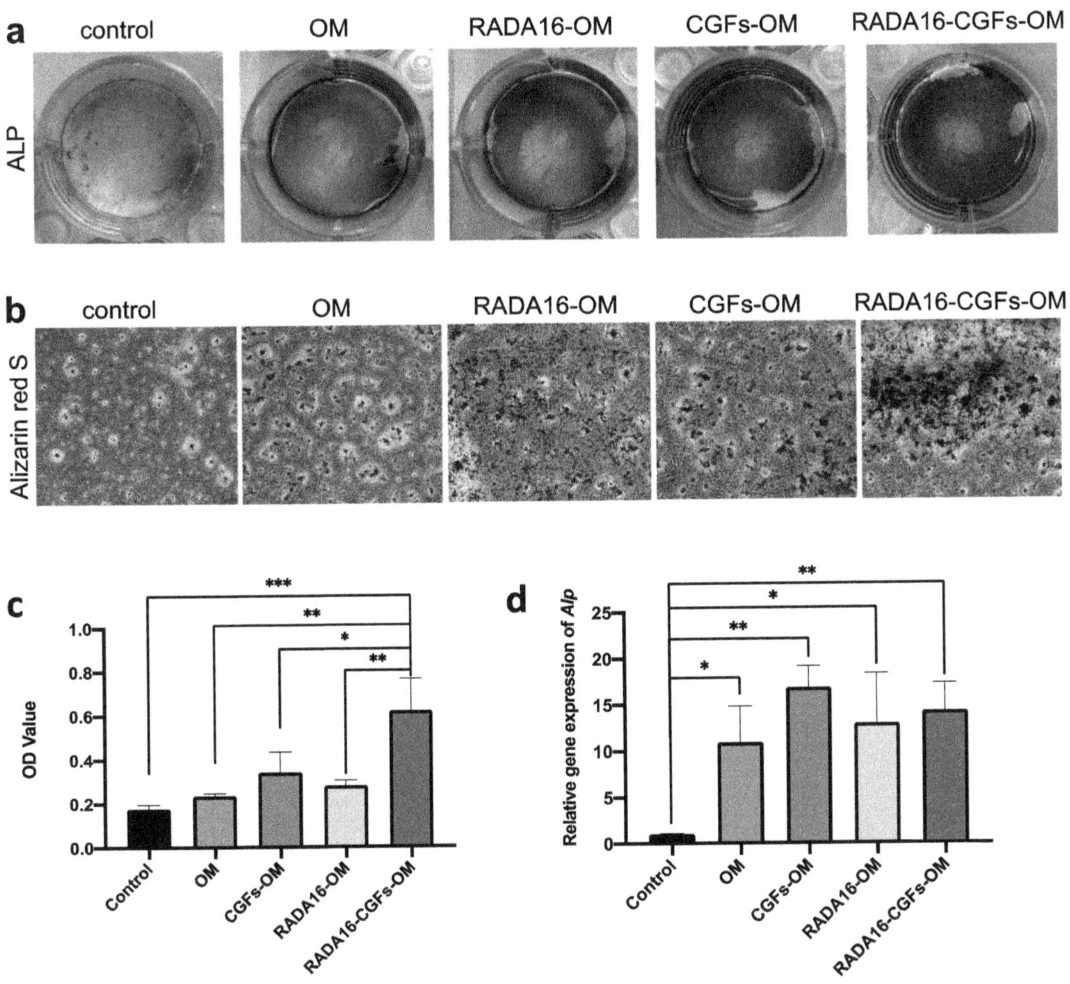

Figure 6. Effect of RADA16-CGFs on mineralization capability. (**a**) ALP staining images after culturing on the different composites for 7 days. (**b**) Optical microscopy images of alizarin red S staining after culturing on the different composites for 21 days. Scale bars, 100 μm. (**c**) Quantitative analysis of calcium nodules on the composites at a wavelength of 562 nm. *, $p < 0.05$; **, $p < 0.005$; ***, $p < 0.001$. (**d**) The gene expression of osteogenic differentiation-related proteins (*Alp*) of MC3T3 cells co-cultured with different composites. *, $p < 0.05$; **, $p < 0.005$.

4. Discussion

Alveolar clefts are common congenital bony defects extending over the alveolar and process toward the hard palate and nasal cavity [28]. To reconstruct the maxillary arch, the autogenous bone from the iliac bone is one of the most common choices in the clinic, which can help induce the formation of new bone. However, the process is complex and risky and will cause trauma to the patient when collecting the bone [29]. The other choice is commercial bone grafting materials, such as Bio-OSS, which can serve as scaffolds for cell migration and vessel regeneration [30], but osteoinduction is not as good as autogenous bone.

To solve the aforementioned problem, blood-derived growth factors are explored and have already been used in the clinic [31], and CGFs are the most widely used concentrations. CGFs can release growth factors locally to enhance tissue regeneration, such as bone remodeling and wound healing [17,18], which are clinically applied to increase the success rate of bone grafting and implant treatment. However, the rapid release of growth factors from CGFs does not match the complex process of bone formation [27], so the potential of CGFs is not fully utilized, and the growth factors can be wasted after being released. Thus, to make the utmost of the CGFs, figuring out a maneuver to slow down and extend the release process should be the first choice. In this study, we collected all the samples for fabricating the CGFs from the same six healthy people who were 20–30 years old, as the CGFs of different people might have different effects.

RADA16, a widely used SAP that can form a stable nanofiber scaffold structure similar to the extracellular matrix (ECM), has been proven to be able to serve as a good environment for cell migration, proliferation, and differentiation [23], and can support the surrounding tissue regeneration [32]. Meanwhile, it can provide the possibility of slow and sustained release of growth factors due to its stable network structure. Thus, we hypothesized that the RADA16 nanofiber scaffold hydrogel could provide a suitable structure and environment for controlling the growth factors released from CGFs.

The SEM images of the RADA16 scaffold showed a stable and polyporous nanostructure.

We first tested whether the growth factors released from CGFs could be slowed down after being wrapped with RADA16 nanofiber scaffold hydrogel. VEGF can induce vessel regeneration, endothelial cell migration, and proliferation, which can help enhance osteogenesis during bone repairing [33]. PDGF can speed tissue recovery and induce osteoblasts and endothelial cell proliferation, which promote bone regeneration [34,35]. Therefore, we chose these two growth factors for measuring release speed. According to a publication [9,27], in pure CGFs, the speed of VEGF and PDGF release showed an increasing pattern from 1 d to 10 d. In our study, after being wrapped by RADA16, the RADA16-CGFs showed stable releasing patterns. For PDGF-BB, the releasing speed increased at 5 d and 7 d, but at 10 d, the speed again slowed down, which is different from the pure CGFs; for VEGF, the releasing speed was always the same from 1 h to 10 d. Thereafter, we confirmed the sustained release function of RADA16 application to CGFs. The assembly of RADA16 with a stable three-dimensional structure can lead to the slow and sustained release of growth factors from CGFs. RADA16 can form three-dimensional pores with similar diameters, providing an ideal network structure for a three-dimensional cell culture. At the same time, this pore-size regime provides the potential for the release of functional proteins [36–38].

The RADA16 nanofiber scaffold hydrogel can provide a suitable microenvironment for cell proliferation and functional protein action as it can highly simulate the ECM structure [37]. During the sustained release process, the bioactivity of CGFs can be maintained. In our study, the morphology of MC3T3-E1 cells was not affected by the application of CGFs and RADA16, in which there was no obvious change in morphology, cell apoptosis, or degenerescence, showing that RADA16 and CGFs were atoxic and safe for cells. The measurement of cell adhesion and CCK-8 testing further confirmed that CGFs, RADA16, and RADA16-CGFs enhanced cell proliferation. In the cell adhesion experiment, cells in the RADA16-CGF group started to be more than the control 60 min after cell seeding, and the CGF group performed the best at 10 min, which again demonstrated the sustained release property of RADA16. Meanwhile, the outcomes revealed that the RADA16 nanofiber scaffold hydrogel could serve better as a cultural environment by providing a 3D polyporous nanostructure [39]. CCK-8 testing also confirmed that there was no cytotoxicity of CGFs or RADA16, which was the same as the published results [40,41].

Finally, we assessed the effects of RADA16-CGFs on the mineralization capability of MC3T3-E1 cells. ALP staining was applied to assess the early stage (7 d) of the osteogenesis induction of RADA16-CGFs on MC3T3-E1 cells, while ARS staining was used to measure the mineralized nodule at 21 d. At both 7 d and 21 d, CGFs and RADA16 alone can help promote the mineralization of MC3T3-E1 cells, and RADA16-CGFs showed the most enhancement. The enhancement was further confirmed by q-PCR of ALP gene expression. Thus, we found that with the help of the RADA16 nanofiber scaffold hydrogel, the function of CGFs for promoting osteogenesis could be enhanced, which showed great potential for clinical application.

In conclusion, we provide a new method for endowing the sustained release of CGFs, which can help maximize the function of CGFs in osteoinduction. The application of the atoxic RADA16 nanofiber scaffold hydrogel with CGFs can be a new therapeutic strategy for the treatment of alveolar bone loss and other problems that require tissue regeneration. However, the limitation of our study is the lack of in vivo experiments, which should be carried out in the future.

Author Contributions: R.Y., J.C. and D.W. contributed to the collection of data. R.Y and J.C. analyzed the data. R.Y., Y.X., and G.O. contributed to writing and revising the paper. Y.X. and G.O. supervised the research. All authors have read and agreed to the published version of the manuscript.

Funding: This research was supported by Sichuan Science and Technology Program (Nos. 2019YJ0143 and 2021YFS0085), Research and Development Program of West China Hospital of Stomatology Sichuan University (No. RD-02-202114), and Research Funding from West China School/Hospital of Stomatology Sichuan University (No. RCDWJS2021-13).

Institutional Review Board Statement: The study was reviewed and approved by the Ethics Committee of West China Hospital of Stomatology, Sichuan University.

Data Availability Statement: Raw data supporting the conclusion of this paper will be provided by the author without improper reservation.

Conflicts of Interest: The authors have no other funding, financial relationships, or conflict of interest to disclose.

References

1. McCrary, H.; Skirko, J.R. Bone Grafting of Alveolar Clefts. *Oral Maxillofac. Surg. Clin. N. Am.* **2021**, *33*, 231–238. [CrossRef] [PubMed]
2. Kim, J.; Jeong, W. Secondary bone grafting for alveolar clefts: Surgical timing, graft materials, and evaluation methods. *Arch. Craniofacial Surg.* **2022**, *23*, 53–58. [CrossRef] [PubMed]
3. Tadic, D.; Epple, M. A thorough physicochemical characterisation of 14 calcium phosphate-based bone substitution materials in comparison to natural bone. *Biomaterials* **2004**, *25*, 987–994. [CrossRef] [PubMed]
4. Baldwin, P.; Li, D.J.; Auston, D.A.; Mir, H.S.; Yoon, R.S.; Koval, K.J. Autograft, Allograft, and Bone Graft Substitutes: Clinical Evidence and Indications for Use in the Setting of Orthopaedic Trauma Surgery. *J. Orthop. Trauma* **2019**, *33*, 203–213. [CrossRef] [PubMed]
5. Simonpieri, A.; Del Corso, M.; Sammartino, G.; Ehrenfest, D.M.D. The Relevance of Choukroun's Platelet-Rich Fibrin and Metronidazole during Complex Maxillary Rehabilitations Using Bone Allograft. Part I: A New Grafting Protocol. *Implant. Dent.* **2009**, *18*, 102–111. [CrossRef] [PubMed]
6. Zheng, C.; Chen, J.; Liu, S.; Jin, Y. Stem cell-based bone and dental regeneration: A view of microenvironmental modulation. *Int. J. Oral Sci.* **2019**, *11*, 1–15. [CrossRef] [PubMed]
7. Lei, L.; Yu, Y.; Han, J.; Shi, D.; Sun, W.; Zhang, D.; Chen, L. Quantification of growth factors in advanced platelet-rich fibrin and concentrated growth factors and their clinical efficacy as adjunctive to the GTR procedure in periodontal intrabony defects. *J. Periodontol.* **2020**, *91*, 462–472. [CrossRef]
8. Masuki, H.; Okudera, T.; Watanabe, T.; Suzuki, M.; Nishiyama, K.; Okudera, H.; Nakata, K.; Uematsu, K.; Su, C.-Y.; Kawase, T. Growth factor and pro-inflammatory cytokine contents in platelet-rich plasma (PRP), plasma rich in growth factors (PRGF), advanced platelet-rich fibrin (A-PRF), and concentrated growth factors (CGF). *Int. J. Implant. Dent.* **2016**, *2*, 1–6. [CrossRef] [PubMed]

9. Hong, S.; Li, L.; Cai, W.; Jiang, B. The potential application of concentrated growth factor in regenerative endodontics. *Int. Endod. J.* **2019**, *52*, 646–655. [CrossRef]
10. Boonyagul, S.; Banlunara, W.; Sangvanich, P.; Thunyakitpisal, P. Effect of acemannan, an extracted polysaccharide from Aloe vera, on BMSCs proliferation, differentiation, extracellular matrix synthesis, mineralization, and bone formation in a tooth extraction model. *Odontology* **2014**, *102*, 310–317. [CrossRef]
11. Honda, H.; Tamai, N.; Naka, N.; Yoshikawa, H.; Myoui, A. Bone tissue engineering with bone marrow-derived stromal cells integrated with concentrated growth factor in Rattus norvegicus calvaria defect model. *J. Artif. Organs* **2013**, *16*, 305–315. [CrossRef] [PubMed]
12. Takeda, Y.; Katsutoshi, K.; Matsuzaka, K.; Inoue, T. The Effect of Concentrated Growth Factor on Rat Bone Marrow Cells In Vitro and on Calvarial Bone Healing In Vivo. *Int. J. Oral Maxillofac. Implant.* **2015**, *30*, 1187–1196. [CrossRef] [PubMed]
13. Park, H.-C.; Kim, S.-G.; Oh, J.-S.; You, J.-S.; Kim, J.-S.; Lim, S.-C.; Jeong, M.-A.; Kim, J.-S.; Jung, C.; Kwon, Y.-S.; et al. Early Bone Formation at a Femur Defect Using CGF and PRF Grafts in Adult Dogs: A Comparative Study. *Implant. Dent.* **2016**, *25*, 387–393. [CrossRef] [PubMed]
14. Kim, T.-H.; Kim, S.-H.; Sándor, G.K.; Kim, Y.-D. Comparison of platelet-rich plasma (PRP), platelet-rich fibrin (PRF), and concentrated growth factor (CGF) in rabbit-skull defect healing. *Arch. Oral Biol.* **2014**, *59*, 550–558. [CrossRef]
15. Hong, S.; Chen, W.; Jiang, B. A Comparative Evaluation of Concentrated Growth Factor and Platelet-rich Fibrin on the Proliferation, Migration, and Differentiation of Human Stem Cells of the Apical Papilla. *J. Endod.* **2018**, *44*, 977–983. [CrossRef]
16. Yu, B.; Wang, Z. Effect of concentrated growth factors on beagle periodontal ligament stem cells in vitro. *Mol. Med. Rep.* **2014**, *9*, 235–242. [CrossRef]
17. Calabriso, N.; Stanca, E.; Rochira, A.; Damiano, F.; Giannotti, L.; Stanca, B.D.C.; Massaro, M.; Scoditti, E.; Demitri, C.; Nitti, P.; et al. Angiogenic Properties of Concentrated Growth Factors (CGFs): The Role of Soluble Factors and Cellular Components. *Pharmaceutics* **2021**, *13*, 635. [CrossRef]
18. Wang, L.; Wan, M.; Li, Z.; Zhong, N.; Liang, D.; Ge, L. A comparative study of the effects of concentrated growth factors in two different forms on osteogenesis in vitro. *Mol. Med. Rep.* **2019**, *20*, 1039–1048. [CrossRef]
19. Yokoi, H.; Kinoshita, T.; Zhang, S. Dynamic reassembly of peptide RADA16 nanofiber scaffold. *Proc. Natl. Acad. Sci. USA* **2005**, *102*, 8414–8419. [CrossRef]
20. Kulkarni, D.; Musale, S.; Panzade, P.; Paiva-Santos, A.C.; Sonwane, P.; Madibone, M.; Choundhe, P.; Giram, P.; Cavalu, S. Surface Functionalization of Nanofibers: The Multifaceted Approach for Advanced Biomedical Applications. *Nanomaterials* **2022**, *12*, 3899. [CrossRef]
21. Semino, C.E. Self-assembling Peptides: From Bio-inspired Materials to Bone Regeneration. *J. Dent. Res.* **2008**, *87*, 606–616. [CrossRef] [PubMed]
22. Koutsopoulos, S. Self-assembling peptide nanofiber hydrogels in tissue engineering and regenerative medicine: Progress, design guidelines, and applications. *J. Biomed. Mater. Res. Part A* **2016**, *104*, 1002–1016. [CrossRef]
23. Horii, A.; Wang, X.; Gelain, F.; Zhang, S. Biological Designer Self-Assembling Peptide Nanofiber Scaffolds Significantly Enhance Osteoblast Proliferation, Differentiation and 3-D Migration. *PLoS ONE* **2007**, *2*, e190. [CrossRef] [PubMed]
24. Yang, H.; Hong, N.; Liu, H.; Wang, J.; Li, Y.; Wu, S. Differentiated adipose-derived stem cell cocultures for bone regeneration in RADA16-I in vitro. *J. Cell. Physiol.* **2018**, *233*, 9458–9472. [CrossRef] [PubMed]
25. Tian, H.; Guo, A.; Li, K.; Tao, B.; Lei, D.; Deng, Z. Effects of a novel self-assembling peptide scaffold on bone regeneration and controlled release of two growth factors. *J. Biomed. Mater. Res. Part A* **2022**, *110*, 943–953. [CrossRef]
26. Takeuchi, T.; Bizenjima, T.; Ishii, Y.; Imamura, K.; Suzuki, E.; Seshima, F.; Saito, A. Enhanced healing of surgical periodontal defects in rats following application of a self-assembling peptide nanofibre hydrogel. *J. Clin. Periodontol.* **2016**, *43*, 279–288. [CrossRef]
27. Yu, M.; Wang, X.; Liu, Y.; Qiao, J. Cytokine release kinetics of concentrated growth factors in different scaffolds. *Clin. Oral Investig.* **2019**, *23*, 1663–1671. [CrossRef]
28. Campbell, S.; Lees, C.; Moscoso, G.; Hall, P. Ultrasound antenatal diagnosis of cleft palate by a new technique: The 3D 'reverse face' view. *Ultrasound Obstet. Gynecol.* **2005**, *25*, 12–18. [CrossRef]
29. Sakkas, A.; Wilde, F.; Heufelder, M.; Winter, K.; Schramm, A. Autogenous bone grafts in oral implantology—Is it still a "gold standard"? A consecutive review of 279 patients with 456 clinical procedures. *Int. J. Implant. Dent.* **2017**, *3*, 1–17. [CrossRef]
30. Tapety, F.I.; Amizuka, N.; Uoshima, K.; Nomura, S.; Maeda, T. A histological evaluation of the involvement of Bio-OssR in osteoblastic differentiation and matrix synthesis. *Clin. Oral Implant. Res.* **2004**, *15*, 315–324. [CrossRef]
31. Burnouf, T.; Goubran, H.A.; Chen, T.-M.; Ou, K.-L.; El-Ekiaby, M.; Radosevic, M. Blood-derived biomaterials and platelet growth factors in regenerative medicine. *Blood Rev.* **2013**, *27*, 77–89. [CrossRef] [PubMed]
32. Wang, J.; Zheng, J.; Zheng, Q.; Wu, Y.; Wu, B.; Huang, S.; Fang, W.; Guo, X. FGL-functionalized self-assembling nanofiber hydrogel as a scaffold for spinal cord-derived neural stem cells. *Mater. Sci. Eng. C* **2015**, *46*, 140–147. [CrossRef] [PubMed]
33. Hu, K.; Olsen, B.R. The roles of vascular endothelial growth factor in bone repair and regeneration. *Bone* **2016**, *91*, 30–38. [CrossRef] [PubMed]
34. Kaigler, D.; Avila, G.; Wisner-Lynch, L.; Nevins, M.L.; Nevins, M.; Rasperini, G.; E Lynch, S.; Giannobile, W.V. Platelet-derived growth factor applications in periodontal and peri-implant bone regeneration. *Expert Opin. Biol. Ther.* **2011**, *11*, 375–385. [CrossRef] [PubMed]

35. Shah, P.; Keppler, L.; Rutkowski, J. A Review of Platelet Derived Growth Factor Playing Pivotal Role in Bone Regeneration. *J. Oral Implant.* **2014**, *40*, 330–340. [CrossRef]
36. Gelain, F.; Unsworth, L.D.; Zhang, S. Slow and sustained release of active cytokines from self-assembling peptide scaffolds. *J. Control. Release* **2010**, *145*, 231–239. [CrossRef] [PubMed]
37. Zhou, A.; Chen, S.; He, B.; Zhao, W.; Chen, X.; Jiang, D. Controlled release of TGF-beta 1 from RADA self-assembling peptide hydrogel scaffolds. *Drug Des. Devel. Ther.* **2016**, *10*, 3043–3051. [CrossRef]
38. Hunt, N.; Grover, L.M. Cell encapsulation using biopolymer gels for regenerative medicine. *Biotechnol. Lett.* **2010**, *32*, 733–742. [CrossRef]
39. Wang, R.; Wang, Z.; Guo, Y.; Li, H.; Chen, Z. Design of a RADA16-based self-assembling peptide nanofiber scaffold for biomedical applications. *J. Biomater. Sci. Polym. Ed.* **2019**, *30*, 713–736. [CrossRef]
40. Sankar, S.; O'neill, K.; D'arc, M.B.; Rebeca, F.; Buffier, M.; Aleksi, E.; Fan, M.; Matsuda, N.; Gil, E.S.; Spirio, L. Clinical Use of the Self-Assembling Peptide RADA16: A Review of Current and Future Trends in Biomedicine. *Front. Bioeng. Biotechnol.* **2021**, *9*, 679525. [CrossRef]
41. Tabatabaei, F.; Aghamohammadi, Z.; Tayebi, L. In vitro and in vivo effects of concentrated growth factor on cells and tissues. *J. Biomed. Mater. Res. Part A* **2020**, *108*, 1338–1350. [CrossRef] [PubMed]

Journal of
Functional Biomaterials

Article

Cytotoxicity Induced by Black Phosphorus Nanosheets in Vascular Endothelial Cells via Oxidative Stress and Apoptosis Activation

Hao Dong [1,†], Yin Wen [1,†], Jiating Lin [1], Xianxian Zhuang [1], Ruoting Xian [1], Ping Li [1,*] and Shaobing Li [1,2,3,*]

1 Center of Oral Implantology, Stomatological Hospital, School of Stomatology, Southern Medical University, Guangzhou 510280, China; donghaoct3@163.com (H.D.); wennyinn@gmail.com (Y.W.)
2 First Clinical Medical College, Xinjiang Medical University, Urumqi 830011, China
3 The First People's Hospital of Kashgar Region, Kashgar 844000, China
* Correspondence: ping_li_88@smu.edu.cn (P.L.); lishaobing@smu.edu.cn (S.L.)
† These authors contributed equally to this work.

Abstract: Black phosphorus (BP), an emerging two-dimensional material with unique optical, thermoelectric, and mechanical properties, has been proposed as bioactive material for tissue engineering. However, its toxic effects on physiological systems remain obscure. The present study investigated the cytotoxicity of BP to vascular endothelial cells. BP nanosheets (BPNSs) with a diameter of 230 nm were fabricated via a classical liquid-phase exfoliation method. Human umbilical vein endothelial cells (HUVECs) were used to determine the cytotoxicity induced by BPNSs (0.31–80 μg/mL). When the concentrations were over 2.5 μg/mL, BPNSs adversely affected the cytoskeleton and cell migration. Furthermore, BPNSs caused mitochondrial dysfunction and generated excessive intercellular reactive oxygen species (ROS) at tested concentrations after 24 h. BPNSs could influence the expression of apoptosis-related genes, including the P53 and BCL-2 family, resulting in the apoptosis of HUVECs. Therefore, the viability and function of HUVECs were adversely influenced by the concentration of BPNSs over 2.5 μg/mL. These findings provide significant information for the potential applications of BP in tissue engineering.

Keywords: nanomaterials; black phosphorus; cytotoxicity; vascular endothelial cell; reactive oxygen species; apoptosis; biomedical applications

Citation: Dong, H.; Wen, Y.; Lin, J.; Zhuang, X.; Xian, R.; Li, P.; Li, S. Cytotoxicity Induced by Black Phosphorus Nanosheets in Vascular Endothelial Cells via Oxidative Stress and Apoptosis Activation. *J. Funct. Biomater.* **2023**, *14*, 284. https://doi.org/10.3390/jfb14050284

Academic Editors: Matt Kipper and Jarosław Jakubowicz

Received: 15 March 2023
Revised: 25 April 2023
Accepted: 15 May 2023
Published: 20 May 2023

1. Introduction

Maxillofacial bone defects are a common pathology that often leads to tooth loss, masticatory difficulties, and even facial deformities [1]. Current clinical treatments, such as bone grafting or bone transplantation, are associated with postoperative complications. Therefore, the safer and more effective therapeutic approaches are required [2]. Black phosphorus (BP) as a novel two-dimensional nanomaterial produces phosphate ions upon biodegradation, which serve as substrates for osteogenesis and facilitate in situ treatment of bone defects [3–5]. This approach provides a new strategy for the treatment of bone tissue diseases. BP has a similar lamellar structure to graphene [6,7]; its regular ribbed surface shape makes BP an excellent vector for drug delivery [8,9] and gene editing [10]. Scaffolds doped with BP nanoparticles have a higher mechanical strength [11,12]. Based on its excellent photonic properties, BP has been investigated for cancer treatment using photothermal [13,14] and photodynamic [15] therapies. Therefore, a multifunctional disease-therapeutic nanoplatform based on BP could be developed in the treatment of maxillofacial bone defects [16].

Phosphorus is a widely present element in the human body. It participates in almost all physiological and chemical reactions and plays a vital role in maintaining the physiological functions of the human body [16]. It is generally believed that the degradation of BP in vivo produces non-toxic phosphate [17,18]. However, some previous studies did not support this conclusion. BP nanosheets (BPNSs) showed obvious cytotoxicity to L929 fibroblasts at concentrations above 4 μg/mL [19], and BP quantum dots (BPQDs) significantly inhibited the growth of Beas-2E cells at a concentration of 5 μg/mL [20]. A study suggests that the toxicity of BP to organisms is related to its concentration, size [21], and surface modification [22]. Therefore, a better understanding of the biotoxicity of BP is required to facilitate the tissue regeneration. After nanomaterials enter the body, they pass through the blood circulation to reach the target [13]. Vascular endothelial cells constitute the inner wall of the blood vessels. Previous studies have shown that nanoparticles can be transferred into the blood [23] so the BP entering the body inevitably comes into contact with vascular endothelial cells. At the same time, the process of osteogenesis relies heavily on the infiltration of early-stage vascular endothelial cells, which provide nutrients to local tissues and remove metabolic waste through neovascularization [24]. However, there are few studies on the toxicity of BP during the process of angiogenesis.

In this study, the effects of BPNSs on the metabolic activity, cell migration, and cell apoptosis and its related gene expression of human umbilical vein endothelial cells (HUVECs) were investigated. The BPNSs were prepared using the classic liquid-phase exfoliation method. HUVECs were used to simulate the vascular structure in vitro [25]. The first objective of this study was to determine the cytotoxic effects of BPNSs on vascular endothelial cells. The second objective was to investigate the intracellular response and the underlying mechanisms of BPNSs-induced cytotoxicity.

2. Materials and Methods

2.1. The Synthesis of BPNSs

BPNSs were fabricated at a low temperature using a classical liquid-phase exfoliation method. BP crystals (50 mg) (99.998%, XFNANO, Nanjing, China) were mixed with 100 mL of N-methyl-2-pyrrolidone (NMP, Electronic grade 99.9%, Aladdin, Shanghai, China). The mixture was treated using an ultrasonic cell disruptor (JY92-IIDN, Xinzhi, Ningbo, China) in an ice water mixture for 15 h (21.0–22.5 kHz, working for 5 s and paused for 10 s). After a brown suspension was formed, unstripped BP crystals were removed by centrifugation at 4500 rpm for 30 min at 4 °C, and the resulting supernatant was collected. Next, the NMP was eliminated through centrifugation at 15,000 rpm for 30 min at 4 °C. The BPNSs obtained were then subjected to three rounds of centrifugation using absolute ethanol and deionized ultrapure water in sequence. After quantification of the manufactured BPNS powder by lyophilization, it was diluted in the culture medium to different concentrations (0.31, 0.63, 1.25, 2.5, 5, 10, 20, 40 and 80 μg/mL).

2.2. Characterizations of BPNSs

Multiple microscopy techniques were used to assess the surface morphology and thickness of BPNSs. Atomic force microscopy (AFM) was employed to perform a vacuum-based characterization of the surface morphology and thickness of the BPNSs at room temperature, utilizing the Dimension Icon system (Bruker, Karlsruhe, Germany). Meanwhile, scanning electron microscopy (SEM) was used to examine the surface morphology of the BPNSs by depositing them onto an aluminum foil and drying at 60 °C prior to imaging. The imaging was carried out under high vacuum with an acceleration voltage of 10 kV using the Sigma 300 system (Zeiss, Oberkochen, Germany). Lastly, transmission electron microscopy (TEM) was employed to determine the elemental compositions and morphologies of the BPNSs utilizing the FEI Tecnai F20 TEM D545 system (FEI, Hillsboro, OR, USA). The Raman spectra of the BPNSs were obtained at room temperature using Raman spectroscopy (LabRAM HR, HORIBA, Montpellier, France) with an excitation wavelength of 532 nm. The particle size distribution and polydispersity (PDI) in water were determined using a Zetasizer 3000 HS nanosizer (Malvern Instruments, Malvern, UK).

2.3. Cell Culture

The human umbilical vein endothelial cells (HUVECs, National Infrastructure of Cell Line Resource, Beijing, China) were cultured in Dulbecco's modified Eagle's medium (DMEM, Gibco, Waltham, MA, USA), supplemented with 10% (v/v) heat-inactivated fetal bovine serum (FBS, Gibco, Waltham, MA, USA) and antibiotics (100 IU/mL penicillin and 100 μg/mL streptomycin) (P/S, Gibco, Waltham, MA, USA) under standard conditions of 37 °C and 5% CO_2 in a humidified atmosphere.

2.4. Cell Uptake

Cellular uptake assays were performed to verify the intracellular entry of the BPNSs. BPNSs were labeled with fluorescein isothiocyanate (FITC, Thermo Fisher, Waltham, MA, USA), and FITC-labelled BPNSs (10 μg/mL) were exposed to the cells for 0.5, 1, 2 and 4 h. The treated cells were stained with 4′,6-diamidino-2-phenylindole dihydrochloride (DAPI, SolarBio Technologies, Beijing, China) and images were captured under a laser confocal microscope (STELLARIS 5, Leica, Weztlar, Germany). The relative uptake levels were calculated as the fluorescence intensity. Five fields of views were captured per well and fluorescence images were analyzed using the ImageJ 1.46 software.

2.5. Cell Morphology and Metabolic Activity

Morphological changes in HUVECs were observed to evaluate the cytotoxicity of BPNSs. HUVECs were treated with various concentrations of BPNSs (ranging from 0 to 80 μg/mL) in the cell culture medium for 24 h, and images were captured using an inverted microscope (DMIL LED, Leica, Weztlar, Germany). The metabolic activity of HUVECs was measured by using the CCK-8 assay (Dojindo, Kumamoto, Japan). First, 1×10^4 HUVECs were seeded in a 96-well plate and incubated for 24 h, and five wells were set in each experimental group for auxiliary purposes. Subsequently, the cells were treated to a culture medium with varying concentrations of BPNSs for 24 h. After the cell culture period, the cells were incubated with the CCK-8 solution at 37 °C for 1 h in the dark. Absorbance at 450 nm was measured using a microplate reader (iMark, Bio-Rad, Hercules, CA, USA). To calculate cell metabolic activity, the absorbance values of the cells treated with BPNSs were expressed as a percentage relative to the negative control groups (without BPNSs), as previously reported [26].

2.6. Cell Migration Ability

The cell migration ability was detected using a scratch test to reflect the cytotoxicity of BPNSs. According to a classic protocol, plate scratches were created using sterile blades at the bottom of 6-well plates before the cells were seeded. HUVECs were seeded in plates at a density of 5×10^5 cells/well and incubated for 24 h. Then, cell scratches perpendicular to the plate scratches were made using a 1 mL pipette tip. BPNSs were diluted to different concentrations and added to the cells for 24 h. Images of the intersection point of the plate scratch and cell scratch in each well were taken at 0 and 24 h. The area of cell migration was calculated as the difference between two time points.

2.7. Cytoskeleton Staining

Morphological changes in the cytoskeleton of HUVECs were observed by phalloidin staining (Abcam, Cambridge, UK). HUVECs (5×10^4 cells/well) were seeded into 24-well plates lined with climbing slides and incubated for 24 h. Different concentrations of BPNSs (1.25, 5 and 20 μg/mL) were added to the cells and incubated for 24 h. The treated HUVECs were stained with phalloidin-iFluor 488 (Abcam, Cambridge, UK) and DAPI for microscopic fluorescence images.

2.8. Reactive Oxygen Species Test

To determine the intracellular reactive oxygen species (ROS) levels in HUVECs, a ROS assay kit (Meilunbio, Dalian, China) was utilized following the manufacturer's guidelines.

Specifically, HUVECs were seeded into 12-well plates at a density of 1×10^5 cells per well and incubated for 24 h prior to the addition of BPNSs. The cells were then treated with BPNSs for another 24 h before analysis. Next, the previous culture medium was replaced with a medium containing 1% of the fluorescent probe (2,7-dichlorodihydrofluorescein diacetate, DCFH-DA). After incubation for 1 h at 37 °C in the dark, the cells were washed with a culture medium three times. Fluorescence images were captured immediately, utilizing the DMi8 inverted fluorescence microscope under FITC filters (Leica, Weztlar, Germany). The relative ROS levels were calculated as the fluorescence fold relative to the control group. Five fields of views were captured per well and fluorescence images were analyzed using the ImageJ 1.46 software.

2.9. Mitochondrial Membrane Potential Detection

Mitochondrial membrane potential ($\Delta\Psi m$) was detected using the JC-1 assay kit (Meilunbio, Dalian, China). HUVECs were placed in 12-well plates (1×10^5 cells/well) and incubated for 24 h. Afterward, the initial medium was removed, and replaced with a fresh medium containing various concentrations of BPNSs for an additional 24 h. Cells were washed twice with PBS and then stained with a medium containing 0.5% of JC-1 for 30 min at 37 °C in the dark. Cells were washed twice in a stain buffer and covered with the culture medium. Finally, the cells were immediately observed under a confocal microscope. Images were taken at 490 nm and 525 nm excitation wavelengths, and 530 nm and 590 nm emission wavelengths. Changes in mitochondrial membrane potential were expressed as the ratio of red-to-green fluorescence intensity. Five fields of views were captured per well and fluorescence images were analyzed using the ImageJ 1.46 software.

2.10. Cell Apoptosis Test

Cell apoptosis rate was detected using an apoptosis assay kit (Meilunbio, Dalian, China). According to the manufacturer's instructions, HUVECs were placed in 12-well plates (1×10^5 cells/well) and incubated for 24 h. Different concentrations of BPNSs (1.25, 5 and 20 μg/mL) were added to the cells and incubated for 24 h. The treated HUVECs were collected and then washed twice with PBS. Cells were stained with a medium containing 1% of the Annexin-V/PI dye for 15 min at 37 °C in the dark and the apoptosis rate was analyzed using flow cytometry (DxFLEX, Beckman, Brea, CA, USA).

2.11. Apoptosis-Related Gene Expression

The reverse transcription-quantitative polymerase chain reaction (RT-qPCR) assay, in accordance with the manufacturer's instructions, was employed to evaluate the expression levels of apoptosis-related genes in HUVECs. HUVECs were cultured in 12-well plates at a density of 1×10^5 cells per well and incubated for 24 h. Different concentrations of BPNSs (1.25, 5 and 20 μg/mL) were added to the cells and incubated for 24 h. The total RNA was extracted from the cells using the Trizol reagent (Accurate Biology, Changsha, China) and converted to cDNA using the Reverse Transcription Mix Kit (Accurate Biology, Changsha, China). Real-time PCR was performed on a LightCycler 96 real-time PCR system (Roche, Basel, Switzerland) using the SYBR Green Premix Kit (Accurate Biology, Changsha, China). The $2^{-\Delta\Delta CT}$ method, with 18S ribosomal RNA (18s rRNA) as the reference gene, was employed to calculate the relative changes in mRNA expression. Three experimental wells were set up for each group. Control groups did not receive BPNS pretreatment. The primers used in this study are listed in Table 1 and were purchased from Sangon Biotech (Shanghai, China).

2.12. Statistical Analysis

The statistical analyses were conducted using GraphPad 9.0 (GraphPad Software, San Diego, CA, USA), and the data are presented as mean ± standard deviation. The significance of the results was determined using one-way ANOVA followed by Tukey's multiple comparison test, with a threshold for statistical significance set at $p < 0.05$. Each cell experiment was independently conducted in triplicate for statistical reliability.

Table 1. Primer sequences employed for real-time polymerase chain reaction (PCR) in this investigation.

Gene	Forward Primer Sequence (5′-3′)	Reverse Primer Sequence (3′-5′)
P53	TGTGACTTGCACGTACTCCC	ACCATCGCTATCTGAGCAGC
BCL-2	GAACTGGGGGAGGATTGTGG	CATCCCAGCCTCCGTTATCC
BAX	GAGCAGCCCAGAGGCG	GGAAAAAGACCTCTCGGGGG
18s rRNA	CAGCCACCCGAGATTGAGCA	TAGTAGCGACGGGCGGTGTG

3. Results

3.1. BPNSs Characterization

A photograph of the as-exfoliated BPNS dispersion is shown in Figure 1a, where the BPNSs dispersion is dark brown and shows good homogeneity and stability. The surface morphology of layered BP is shown in Figure 1b,d. The size of the lateral particles is distributed from 100 to 500 nm. The height profiles in Figure 1c were obtained from the AFM image in Figure 1b, showing that the thickness of BPNSs was from ~3 to 6 nm (5–10 layers) [27,28]. The TEM image in Figure 1e shows a similar diameter as that in the AFM and SEM images. The elemental distributions in Figure 1e were detected, and the main elements in the particle area were phosphorus and oxygen. The Raman spectrum of the BPNSs is shown in Figure 1f. The three prominent peaks can be ascribed to the A_g^1 out-of-plane phonon modes at 359 cm^{-1}, B_{2g} in-plane modes at 435 cm^{-1}, and A_g^2 in-plane modes at 463 cm^{-1}. These peaks are generally considered the unique high-frequency interlayer Raman modes of BP [29]. The particle size distributions of the BPNSs are shown in Figure 1g with a mean size of 239.09 nm. The polydispersity (PDI) of BPNSs was measured as 0.123, indicating excellent dispersion [21].

3.2. Cell Uptake

FITC-labeled BPNSs were constructed and FITC fluorescence was observed in the cytoplasm for different time points, as shown in Figure 2a. The intracellular fluorescence intensity enhanced with an increase in the exposure time. The quantitative analysis also proved this finding (Figure 2b), indicating the accumulation of BPNSs in the cells.

3.3. Cell Morphology and Metabolic Activity

To investigate the cell morphology of BPNSs, HUVECs were treated with various concentrations of BPNSs for 24 h. As shown in Figure 3, cells without BPNSs had a blunt and cobblestone-like morphology. The number and morphology of HUVECs were not significantly different when the BPNSs concentration was less than 2.5 μg/mL. The number of black particles and aggregated BPs gradually increased with increasing concentrations. When the concentrations were greater than 10 μg/mL, black particles were observed in the cellular region, indicating the cell uptake of BPNSs.

HUVECs were exposed to increasing BPNSs concentrations (0.3–80 μg/mL) for 24 h, and the cellular metabolic activity was evaluated. Figure 4 shows that relative cell metabolic activity decreases with increasing concentrations at 24 h. This result indicated that the cytotoxicity of BP was concentration-dependent. Compared to the control group, the cell metabolic activity was higher than 80% at concentrations lower than 5 μg/mL, while a 50% reduction in cell metabolic activity was observed at a concentration of 40 μg/mL at 24 h.

3.4. Cell Migration Ability

The migration of vascular cells is essential for their proper function. Figure 5a shows the cell scratches at different BPNSs concentrations. At 24 h after scratch formation, the scratch area healed at all concentrations. The migration area of the cells was not significantly different from that of the control group until 1.25 μg/mL (Figure 5b). A significant decrease in migration area was detected at higher concentrations, indicating that BPNSs inhibited the migration of cells. After BPNSs were added to the culture medium for 24 h, the number of dead cells (round and bright in morphology [30]) increased with the concentration.

3.5. Cytoskeleton Staining

The cytoskeleton is an intracellular structure that maintains a balance between centrifugal and centripetal forces within the cell. The cytoskeleton can stabilize the cell membrane by regulating junctional complexes. As shown in Figure 6, more radial actin fibers were found in the control group cells. The cells maintained a similar state when 1.25 μg/mL of BPNSs were introduced. However, disordered fibers emerged at 5 μg/mL. At 20 μg/mL, the cells retained their morphology but had vague spot-like F-actin filaments.

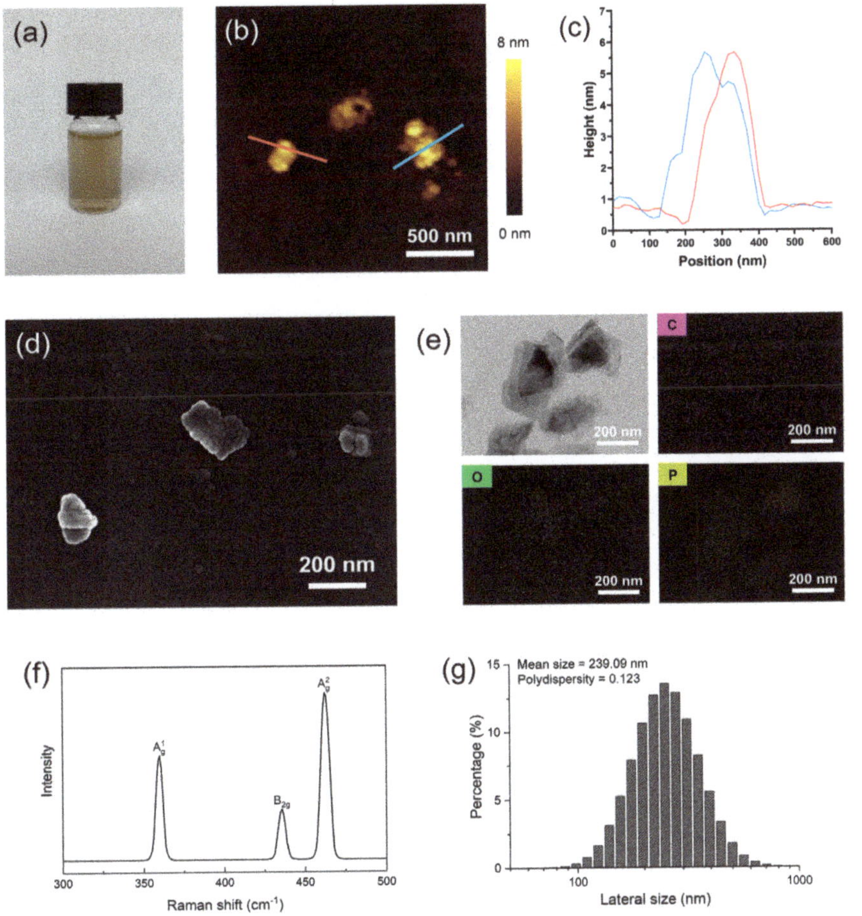

Figure 1. Physicochemical characterizations of the BPNSs. (**a**) Macroscopic image of 10 μg/mL BPNSs dispersed in water; (**b**) AFM image of BPNSs and (**c**) height profiles of BPNSs obtained from the AFM image; (**d**) SEM image of BPNS morphology; (**e**) TEM image and surface elements mapping images of BPNSs; (**f**) Raman spectrum of BPNSs; (**g**) Lateral size distributions and polydispersity of BPNSs in water.

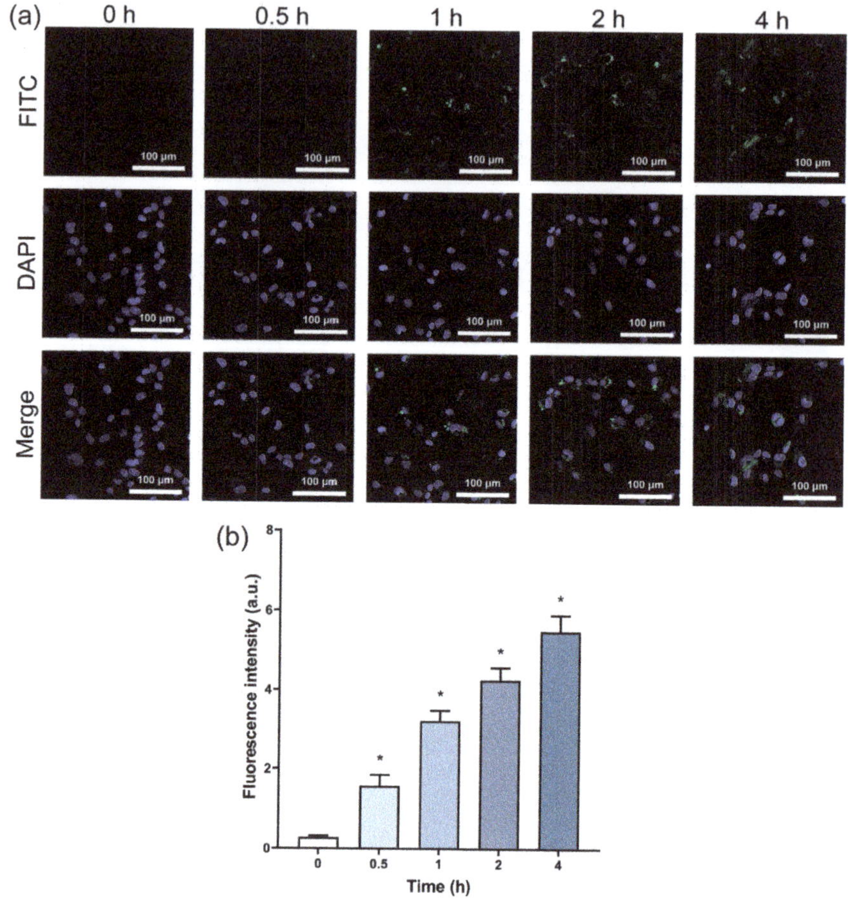

Figure 2. Cell uptake of BPNSs exposed to HUVECs. The HUVEC cells were exposed to the cell culture medium containing 10 μg/mL FITC-labeled BPNSs for different time periods of 0.5, 1, 2 and 4 h. (**a**) Fluorescence images of HUVECs after exposure to 10 μg/mL FITC-labeled BPNSs for 0, 0.5, 1, 2 and 4 h. (**b**) FITC fluorescence intensity was quantitatively analyzed in the cells. * denotes statistical significance with a p-value of less than 0.05 compared to the control.

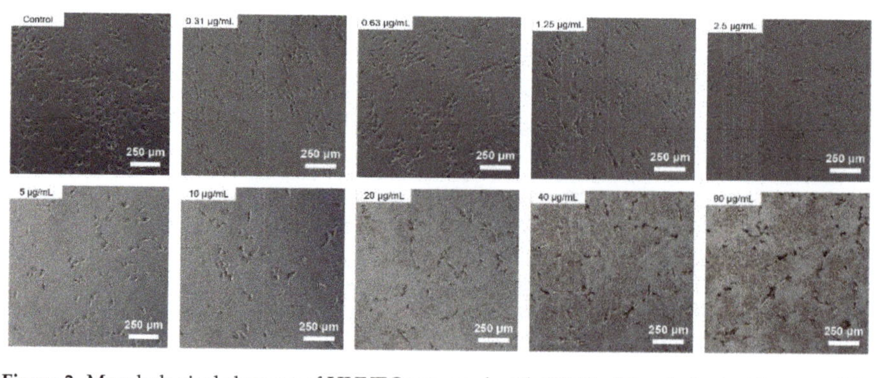

Figure 3. Morphological changes of HUVECs exposed with BPNSs. Morphological images of the HUVEC cells treated with varying concentrations of BPNS (ranging from 0 to 80 μg/mL) in cell culture medium for 24 h.

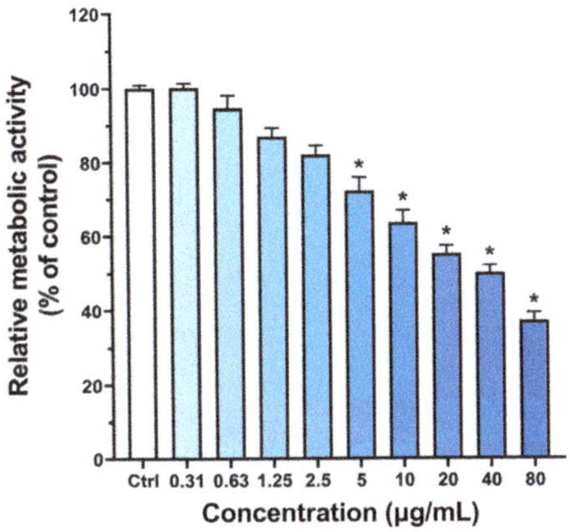

Figure 4. Relative metabolic activity of HUVECs exposed with BPNSs. The relative metabolic activity of HUVECs after treatment with different concentrations of BPNSs (0–80 μg/mL) for 24 h was calculated by the CCK-8 method. * denotes statistical significance with a p-value of less than 0.05 compared to the control.

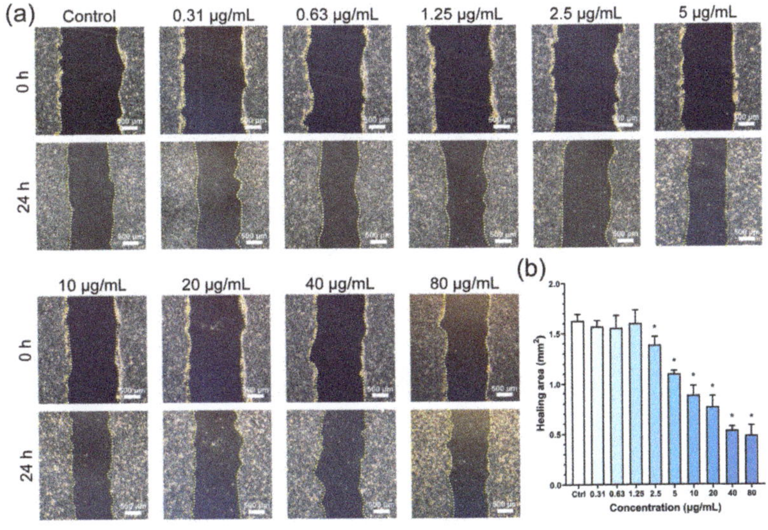

Figure 5. Evaluation of the impact of BPNSs on cellular migration capability. (**a**) Micrographs displaying the cell scratch zone of HUVECs treated with various concentrations of BPNSs for 0 h and 24 h. (**b**) Quantitative analysis of the healing area in the cell scratch region over 24 h. * denotes statistical significance with a p-value of less than 0.05 compared to the control.

3.6. ROS and Mitochondrial Membrane Potential Test

Once inside the cell, the fluorescent probe DCFH-DA was hydrolyzed to DCFH. In the presence of ROS, DCFH was oxidized to generate DCF, which emits green fluorescence. Therefore, the intensity of green fluorescence derived from DCF can be used to gauge the intracellular ROS level. As shown in Figure 7a, green fluorescence in the cells was not obvious at BP concentrations of less than 2.5 μg/mL. However, with the increase in BP concentration, the green fluorescence intensity gradually increased. The highest concentration (80 μg/mL) showed the most vigorous fluorescence intensity, which was 10.5 times higher than that of the control group (Figure 7b).

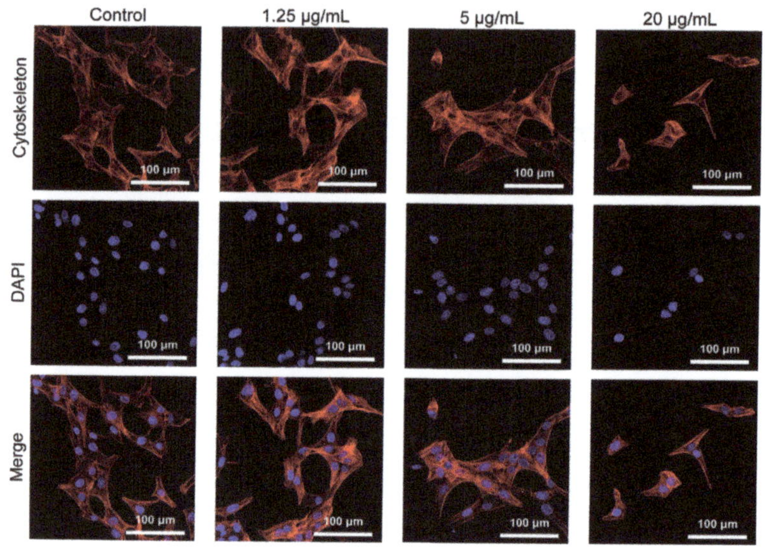

Figure 6. Impact of BPNSs exposure on HUVECs cytoskeleton morphology. Fluorescence microscopy images of phalloidin-stained cytoskeleton and DAPI-stained nuclei of HUVECs incubated with different concentrations (1.25, 5 and 20 μg/mL) of BPNSs for 24 h.

The alteration in mitochondrial membrane potential of cells following 24 h of treatment with BPNSs is illustrated in Figure 7d. Red fluorescence was dominant at concentrations below 1.25 μg/mL; red and green fluorescence was observed within the concentration range of 2.5–20 μg/mL, and green fluorescence was prevalent at 40 and 80 μg/mL. In Figure 7c, the mitochondrial membrane potential was reduced for the groups with BPNSs, and there was a statistical difference between the group with the lowest BPNSs concentration and the control group.

3.7. Cell Apoptosis Test

Elevated levels of ROS and mitochondrial dysfunction are both implicated in the induction of cell apoptosis. Flow cytometry determines the cell state by detecting the fluorescence of cells, and apoptotic cells appear in the two quadrants on the right side. The results of flow cytometry analysis showed an increased number of the experimental group cells in the two quadrants on the right side (Figure 8a), indicative of a greater presence of apoptotic cells. Quantitative analysis further demonstrated a statistically significant increase in the proportion of apoptotic cells within the experimental group compared to the control group (Figure 8b), with significant differences observed at concentrations of 5 and 20 μg/mL.

3.8. Apoptosis-Related Gene Expression

The objective of this study was to evaluate the impact of BPNSs on the expression of apoptosis-associated genes in the HUVEC cells through the RT-PCR analysis, as shown in Figure 9. After 24 h of exposure with BPNSs of varying concentrations, the experimental groups exhibited an incremental rise in the expression of the P53 gene compared to the control group. Additionally, concentrations of 5 and 20 μg/mL showed statistically significant differences from the control group. In contrast, the expression of the BCL-2 gene was significantly higher in the control group than in the experimental group, with a statistically significant difference observed at concentrations of 5 and 20 μg/mL. The pattern of BAX expression followed that of P53, with higher expression levels in the experimental group than the control group, resulting in statistically significant differences at concentrations of 5 and 20 μg/mL.

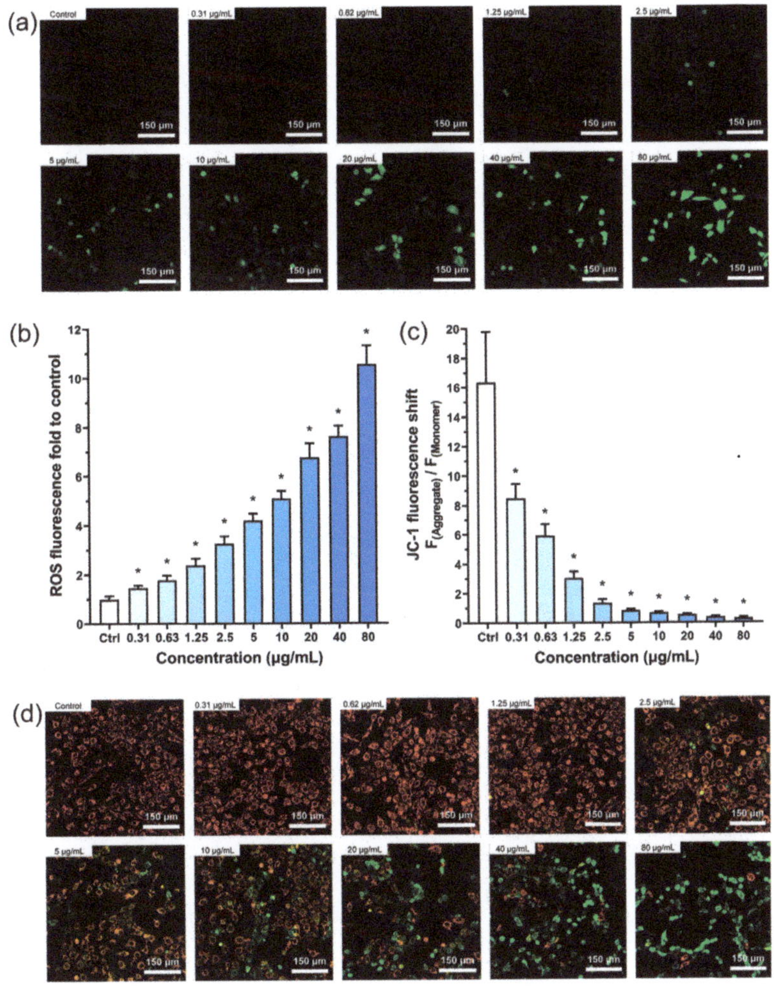

Figure 7. Effects of BPNSs on intracellular ROS levels and mitochondrial membrane potential in HUVECs. (**a**) Fluorescence images were produced by the reaction of ROS with fluorescent probes in the HUVEC cells following 24 h incubation with varying concentrations of BPNSs. (**b**) Quantitative assessment of the relative intracellular ROS levels in HUVECs after 24 h exposure to BPNSs. The relative ROS levels were computed as the fold change in fluorescence intensity compared to the control group. (**c**) Quantitative assessment of the mitochondrial membrane potential changes in HUVECs after 24 h exposure to BPNSs. The relative mitochondrial membrane potential changes were computed as the ratio of intracellular JC-1 red and green fluorescence. (**d**) Following incubation of the HUVEC cells with varying concentrations of BPNSs for 24 h, fluorescent images of mitochondrial membrane potential were obtained using the JC-1 fluorescent probe. * denotes statistical significance with a p-value of less than 0.05 compared to the control.

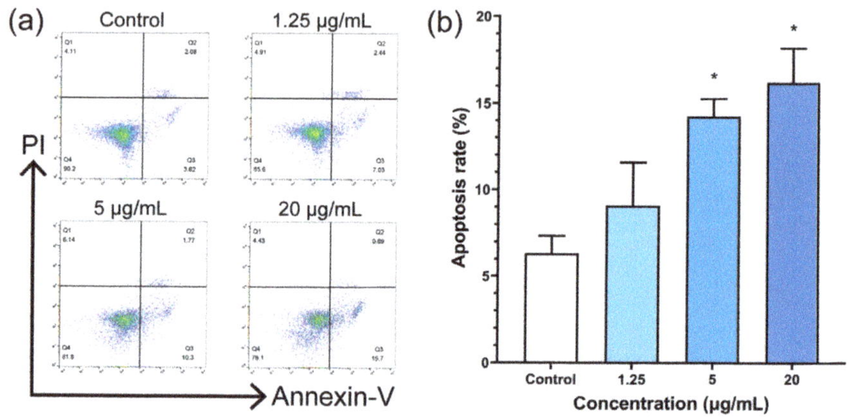

Figure 8. Apoptotic effects of BPNS exposure on HUVECs. (**a**) Flow cytometry analysis of HUVEC apoptosis after exposure to different concentrations of BPNSs for 24 h, and (**b**) corresponding quantitative analysis. * denotes statistical significance with a *p*-value of less than 0.05 compared to the control.

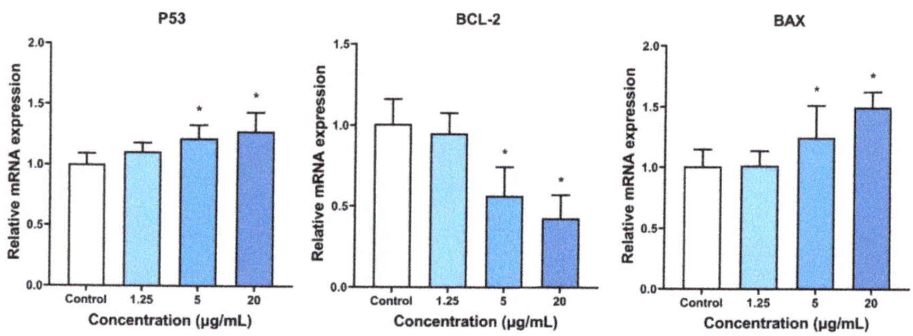

Figure 9. Assessment of the impact of BPNS exposure on apoptosis-related gene expression in HUVECs. RT-PCR analysis was conducted to evaluate the expression levels of P53, BCL-2, and BAX in HUVECs after incubation with different concentrations (1.25, 5 and 20 μg/mL) of BPNSs for 24 h. * denotes statistical significance with a *p*-value of less than 0.05 compared to the control.

4. Discussion

The unique characteristic of a single-element composition and exceptional optoelectronic properties render two-dimensional BP nanomaterials a promising candidate for deployment in craniofacial applications [31]. Biosafety is a key parameter that determines the application of nanomaterials in vivo. Nevertheless, the range of biocompatibility of BP nanomaterials remains unclear [32]. Hence, the current study investigated the cytotoxicity and the underlying mechanisms of black phosphorus nanosheets on vascular endothelial cells. In this study, BPNSs with stable dimensions and properties were obtained via liquid phase exfoliation. Concurrently, the results revealed that exposure of HUVECs with BPNSs at concentrations exceeding 2.5 μg/mL led to a significant reduction in cellular metabolic activity. Our findings indicated that higher concentrations of BPNSs increased the levels of oxidative stress and apoptosis within HUVECs. Additionally, RT-PCR analyses demonstrated that BPNSs had a significant effect on the expression of genes associated with apoptosis.

BPNSs were fabricated using the classical liquid-phase stripping method. Morphological characterization showed that the diameter range of the BPNSs was 100–500 nm, and the height was 3–6 nm, corresponding to 5–10 layers. The BPNSs exhibited good stability and dispersibility, as characterized by Raman spectroscopy and particle-size measurements in Figure 1. The BPNSs in this study had size and properties similar to those of BP nanomaterials in other biomedical applications [5,11].

Nanomaterials can directly enter the cells or interact with them through extracellular contact [33]. To demonstrate the effects of BPNSs on HUVECs, FITC-labelled BPNSs were constructed and the green fluorescence of FITC was observed in the cytoplasm as shown in Figure 2a, implying that BPNSs were internalized by cells. A previous report suggested that the ribbed surface of BP could reduce damage to the cell membrane [34], while the opposite view revealed that BPNSs could damage the integrity of the cell membrane when entering cells [20]. Furthermore, damage to the cell membrane affected the cytoskeleton [35,36], damaging cell morphology and motor function. Disordered myofilaments appeared in the cells when the concentration was higher than 5 μg/mL. Cytoskeletal alterations, as well as excessive intake of BPNSs, could affect cell motility [37]. When the concentration of BPNSs exceeded 2.5 μg/mL, the scratch assay revealed a significant inhibition of cell migration ability (Figure 5b), which could undermine the suitability of BPNSs for tissue engineering applications. Cell viability continued to decrease with an increasing incubation time, further reducing migration ability. Therefore, under the abovementioned actions, BPNSs ultimately decreased cell activity and caused cytotoxicity.

The ROS induced by nanomaterials plays an important role in cytotoxicity [38]. An appropriate level of ROS may be crucial for the physiological functions of biological components [39]. Although the body can balance oxidative stress by processing redox products, excessive ROS can cause various adverse effects [40]. Our results showed a significant increase in intracellular ROS levels after 24 h of exposure to BPNSs at all concentrations in Figure 7b. A negative correlation between ROS levels and cell viability was demonstrated (Figure 10), consistent with the other toxicity studies [41]. Based on previous research, excessive production of ROS has been implicated in the activation of the apoptotic pathway, ultimately resulting in cell death [42], which was also observed in our cell viability and migration ability tests. Mitochondria are the main sites of cellular ROS production. Excessive ROS causes mitochondrial destruction, and mitochondrial fission accelerates ROS accumulation. Such a vicious cycle could amplify the effects of oxidative stress and eventually lead to mitochondrial dysfunction [43]. The results showed that even the lowest concentration of BPNSs (0.31 μg/mL) could decrease mitochondrial membrane potential (Figure 7c), and this effect was more obvious at a higher concentration (up to 80 μg/mL).

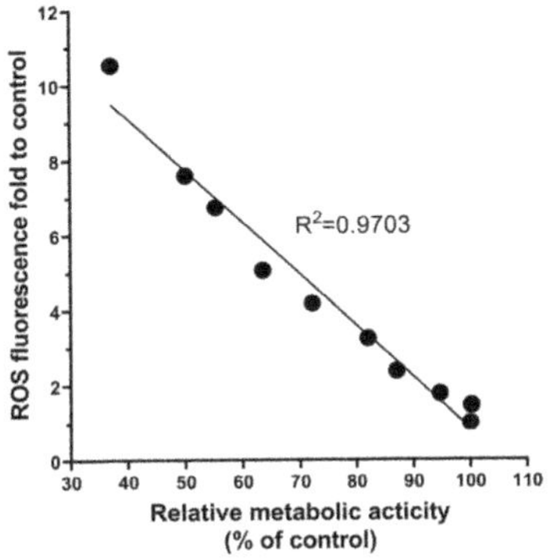

Figure 10. Correlation analysis of BPNSs on the HUVEC cell survival rate and relative intracellular ROS level. The data on relative metabolic activity and ROS levels of HUVECs were used to conduct correlation analysis after 24 h incubation with BPNSs.

Excessive ROS generation exerts direct damage on intracellular biomolecules, resulting in adverse effects on cellular physiological function and the induction of apoptosis [43]. In addition, mitochondrial dysfunction adversely affects intracellular energy metabolism, contributing to the onset of apoptosis [32]. The P53 gene is known to play a pivotal role in regulating cell fate. In response to DNA damage, it can elicit various responses including autophagy, senescence, and apoptosis [44]. Furthermore, P53 has been shown to be causally associated with G1 arrest, which in turn affects embryonic development [45]. Additionally, P53 has been implicated in promoting apoptosis via the regulation of ferroptosis [46]. In response to oxidative stress, the P53 gene-encoded protein could influence the process of apoptosis by mediating mitochondrial membrane potential and activating BAX [47]. Both BCL-2 and BAX are constituents of the BCL-2 gene family, contributing to the intricate regulation of cell apoptosis. In response to apoptotic signals, members of the BCL-2 family become activated and mediate the permeabilization of the outer mitochondrial membrane, a critical step in the process of apoptosis [48]. The protein encoded by the BCL-2 gene restrains apoptosis, whereas the protein encoded by the BAX gene promotes it. These proteins hold opposing effects on each other, affecting the course of apoptosis [49]. The flow cytometry analysis demonstrated a significant increase in apoptosis when the concentration of BPNSs exceeded 5 μg/mL in Figure 8b. Additionally, the RT-PCR results revealed that BPNSs induced the up-regulation of P53 gene expression in cells. Moreover, at concentrations above 5 μg/mL, statistically significant down-regulation of BCL-2 and up-regulation of BAX were observed relative to the control group. As an upstream gene of BCL-2 gene family, the up-regulation of P53 expression promotes cell apoptosis. Therefore, the down-regulation of BCL-2, which inhibits cell apoptosis, leads to the up-regulation of BAX expression that antagonizes BCL-2, indicating that BPNSs promote cell apoptosis. The expression levels of apoptosis-related genes also reflected that BPNSs promoted apoptosis in a concentration-dependent manner. Hence, the cytotoxicity of BPNSs on HUVECs is contingent on several factors, including the integrity of the cell membrane, intracellular oxidative stress levels, and the down-regulation of mitochondrial membrane potential. These factors collectively alter the cellular activity and function by inducing apoptosis (Figure 11).

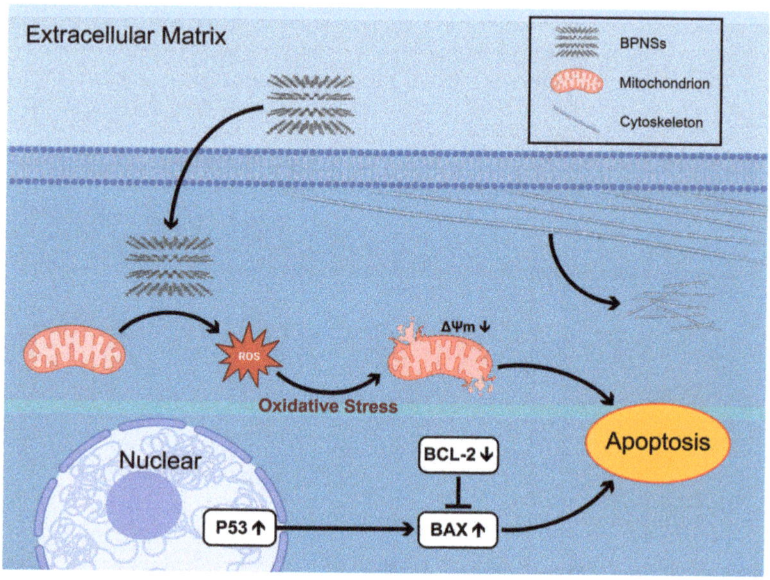

Figure 11. Schematic mechanisms of BPNSs-induced cytotoxicity of HUVECs. The internalization of BPNSs by HUVECs had an impact on the cytoskeletal structure, which was manifested in the form of alterations in cellular morphology and a subsequent decrease in the migration ability. After internalization, BPNSs induced excessive ROS and further led to the dysfunction of mitochondria. These sequential cellular responses promote the progression of cell apoptosis, which further affects cell activity and ultimately leads to reduced cell survival. Abbreviations: BPNSs: Black phosphorus nanosheets; ROS: Reactive oxygen species.

Nevertheless, the size and surface properties of nanomaterials could additionally influence their cytotoxicity. Apart from inducing apoptosis, they could also generate cytotoxicity by affecting the cell cycle and triggering intracellular inflammatory responses. Therefore, more investigations are warranted to gain a comprehensive understanding of the toxicity and to provide a framework for future research based on BP.

5. Conclusions

In this study, the cytotoxicity of BPNSs to HUVECs and its underlying mechanisms were investigated, and the following conclusions were drawn:

1. BPNSs exhibited significant cytotoxicity towards HUVECs at concentrations exceeding 2.5 μg/mL, characterized by inhibited cell metabolic activity, disrupted cytoskeleton, and suppressed cell migration.
2. The cytotoxicity mechanism of BPNSs on HUVECs involves the generation of excessive ROS and mitochondrial dysfunction, ultimately leading to apoptosis.
3. The cytotoxic effect of BPNSs on HUVECs was associated with apoptosis-associated genes, including P53 and the BCL-2 family.

Author Contributions: Conceptualization, S.L. and H.D.; methodology, H.D. and Y.W.; software, Y.W. and R.X.; validation, X.Z. and J.L.; formal analysis, H.D. and Y.W.; investigation, J.L. and H.D.; resources, R.X. and X.Z.; data curation, R.X. and X.Z.; writing—original draft preparation, H.D. and Y.W.; writing—review and editing, P.L. and S.L.; visualization, J.L. and R.X.; supervision, P.L. and S.L.; project administration, S.L.; funding acquisition, S.L. and P.L. All authors have read and agreed to the published version of the manuscript.

Funding: This study was financially supported by the Natural Science Foundation of Xinjiang Uygur Autonomous Region in China [2020D01C004], the National Natural Science Foundation of China (81801008), GuangDong Basic and Applied Basic Research Foundation, China (No. 2021A1515111140), Medical Scientific Research Foundation of Guangdong Province, China (No. A2023369), the Scientific Research Project of Traditional Chinese Medicine Bureau of Guangdong Province (20221267), and the Science Research Cultivation Program of Stomatological Hospital, Southern Medical University (PY2021019 and PY2021003).

Data Availability Statement: Access to the data could be requested by contacting the corresponding author.

Conflicts of Interest: The authors declare no conflict of interest.

References

1. Francisco, I.; Basílio, Â.; Ribeiro, M.P.; Nunes, C.; Travassos, R.; Marques, F.; Pereira, F.; Paula, A.B.; Carrilho, E.; Marto, C.M.; et al. Three-Dimensional Impression of Biomaterials for Alveolar Graft: Scoping Review. *J. Funct. Biomater.* **2023**, *14*, 76. [CrossRef] [PubMed]
2. Romasco, T.; Tumedei, M.; Inchingolo, F.; Pignatelli, P.; Montesani, L.; Iezzi, G.; Petrini, M.; Piattelli, A.; Di Pietro, N. A Narrative Review on the Effectiveness of Bone Regeneration Procedures with Osteobiol (R) Collagenated Porcine Grafts: The Translational Research Experience over 20 Years. *J. Funct. Biomater.* **2022**, *13*, 121. [CrossRef]
3. Qing, Y.A.; Li, R.Y.; Li, S.H.; Li, Y.H.; Wang, X.Y.; Qin, Y.G. Advanced Black Phosphorus Nanomaterials for Bone Regeneration. *Int. J. Nanomed.* **2020**, *15*, 2045–2058. [CrossRef] [PubMed]
4. Li, Z.; Zhang, X.; Ouyang, J.; Chu, D.; Han, F.; Shi, L.; Liu, R.; Guo, Z.; Gu, G.X.; Tao, W.; et al. Ca2+-Supplying Black Phosphorus-Based Scaffolds Fabricated with Microfluidic Technology for Osteogenesis. *Bioact. Mater.* **2021**, *6*, 4053–4064. [CrossRef]
5. Xu, H.; Liu, X.; Park, S.; Terzic, A.; Lu, L. Size-Dependent Osteogenesis of Black Phosphorus in Nanocomposite Hydrogel Scaffolds. *J. Biomed. Mater. Res. A* **2022**, *110*, 1488–1498. [CrossRef] [PubMed]
6. Li, L.; Yu, Y.; Ye, G.J.; Ge, Q.; Ou, X.; Wu, H.; Feng, D.; Chen, X.H.; Zhang, Y. Black Phosphorus Field-Effect Transistors. *Nat. Nanotechnol.* **2014**, *9*, 372–377. [CrossRef]
7. Qu, G.; Xia, T.; Zhou, W.; Zhang, X.; Zhang, H.; Hu, L.; Shi, J.; Yu, X.F.; Jiang, G. Property-Activity Relationship of Black Phosphorus at the Nano-Bio Interface: From Molecules to Organisms. *Chem. Rev.* **2020**, *120*, 2288–2346. [CrossRef]
8. Tao, W.; Zhu, X.; Yu, X.; Zeng, X.; Xiao, Q.; Zhang, X.; Ji, X.; Wang, X.; Shi, J.; Zhang, H.; et al. Black Phosphorus Nanosheets as a Robust Delivery Platform for Cancer Theranostics. *Adv. Mater.* **2017**, *29*, 1603276. [CrossRef]

9. Liu, W.X.; Dong, A.; Wang, B.; Zhang, H. Current Advances in Black Phosphorus-Based Drug Delivery Systems for Cancer Therapy. *Adv. Sci.* **2021**, *8*, 2003033. [CrossRef]
10. Zhou, W.; Cui, H.; Ying, L.; Yu, X.F. Enhanced Cytosolic Delivery and Release of Crispr/Cas9 by Black Phosphorus Nanosheets for Genome Editing. *Angew. Chem. Int. Ed. Engl.* **2018**, *57*, 10268–10272. [CrossRef]
11. Qian, Y.; Yuan, W.E.; Cheng, Y.; Yang, Y.Q.; Qu, X.H.; Fan, C.Y. Concentrically Integrative Bioassembly of a Three-Dimensional Black Phosphorus Nanoscaffold for Restoring Neurogenesis, Angiogenesis, and Immune Homeostasis. *Nano Lett.* **2019**, *19*, 8990–9001. [CrossRef] [PubMed]
12. Huang, K.; Wu, J.; Gu, Z. Black Phosphorus Hydrogel Scaffolds Enhance Bone Regeneration Via a Sustained Supply of Calcium-Free Phosphorus. *ACS Appl. Mater. Interfaces* **2019**, *11*, 2908–2916. [CrossRef] [PubMed]
13. Luo, M.M.; Cheng, W.; Zeng, X.W.; Mei, L.; Liu, G.; Deng, W.B. Folic Acid-Functionalized Black Phosphorus Quantum Dots for Targeted Chemo-Photothermal Combination Cancer Therapy. *Pharmaceutics* **2019**, *11*, 242. [CrossRef]
14. Sutrisno, L.; Chen, H.; Chen, Y.; Yoshitomi, T.; Kawazoe, N.; Yang, Y.; Chen, G. Composite Scaffolds of Black Phosphorus Nanosheets and Gelatin with Controlled Pore Structures for Photothermal Cancer Therapy and Adipose Tissue Engineering. *Biomaterials* **2021**, *275*, 120923. [CrossRef] [PubMed]
15. Zhang, F.; Peng, F.F.; Qin, L.; Yang, D.D.; Li, R.R.; Jiang, S.S.; He, H.Y.; Zhang, P. Ph/near Infrared Dual-Triggered Drug Delivery System Based Black Phosphorus Nanosheets for Targeted Cancer Chemo-Photothermal Therapy. *Colloids Surf. B Biointerfaces* **2019**, *180*, 353–361. [CrossRef]
16. Zhu, Y.; Xie, Z.; Li, J.; Liu, Y.; Li, C.; Liang, W.; Huang, W.; Kang, J.; Cheng, F.; Kang, L.; et al. From Phosphorus to Phosphorene: Applications in Disease Theranostics. *Coord. Chem. Rev.* **2021**, *446*, 214110. [CrossRef]
17. Zhang, S.; Zhang, X.; Lei, L.; Yu, X.F.; Chen, J.; Ma, C.; Wu, F.; Zhao, Q.; Xing, B. Ph-Dependent Degradation of Layered Black Phosphorus: Essential Role of Hydroxide Ions. *Angew. Chem. Int. Ed. Engl.* **2019**, *58*, 467–471. [CrossRef]
18. Zhang, T.; Wan, Y.; Xie, H.; Mu, Y.; Du, P.; Wang, D.; Wu, X.; Ji, H.; Wan, L. Degradation Chemistry and Stabilization of Exfoliated Few-Layer Black Phosphorus in Water. *J. Am. Chem. Soc.* **2018**, *140*, 7561–7567. [CrossRef]
19. Song, S.J.; Shin, Y.C.; Lee, H.U.; Kim, B.; Han, D.W.; Lim, D. Dose- and Time-Dependent Cytotoxicity of Layered Black Phosphorus in Fibroblastic Cells. *Nanomaterials* **2018**, *8*, 408. [CrossRef]
20. Ruan, F.; Liu, R.; Wang, K.; Zeng, J.; Zuo, Z.; He, C.; Zhang, Y. Cytotoxicity of Black Phosphorus Quantum Dots on Lung-Derived Cells and the Underlying Mechanisms. *J. Hazard. Mater.* **2021**, *402*, 122875. [CrossRef]
21. Zhang, X.; Zhang, Z.; Zhang, S.; Li, D.; Ma, W.; Ma, C.; Wu, F.; Zhao, Q.; Yan, Q.; Xing, B. Size Effect on the Cytotoxicity of Layered Black Phosphorus and Underlying Mechanisms. *Small* **2017**, *13*. [CrossRef] [PubMed]
22. Chen, D.; Tong, Z.F.; Xiong, Z.; Zhang, X.; Zhao, Q.; Zhang, S. Environmental Stability and Cytotoxicity of Layered Black Phosphorus Modified with Polyvinylpyrrolidone and Zeolitic Imidazolate Framework-67. *Sci. Total Environ.* **2021**, *790*, 148105. [CrossRef] [PubMed]
23. Du, Z.; Zhao, D.; Jing, L.; Cui, G.; Jin, M.; Li, Y.; Liu, X.; Liu, Y.; Du, H.; Guo, C.; et al. Cardiovascular Toxicity of Different Sizes Amorphous Silica Nanoparticles in Rats after Intratracheal Instillation. *Cardiovasc. Toxicol.* **2013**, *13*, 194–207. [CrossRef] [PubMed]
24. Shineh, G.; Patel, K.; Mobaraki, M.; Tayebi, L. Functional Approaches in Promoting Vascularization and Angiogenesis in Bone Critical-Sized Defects Via Delivery of Cells, Growth Factors, Drugs, and Particles. *J. Funct. Biomater.* **2023**, *14*, 99. [CrossRef]
25. Cao, Y.; Gong, Y.; Liu, L.; Zhou, Y.; Fang, X.; Zhang, C.; Li, Y.; Li, J. The Use of Human Umbilical Vein Endothelial Cells (Huvecs) as an in Vitro Model to Assess the Toxicity of Nanoparticles to Endothelium: A Review. *J. Appl. Toxicol.* **2017**, *37*, 1359–1369. [CrossRef] [PubMed]
26. Li, P.; Schille, C.; Schweizer, E.; Kimmerle-Müller, E.; Rupp, F.; Heiss, A.; Legner, C.; Klotz, U.E.; Geis-Gerstorfer, J.; Scheideler, L. Selection of Extraction Medium Influences Cytotoxicity of Zinc and Its Alloys. *Acta Biomater.* **2019**, *98*, 235–245. [CrossRef] [PubMed]
27. Xia, F.; Wang, H.; Jia, Y. Rediscovering Black Phosphorus as an Anisotropic Layered Material for Optoelectronics and Electronics. *Nat. Commun.* **2014**, *5*, 4458. [CrossRef] [PubMed]
28. Ling, X.; Wang, H.; Huang, S.; Xia, F.; Dresselhaus, M.S. The Renaissance of Black Phosphorus. *Proc. Natl. Acad. Sci. USA* **2015**, *112*, 4523–4530. [CrossRef]
29. Huang, S.; Ling, X. Black Phosphorus: Optical Characterization, Properties and Applications. *Small* **2017**, *13*, 1700823. [CrossRef]
30. Liu, X.; Xie, J.; Yang, L.; Li, Y.; He, Y.; Liu, Z.; Zhang, Y.; Su, G. Bone Marrow Mesenchymal Stem Cells Enhance Autophagy and Help Protect Cells under Hypoxic and Retinal Detachment Conditions. *J. Cell. Mol. Med.* **2020**, *24*, 3346–3358. [CrossRef]
31. Jiang, H.; Xia, Q.; Liu, D.; Ling, K. Calcium-Cation-Doped Polydopamine-Modified 2d Black Phosphorus Nanosheets as a Robust Platform for Sensitive and Specific Biomolecule Sensing. *Anal. Chim. Acta* **2020**, *1121*, 1–10. [CrossRef]
32. Zhang, X.; Donskyi, I.S.; Tang, W.; Deng, S.; Liu, D.; Zhang, S.; Zhao, Q.; Xing, B. Biological Effects of Black Phosphorus Nanomaterials on Mammalian Cells and Animals. *Angew. Chem. Int. Ed.* **2022**, *62*, e202213336. [CrossRef] [PubMed]
33. Zhang, Y.; Ali, S.F.; Dervishi, E.; Xu, Y.; Li, Z.; Casciano, D.; Biris, A.S. Cytotoxicity Effects of Graphene and Single-Wall Carbon Nanotubes in Neural Phaeochromocytoma-Derived Pc12 Cells. *ACS Nano* **2010**, *4*, 3181–3186. [CrossRef] [PubMed]
34. Ma, X.; Zhu, X.; Huang, C.; Li, Z.; Fan, J. Molecular Mechanisms Underlying the Role of the Puckered Surface in the Biocompatibility of Black Phosphorus. *Nanoscale* **2021**, *13*, 3790–3799. [CrossRef]
35. Snyder, R.J.; Hussain, S.; Rice, A.B.; Garantziotis, S. Multiwalled Carbon Nanotubes Induce Altered Morphology and Loss of Barrier Function in Human Bronchial Epithelium at Noncytotoxic Doses. *Int. J. Nanomed.* **2014**, *9*, 4093–4105. [CrossRef]

36. Wu, J.; Zhu, Z.; Liu, W.; Zhang, Y.; Kang, Y.; Liu, J.; Hu, C.; Wang, R.; Zhang, M.; Chen, L.; et al. How Nanoparticles Open the Paracellular Route of Biological Barriers: Mechanisms, Applications, and Prospects. *ACS Nano* **2022**, *16*, 15627–15652. [CrossRef] [PubMed]
37. Fojtu, M.; Balvan, J.; Raudenska, M.; Vicar, T.; Bousa, D.; Sofer, Z.; Masarik, M.; Pumera, M. Black Phosphorus Cytotoxicity Assessments Pitfalls: Advantages and Disadvantages of Metabolic and Morphological Assays. *Chemistry* **2019**, *25*, 349–360. [CrossRef] [PubMed]
38. Mu, X.; Wang, J.-Y.; Bai, X.; Xu, F.; Liu, H.; Yang, J.; Jing, Y.; Liu, L.; Xue, X.; Dai, H.; et al. Black Phosphorus Quantum Dot Induced Oxidative Stress and Toxicity in Living Cells and Mice. *ACS Appl. Mater. Interfaces* **2017**, *9*, 20399–20409. [CrossRef]
39. Schieber, M.; Chandel, N.S. Ros Function in Redox Signaling and Oxidative Stress. *Curr. Biol.* **2014**, *24*, R453–R462. [CrossRef] [PubMed]
40. Wang, H.; Fu, X.; Shi, J.; Li, L.; Sun, J.; Zhang, X.; Han, Q.; Deng, Y.; Gan, X. Nutrient Element Decorated Polyetheretherketone Implants Steer Mitochondrial Dynamics for Boosted Diabetic Osseointegration. *Adv. Sci.* **2021**, *8*, e2101778. [CrossRef]
41. Song, W.; Zhang, J.; Guo, J.; Zhang, J.; Ding, F.; Li, L.; Sun, Z. Role of the Dissolved Zinc Ion and Reactive Oxygen Species in Cytotoxicity of Zno Nanoparticles. *Toxicol. Lett.* **2010**, *199*, 389–397. [CrossRef] [PubMed]
42. Lv, R.; Yang, D.; Yang, P.; Xu, J.; He, F.; Gai, S.; Li, C.; Dai, Y.; Yang, G.; Lin, J. Integration of Upconversion Nanoparticles and Ultrathin Black Phosphorus for Efficient Photodynamic Theranostics under 808 Nm near-Infrared Light Irradiation. *Chem. Mater.* **2016**, *28*, 4724–4734. [CrossRef]
43. Scheibye-Knudsen, M.; Fang, E.F.; Croteau, D.L.; Wilson, D.M., 3rd; Bohr, V.A. Protecting the Mitochondrial Powerhouse. *Trends Cell. Biol.* **2015**, *25*, 158–170. [CrossRef] [PubMed]
44. Gupta, S.; Silveira, D.A.; Mombach, J.C.M. Towards DNA-Damage Induced Autophagy: A Boolean Model of P53-Induced Cell Fate Mechanisms. *DNA Repair* **2020**, *96*, 102971. [CrossRef] [PubMed]
45. Moallem, S.A.; Hales, B.F. The Role of P53 and Cell Death by Apoptosis and Necrosis in 4-Hydroperoxycyclophosphamide-Induced Limb Malformations. *Development* **1998**, *125*, 3225–3234. [CrossRef]
46. Kang, R.; Kroemer, G.; Tang, D. The Tumor Suppressor Protein P53 and the Ferroptosis Network. *Free Radic. Biol. Med.* **2019**, *133*, 162–168. [CrossRef]
47. Liu, J.; Wang, X.; Zheng, M.; Luan, Q. Oxidative Stress in Human Gingival Fibroblasts from Periodontitis Versus Healthy Counterparts. *Oral Dis.* **2021**, *29*, 1214–1225. [CrossRef]
48. Peña-Blanco, A.; García-Sáez, A.J. Bax, Bak and Beyond—Mitochondrial Performance in Apoptosis. *FEBS J.* **2018**, *285*, 416–431. [CrossRef]
49. Spitz, A.Z.; Gavathiotis, E. Physiological and Pharmacological Modulation of Bax. *Trends Pharmacol. Sci.* **2022**, *43*, 206–220. [CrossRef]

Journal of
Functional Biomaterials

MDPI

Review

The Use of Warm Air for Solvent Evaporation in Adhesive Dentistry: A Meta-Analysis of In Vitro Studies

Rim Bourgi [1,2], Louis Hardan [1,†], Carlos Enrique Cuevas-Suárez [3], Francesco Scavello [4,*], Davide Mancino [2,5,6], Naji Kharouf [2,5,*] and Youssef Haikel [2,5,6,†]

1 Department of Restorative Dentistry, School of Dentistry, Saint-Joseph University, Beirut 1107 2180, Lebanon; rim.bourgi@net.usj.edu.lb (R.B.); louis.hardan@usj.edu.lb (L.H.)
2 Department of Biomaterials and Bioengineering, INSERM UMR_S 1121, University of Strasbourg, 67000 Strasbourg, France; mancino@unistra.fr (D.M.); youssef.haikel@unistra.fr (Y.H.)
3 Dental Materials Laboratory, Academic Area of Dentistry, Autonomous University of Hidalgo State, San Agustín Tlaxiaca 42160, Mexico; cecuevas@uaeh.edu.mx
4 IRCCS Humanitas Research Hospital, 20089 Rozzano, Milan, Italy
5 Department of Endodontics and Conservative Dentistry, Faculty of Dental Medicine, University of Strasbourg, 67000 Strasbourg, France
6 Pôle de Médecine et Chirurgie Bucco-Dentaire, Hôpital Civil, Hôpitaux Universitaire de Strasbourg, 67000 Strasbourg, France
* Correspondence: francesco.scavello@humanitasresearch.it (F.S.); dentistenajikharouf@gmail.com (N.K.); Tel.: +33-66752-2841 (N.K.)
† These authors contributed equally to this work.

Abstract: Any excess solvent from dental adhesive systems must be eliminated prior to material photopolymerization. For this purpose, numerous approaches have been proposed, including the use of a warm air stream. This study aimed to investigate the effect of different temperatures of warm air blowing used for solvent evaporation on the bond strength of resin-based materials to dental and nondental substrates. Two different reviewers screened the literature in diverse electronic databases. In vitro studies recording the effect of warm air blowing to evaporate solvents of adhesive systems on the bond strength of resin-based materials to direct and indirect substrates were included. A total of 6626 articles were retrieved from all databases. From this, 28 articles were included in the qualitative analysis, and 27 remained for the quantitative analysis. The results of the meta-analysis for etch-and-rinse adhesives revealed that the use of warm air for solvent evaporation was statistically significantly higher ($p = 0.005$). For self-etch adhesives and silane-based materials, this effect was observed too ($p < 0.001$). The use of a warm air stream for solvent evaporation enhanced the bonding performance of alcohol-/water-based adhesive systems for dentin. This effect seems to be similar when a silane coupling agent is submitted to a heat treatment before the cementation of a glass-based ceramic.

Keywords: dentine; silane; solvent-based; total-etch; warm air

Citation: Bourgi, R.; Hardan, L.; Cuevas-Suárez, C.E.; Scavello, F.; Mancino, D.; Kharouf, N.; Haikel, Y. The Use of Warm Air for Solvent Evaporation in Adhesive Dentistry: A Meta-Analysis of In Vitro Studies. *J. Funct. Biomater.* **2023**, *14*, 285. https://doi.org/10.3390/jfb14050285

Academic Editors: Ping Li, Guojiang Wan, Shulan Xu and An Li

Received: 20 April 2023
Revised: 15 May 2023
Accepted: 17 May 2023
Published: 20 May 2023

1. Introduction

Adhesive systems contain resin monomers with hydrophilic and hydrophobic characteristics, polymerization modulators, and a high concentration of solvents [1]. Organic solvents act as diluting agents that improve wetting and the infiltration of resin monomers into the dentinal surface [2]. It is crucial that any excess solvent must be eliminated from the dental substrate by means of air drying prior to the photopolymerization of the applied adhesive, in addition to allowing time between these two processes [3]. A presence of residual solvents might affect the polymerization of monomers and hinder the integrity of the bond, creating unwanted pathways for voids inside the adhesive interface and causing a decrease in the bond strength [4,5].

Various approaches favoring solvent evaporation have been proposed. These include an increased adhesive application time [6], multiple adhesive layers [7], delayed light curing [8], vigorous adhesive rubbing [9], longer exposure duration of bonding systems [10], and extended air drying [11]. Additionally, using a warm air stream was found to sufficiently evaporate solvents included in an adhesive system [12]. Authors have experimented on external sources for heating before photopolymerization, ranging within biologically acceptable limits [13–15], leading to immediate monomer conversion gains while reducing the concentration of the final solvent in the adhesive system [16].

The application of a warm air stream was found to adequately evaporate solvents from an adhesive system [12]; this might clinically increase the dentin bond strength. Authors have experimented on external sources of heating before photopolymerization, ranging within biologically acceptable limits [12–15,17,18], resulting in immediate monomer conversion gains while reducing the concentration of the final solvent in the adhesive system [16]. Furthermore, this approach enhances resin infiltration into the decalcified dentin, consequently improving dentinal bond strength. It is important to mention that a warm air stream within the thermal tolerance zone of the dentin pulp organ (29–56 °C) will not yield any undesirable pathological effects on the dentin pulp organ and dentin will respond physiologically to warm air blowing [19–21]. A previous meta-analysis proposed that the dentin bond strength of adhesive systems might be improved by using warm air streams for solvent evaporation [22]. Several air temperatures for solvent evaporation, comprising 37 °C, 38 °C, 50 °C, 60 °C, and 80 °C, were recognized in the aforementioned review. Approximately 40 °C and 60 °C warm-air-drying temperatures were considered efficient for improving solvent evaporation. Nonetheless, the 60 °C temperature was more favorable in terms of stable bond strength and reduced bond degradation [22]. Fittingly, a better description of the gold-standard temperature for air drying should be of great attention.

The use of warm air for an adequate evaporation of solvents has been also explored for silane-based products [23]. In a clinical situation, silanes reduce the contact angle, leaving a film with a thickness ranging between approximately 10 and 50 nm [24]. The outcomes of a study support the theory that silane is capable of raising the surface wettability, causing the formation of chemical bridges with substrates covered by hydroxyl groups (OH) (e.g., glass or quartz fibers) [25]. Furthermore, it has been conveyed that the silane coupling agent significantly improves the bond strength between the fiber posts and the composite core build-up materials [26].

Several layers or a thick film of silane lead to internal cohesive tendencies, and eventually bond failure. Due to this, clinically, an ultrathin silane layer is required to enhance bond strength [27]. An approach to improving its implementation includes thermal drying with temperatures ranging between 50 and 100 °C of silanes once applied to substrates. This permits the evaporation of vehicles, thus accelerating the condensation on the surface and promoting the efficient formation of covalent bonds [28]. However, silane heat treatment at high temperatures (70–80 °C) may not be practical for clinical practices, but a warm air stream (38 °C) acceptable to patients can be used to aid in solvent evaporation and the products of the reaction on the silane-treated surface, leading to a dried surface [29].

The purpose of this manuscript was to investigate the effect of different temperatures of warm air blowing used for solvent evaporation on the bond strength of resin-based materials to dental and nondental substrates. Accordingly, the null hypothesis of the present systematic review and meta-analysis was that the use of a different temperature of warm air stream for solvent evaporation does not affect the bond strength of resin-based materials to dental and nondental substrates.

2. Materials and Methods

2.1. Protocol and Registration

This systematic review was registered in the Open Science Framework under the identifier DOI 10.17605/OSF.IO/JUQT5 and it respected the suggestions of the Preferred Reporting Items for Systematic Reviews and Meta-Analysis (PRISMA) statement.

2.2. Information Sources and Search Strategy

The search strategy (Table 1) was firstly described for the MEDLINE database using keywords for each concept of the following PICOT strategy: population, permanent enamel and dentin, and indirect substrates; intervention: application of warm air for solvent evaporation; control, application of room-temperature air; outcome, bond strength; and type of studies, in vitro studies. The MEDLINE search strategy was adapted to other electronic databases, including Scielo, Web of Science, Scopus, and Embase. Additionally, the first 100 results of Google Scholar were also consulted.

Table 1. Search strategy used.

Search	Keywords
# 1	Dental Bonding OR Self-Cured Dental Bonding OR Chemical-Curing of Dental Adhesives OR Chemical Curing of Dental Adhesives OR Dentin-Bonding Agents OR dental primer OR Dental Materials OR Dental Material OR dental resin OR Dental Resins OR bonding interface OR adhesive OR Dentin-Bonding Agents OR Dentin Bonding Agents
# 2	warm air OR temperature air OR air-blowing OR air-drying OR air-stream
# 3	# 1 and # 2

2.3. Selection Process and Data Collection Process

After running a search strategy, an online software program (Rayyan, Qatar Computing Research Institute, HBKU, Doha, Qatar) was used to store the files from all databases, and for duplicate detection. The same software program was used for evaluating the title and abstract of the articles. This phase was carried out by two independent reviewers to check whether they encountered the following inclusion criteria: (1) in vitro studies recording the effect of the use of different temperatures of warm air blowing to evaporate solvents from an adhesive system on the bond strength of resin-based materials to enamel, dentin, glass-based ceramics, oxide-based ceramics, metal alloys, and indirect composites; (2) evaluated the bond strength of adhesive systems to the aforementioned substrates with 2 antagonists: composite resin or resin-based cement; (3) comprised a control group; (4) included mean and standard deviation (SD) data in MPa on shear, microshear, microtensile, and tensile bond tests; and (5) available in the English, Spanish, or Portuguese language. Case series, case reports, pilot studies, and reviews were omitted.

Each adequate article received a study identification, merging the last name of the first author with the year of publication. The same two referees reviewed and categorized data, such as the material tested, the solvent contained in the material, the temperature of warm air stream used, the substrate tested, and the bond strength test.

2.4. Quality Assessment

The methodological quality of, respectively, integrated manuscripts was evaluated independently by two authors (L.H. and R.B.). The risk of bias in individual articles was considered via the description of the following factors: specimens' randomization, single operator, operator blinded, control group, standardized specimens, failure mode, manufacturer's instructions, sample size calculation, and coefficient of variation. If the article included the factor, the study received a "Yes" for that specific parameter. In the case of missing data, the parameter received a "No". The risk of bias was classified regarding the sum of "Yes" answers received: 1 to 3 denoted a high bias, 4 to 6 medium, and 7 to 9 showed a low risk of bias.

2.5. Statistical Analysis

A meta-analysis was executed by means of a software program (Review Manager version 5.3.5; The Cochrane Collaboration, Copenhagen, Denmark). A random-effect model was used, and estimates were acquired by comparing the standardized mean difference between the bond strength values for the groups where a warm air stream was used against the control group in which a room-temperature air stream was considered. The bond strength from etch-and-rinse (ER), self-etch (SE), and silane-based materials was analyzed separately. For each analysis, subgroups were formed considering the type of solvent used for the adhesive formulation. A p-value < 0.05 was considered statistically significant. The heterogeneity was designed by means of the Cochran Q test and the inconsistency I^2 test.

3. Results

A total of 6626 articles were retrieved from all databases. Figure 1 shows the flowchart that summarizes the study selection according to the PRISMA statement. A total of 4867 documents were screened by title and abstract after duplicate removal. From these, 4831 were excluded after the initial screening. Thirty-six studies were chosen for full-text reading, and one article was included after hand searching the references from these documents. After reading the full text, nine articles were excluded due to several reasons: in (4), the bond strength was not evaluated [18,30–32]; in (4), a warm air stream was not tested [33–36]; and (1) was a thesis [37]. Then, a total of 28 articles were included in the qualitative analysis [12–14,16,20,21,23,24,29,38–56]. One study was excluded from the meta-analysis because there were no more studies for comparison [53].

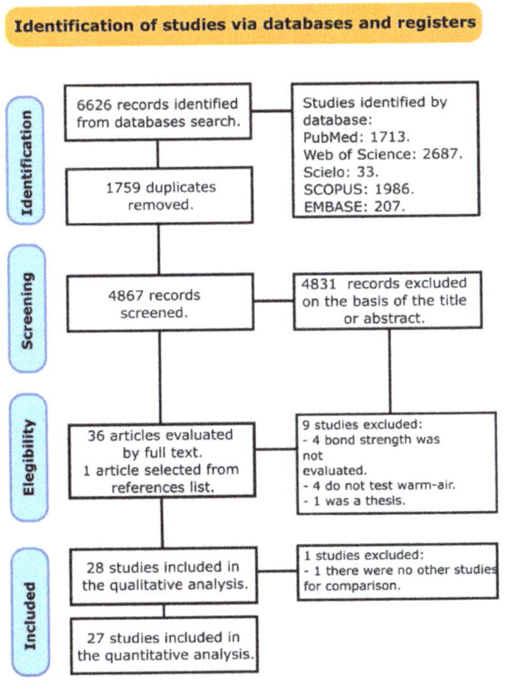

Figure 1. Study selection according to the PRISMA statement.

The features of the included articles are recapitulated in Tables 2 and 3. Numerous warm-air temperatures, including 37 °C, 38 °C, 50 °C, 60 °C, and 80 °C, were acknowledged in the present review for solvent evaporation. Most of the included articles assessed the bond strength of both SE and ER adhesive systems. Universal adhesives were found too; however, due their composition, these adhesives were considered as SE adhesives. Different ceramic primers were evaluated too; all of them had a silane coupling agent within its composition. All studies immediately evaluated the bond strength, but only two tested this property after some aging. Microtensile bond strength was the most commonly used test to evaluate bond strength; other tests such as shear and push-out were used too.

Table 2. Demographic and study design data of the included studies regarding direct substrates.

Study	Year	Material Tested and Category of the Material	Solvent Contained within the Material	Temperature of Warm Air Stream	Substrate Tested	Bond Strength Test
Al-Salamony [38]	2020	Optibond XTR (Kerr Co., Orange, CA, USA): Two step self-etch adhesive	Water, ethanol, and acetone	39 and 50 °C	Human dentin (Permanent)	Microtensile bond strength (μTBS)
		Optibond All-In-One (Kerr Co., Orange, CA, USA): One step self-etch adhesive	Water, ethanol, and acetone			
Carvalho [41]	2016	Adper Single Bond 2 (3M ESPE, St. Paul, MN, USA): Two-step total-etch adhesive system	Ethanol, water	50 °C	Human dentin (Permanent)	μTBS
Chen [42]	2021	Adhese Universal Vivapen (Ivoclar Vivadent, Schaan, Liechtenstein)	Ethanol, water	60 ± 2 °C	Human dentin (Permanent)	μTBS
		Gluma Bond Universal (Heraeus Kulzer, Hanau, Germany)	Acetone, water			
		All Bond Universal® (Bisco, Schaumburg, USA)	Ethanol, water			
		Single Bond Universal (3M ESPE, St. Paul, MN, USA)	Ethanol, water			
		Clearfil Universal Bond (Kuraray Noritake Dental Inc., Osaka, Japan)	Ethanol, water			
Garcia [47]	2009	Clearfil SE Bond (Kuraray Co. Inc., Osaka, Japan): Two-step self-etch	Ethanol, water	38 °C	Human dentin (Permanent)	μTBS
		Clearfil SE Protect (Kuraray Noritake Dental Inc.). Two-step self-etch	Water			
		Adper Prompt-L-Pop (3M ESPE, St. Paul, MN, USA): One-step self-etch	Water			
		Xeno III (Dentsply, Konstanz, Germany): Two-step self-etch	Ethanol, water			
Klein-Júnior [13]	2008	Adper Single Bond 2 (3M ESPE, St. Paul, MN, USA): Two-step etch-and-rinse	Ethanol, water	60 ± 2 °C	Human dentin (Permanent)	μTBS
		Prime & Bond 2.1 (Dentsply, Konstanz, Germany): Two-step etch-and-rinse	Acetone			
Marsiglio [21]	2012	Adper Scotchbond Multi-Purpose (3M ESPE, St. Paul, MN, USA): Three- step etch- and-rinse	Water	38 °C	Human dentin (Permanent)	μTBS
		Adper Single Bond 2 (3M ESPE, St. Paul, MN, USA): Two-step etch-and-rinse	Ethanol, water			
		Prime & Bond 2.1 (Dentsply, Mildford, Germany): Two-step etch-and-rinse	Acetone			
Moura [16]	2014	Adper SE Plus (3M ESPE; St. Paul, MN, USA): Two-step self-etch	Water	60 ± 2 °C	Human dentin (Permanent)	μTBS
		Clearfil 3S Bond (Kuraray Medical Inc, Tokyo, Japan): One-step self-etch	Water, ethanol			
		OptiBond All in-one (Kerr Co., Orange, CA, USA): One-step self-etch	Water, ethanol, and acetone			
		Silorane (3M ESPE; St. Paul, MN, USA): Two-step self-etch	Water, ethanol			

Table 2. *Cont.*

Study	Year	Material Tested and Category of the Material	Solvent Contained within the Material	Temperature of Warm Air Stream	Substrate Tested	Bond Strength Test
Ogura [20]	2012	Adper Easy Bond (3M ESPE, St. Paul, MN, USA): One-step self-etch	Ethanol, water	37 °C	Bovine dentin	Shear bond strength (SBS)
		Clearfil tri-S Bond (Kuraray Medical Inc., Tokyo, Japan): One-step self-etch	Ethanol, water			
		G-Bond Plus (GC Corp., Tokyo, Japan): One-step self-etch	Acetone, water			
Reis [12]	2010	Adper Single Bond 2 (3M ESPE, St. Paul, MN, USA): Two-step etch-and-rinse	Ethanol, water	60 ± 2 °C	Human dentin (Permanent)	μTBS
		Prime & Bond 2.1 (Dentsply, Mildford, Germany): Two-step etch-and-rinse	Acetone			
Riad [52]	2022	Adper Single Bond 2 (3M ESPE, St. Paul, MN, USA): Two-step etch-and-rinse	Ethanol, water	50 °C	Human dentin (Permanent)	SBS
		Single Bond Universal (3M ESPE, St. Paul, MN, USA): Universal adhesive applied in a self-etch mode	Ethanol, water			
Shiratsuchi [53]	2013	Bond Force (Tokuyama Dental, Tokyo, Japan): One-step self-etch	Isopropanol, water	37 °C	Bovine enamel	SBS
		Clearfil tri-S Bond (Kuraray Medical Inc., Tokyo, Japan): One-step self-etch	Ethanol, water			
		G-Bond Plus (GC Corp., Tokyo, Japan): One-step self-etch	Acetone, water			
Taguchi [14]	2018	Clearfil Bond SE ONE (Kuraray Noritake Dental Inc., Tokyo, Japan): One-step self-etch	Water, ethanol	80 ± 1 °C	Human dentin (Permanent)	μTBS
		Unifil Core EM Self-etch Bond (GC Corp., Tokyo, Japan): One-step self-etch	Water, ethanol			
		Estelink (Tokuyama Dental Corp., Tokyo, Japan): One-step self-etch	Water, ethanol, and acetone			
		Beauti Dual bond EX (Shofu Inc., Kyoto, Japan): One-step self-etch	Water, ethanol, and acetone			
Yonekura [55]	2020	Scotchbond Universal (3M ESPE, St. Paul, MN, USA): Universal adhesive used in a self-etch mode	Water, ethanol	60 ± 1 °C	Human dentin (Permanent)	μTBS
		Clearfil Bond SE ONE (Kuraray Noritake Dental Inc., Tokyo, Japan): One-step self-etch	Water, ethanol			
		Unifil Core EM Self-etch Bond (GC Corp. Tokyo, Japan): One-step self-etch	Water, acetone			
		Estelink (Tokuyama Dental Corp., Tokyo, Japan): One-step self-etch	Water, ethanol, and acetone			
Zimmer [56]	2022	Single Bond Universal (3M Oral Care, St. Paul, MN, USA): Universal adhesive system	Ethanol, water	50 °C	Human dentin (Permanent)	μTBS

The outcomes of the meta-analysis for ER adhesives are accessible in Figure 2. The overall analysis revealed that the use of warm air for solvent evaporation was statistically significantly higher ($p = 0.005$). This effect was observed only for the adhesives based in water/alcohol ($p = 0.03$), while for acetone-based adhesive systems, the use of warm air did not increase the bond strength ($p = 0.15$).

Table 3. Demographic and study design data of the included studies regarding indirect substrates.

Study	Year	Material Tested and Category of the Material	Solvent Contained within the Material	Temperature of Warm Air Stream	Substrate Tested	Bond Strength Test
Baratto [39]	2015	Silano (Dentsply, Santiago, Chile) Silane (DMG, Hamburg, Germany): Silanes	Ethanol	50 ± 5 °C, 80 °C	Heat-pressed lithium disilicate glass-ceramic discs (IPS e.max Press; Ivoclar Vivadent AG, Schaan, Liechtenstein)	SBS
Carvalho [40]	2015	Clearfil Ceramic Primer (Kuraray Medical Inc., Tokyo, Japan): Silane	Ethanol	50 ± 5 °C	Ceramic (Vacumat, Vita Zahnfabrik)	μTBS
Colares [43]	2013	RelyX Ceramic Primer (3M ESPE, St. Paul, MN, USA): Silane	Ethyl alcohol, water	45 ± 5 °C	Crystallized lithium-disilicate-based glass blocks (IPSe.max CAD; Ivoclar Vivadent)	μTBS
Cotes [44]	2013	RelyX Ceramic Primer (3M ESPE; St. Paul, MN, USA): Silane	Ethyl alcohol, water	50 ± 5 °C	Ceramic block (VITA Zahnfabrik; Bad Säckingen, Germany)	μTBS
Fabianelli [46]	2010	Monobond-S (Ivoclar-Vivadent, Schaan, Liechtenstein): Pre-hydrolyzed silane coupling agent	Ethanol	100 °C	Leucite-reinforced ceramic blocks IPS Empress (Ivoclar Vivadent, Schaan, Liechtenstein)	μTBS
Kim [48]	2013	Porcelain Liner M (Sun Medical, Moriyama City, Japan): Two-component silane coupling agent	Ethanol	38 °C	Glass-fiber post (FRC Postec Plus (Ivoclar Vivadent AG, Schaan, Liechtenstein; and D.T. Light Post, BISCO Inc., Schaumburg, IL, USA)	SBS
Melo-Silva [49]	2012	Silane (3M ESPE, St. Paul, MN, USA)	Ethanol	70 °C	Ceramic-type (Y- TZP and feldspar)	SBS
Monticelli [50]	2006	Monobond-S (Ivoclar-Vivadent, Schaan, Liechtenstein): Pre-hydrolyzed silane coupling agent	Ethanol	38 °C	Quartz fiber posts (DT Light Post #2, RTD, St.Egéve, France)	μTBS
		Porcelain Liner M (Sun Medical Co., Ltd., Japan): Two-component silane coupling agent	Ethanol			
		Porcelain Silane (BJM Lab, Or-Yenuda, Israel): Pre-hydrolyzed silane coupling agent	Ethanol			
Novais [51]	2012	Silano (Angelus, Petrópolis, RJ, Brazil)	Ethanol	60 °C	Glass fiber posts (Exacto; Angelus, Londrina, PR, Brazil; size 2, 17.0 mm long × 1.50 mm diameter)	Push-out testing
		Prosil (FGM, Joinville, SC, Brazil)	Ethanol, water			
		RelyX Ceramic Primer (3M ESPE, St. Paul, MN, USA)	Ethyl alcohol, water			
		Silane coupling agent (Dentsply, Petrópolis, RJ, Brazil): Silanes	Ethanol, acetic acid			
Ramón-Leonardo [24]	2020	RelyX Ceramic Primer (3M ESPE, St. Paul, MN, USA)	Ethyl alcohol, water	100 °C	Lithium-disilicate-r-inforced glass ceramic (e.max CAD Ivoclar Vivadent, Schaan, Liechtenstein)	μSBS
		Monobond N (Ivoclar Vivadent, Schaan, Liechtenstein): Silanes	Ethanol			
de Rosatto [45]	2014	Silano (Angelus, Petrópolis, RJ, Brazil)	Ethanol	60 °C	Fiberglass posts (Exacto, Angelus)	Push-out testing
		Prosil (FGM, Joinville, SC, Brazil)	Ethanol, water			
		RelyX Ceramic Primer (3M ESPE, St. Paul, MN, USA)	Ethyl alcohol, water			
		Silane coupling agent (Dentsply, Petrópolis, RJ, Brazil): Silanes	Ethanol, acetic acid			

Table 3. *Cont.*

Study	Year	Material Tested and Category of the Material	Solvent Contained within the Material	Temperature of Warm Air Stream	Substrate Tested	Bond Strength Test
Shen [29]	2004	Monobond-S (Ivoclar-Vivadent, Schaan, Liechtenstein): Pre-hydrolyzed silane coupling agent	Ethanol	45 ±5 °C	Eris (Ivoclar Vivadent, Schaan, Liechtenstein) IPS Empress (Ivoclar Vivadent)	μTBS
Silva [54]	2013	Monobond-S (Ivoclar Vivadent AG, Schaan, Linchtenstein): Pre-hydrolyzed silane coupling agent	Ethanol	100 °C	Monocrystalline alumina premolar brackets (Pure®, OrthoTechnology, Tampa, FL, USA)	SBS
Yanakiev [23]	2017	Monobond Plus (Ivoclar Vivadent, Schaan, Lichtenstein): Silane coupling agent	Ethanol	38 °C, 50 °C, 100 °C, 120 °C	EX-3 veneering ceramic (Kuraray Noritake Dental, Japan)	Tensile bond strength (TBS)

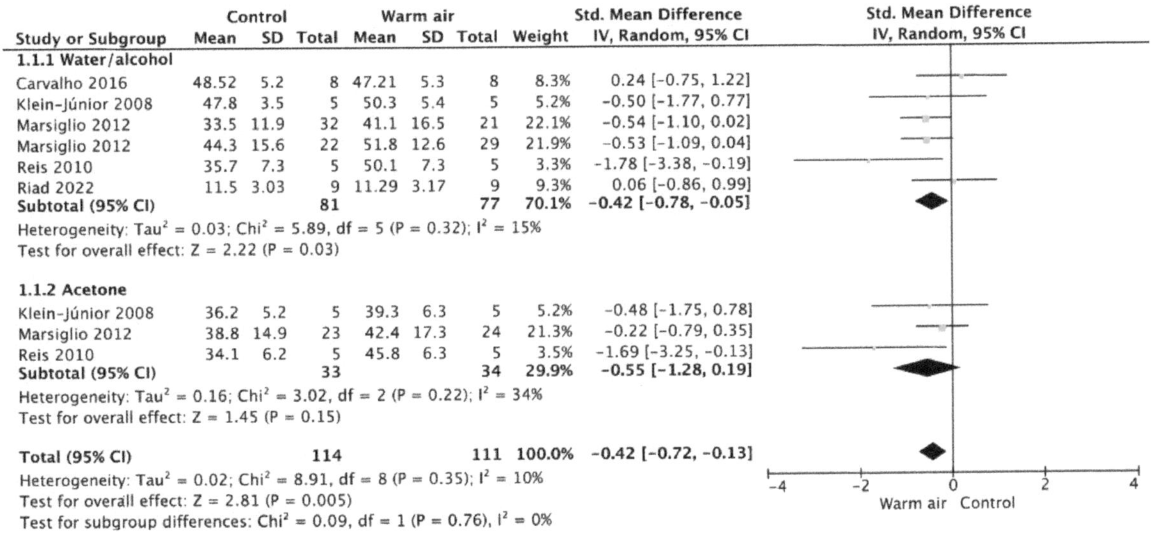

Figure 2. Forest plot showing the effect of warm air stream on the bond strength for etch-and-rinse adhesives.

For SE adhesives (Figure 3), the overall analysis privileged the use of warm air ($p < 0.001$). In the subgroup analysis, the use of a warm air stream allowed statistically significantly higher values for the water- and water/alcohol-based adhesive systems ($p = 0.03$, and $p < 0.0001$). This effect was not perceived for the acetone-based products ($p = 0.24$).

Finally, in Figure 4 the meta-analysis for the silane-based materials is shown. The overall analysis privileged the use of a warm air stream for solvent evaporation ($p = 0.001$). When analyzing the subgroups, it could be observed that this result was only valid when the silane was applied in glass ceramic materials ($p < 0.001$), while for glass fiber posts and oxide-based ceramics, the use of a warm air stream did not represent any advantage ($p = 0.49$, $p = 0.97$).

Table 4 shows the outcomes from the methodological quality evaluation. Most of the included articles recorded between a medium and low risk of bias. Some of the studies failed to report the specimen randomization, operator blinded, single operator, and sample size calculation factors.

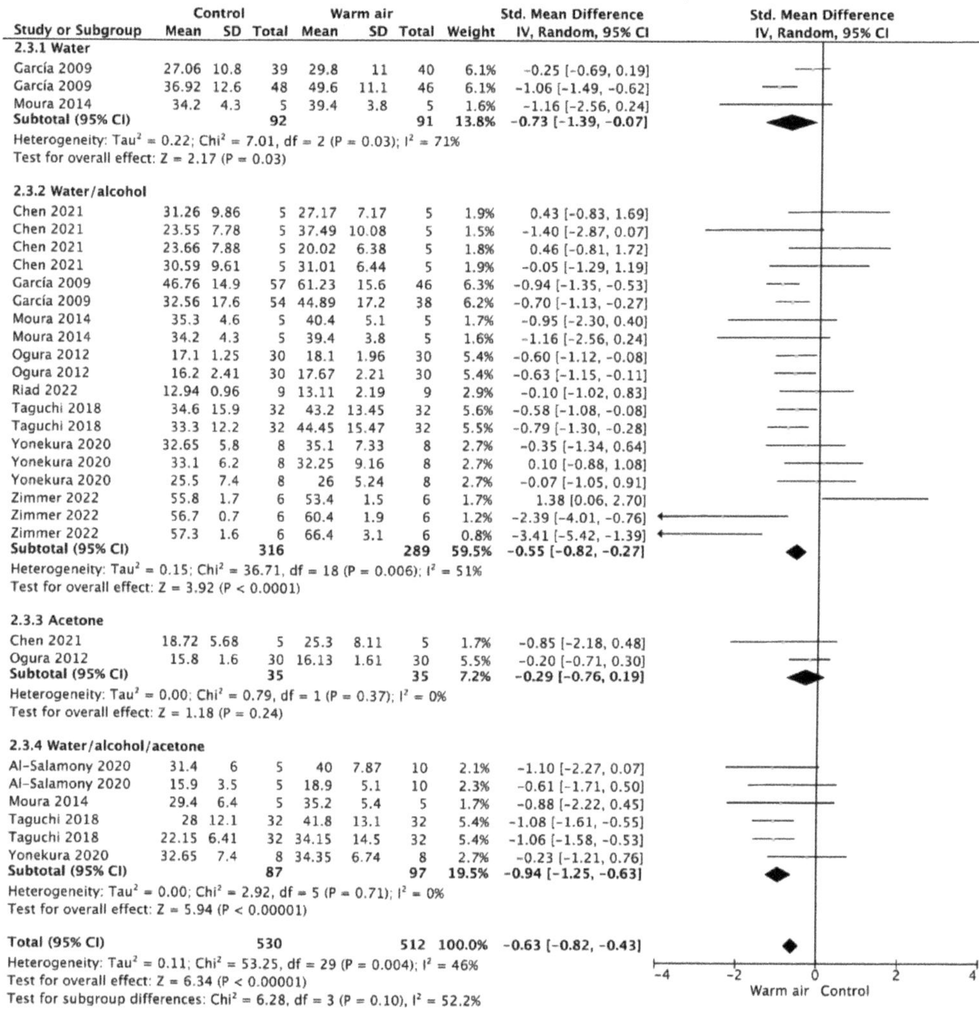

Figure 3. Forest plot showing the effect of warm air stream on the bond strength for self-etch adhesives.

Table 4. Methodological quality assessment.

Study	Specimen Randomization	Single Operator	Operator Blinded	Control Group	Standardized Specimens	Failure Mode	Manufacturer's Instructions	Sample Size Calculation	Coefficient of Variation	Risk of Bias
Al-Salamony [38]	Yes	Yes	No	Yes	Yes	No	Yes	No	Yes	Medium
Baratto [39]	No	No	No	Yes	Yes	Yes	Yes	No	Yes	Medium
Carvalho [40]	Yes	No	No	Yes	Yes	Yes	Yes	Yes	Yes	Low
Carvalho [41]	Yes	No	No	Yes	Yes	Yes	Yes	No	Yes	Medium
Chen [42]	Yes	No	No	Yes	Yes	Yes	Yes	No	Yes	Medium
Colares [43]	Yes	No	No	Yes	No	Yes	Yes	No	Yes	Medium
Cotes [44]	Yes	No	No	Yes	Yes	Yes	Yes	No	Yes	Medium
Fabianelli [46]	Yes	No	No	Yes	Yes	Yes	Yes	No	Yes	Medium
Garcia [47]	Yes	No	No	Yes	Yes	No	Yes	No	Yes	Medium
Klein-Júnior [13]	Yes	Yes	No	Yes	Yes	Yes	Yes	No	Yes	Low
Kim [48]	No	No	No	Yes	Yes	No	Yes	No	Yes	Medium
Marsiglio [21]	Yes	No	No	Yes	Yes	Yes	Yes	No	Yes	Medium
Melo-Silva [49]	No	No	No	Yes	Yes	Yes	Yes	Yes	Yes	Medium
Monticelli [50]	No	No	No	Yes	Yes	Yes	Yes	Yes	Yes	Medium
Moura [16]	No	Yes	No	Yes	Yes	Yes	Yes	No	Yes	Medium
Novais [51]	Yes	No	No	Yes	Yes	Yes	Yes	No	Yes	Medium

Table 4. *Cont.*

Study	Specimen Random-ization	Single Operator	Operator Blinded	Control Group	Standardized Specimens	Failure Mode	Manufacturer's Instructions	Sample Size Calculation	Coefficient of Variation	Risk of Bias
Ogura [20]	No	No	No	Yes	Yes	Yes	Yes	No	Yes	Medium
Ramón-Leonardo [24]	Yes	Yes	No	Yes	Yes	Yes	Yes	No	Yes	Low
Reis [12]	No	Yes	No	Yes	Yes	Yes	Yes	No	Yes	Medium
Riad [52]	Yes	No	No	Yes	Yes	Yes	Yes	Yes	Yes	Low
de Rosatto [45]	No	No	No	Yes	Yes	Yes	Yes	No	Yes	Medium
Shen [29]	No	No	No	Yes	Yes	Yes	Yes	No	Yes	Medium
Shiratsuchi [53]	Yes	No	No	Yes	No	Yes	Yes	No	Yes	Medium
Silva [54]	Yes	Yes	No	Yes	Yes	No	Yes	No	Yes	Medium
Taguchi [14]	No	No	No	Yes	Yes	Yes	Yes	No	Yes	Medium
Yanakiev [23]	No	No	No	Yes	Yes	No	Yes	No	Yes	Medium
Yonekura [55]	No	No	No	Yes	Yes	Yes	Yes	No	Yes	Medium
Zimmer [56]	No	No	No	Yes	Yes	Yes	Yes	No	Yes	Medium

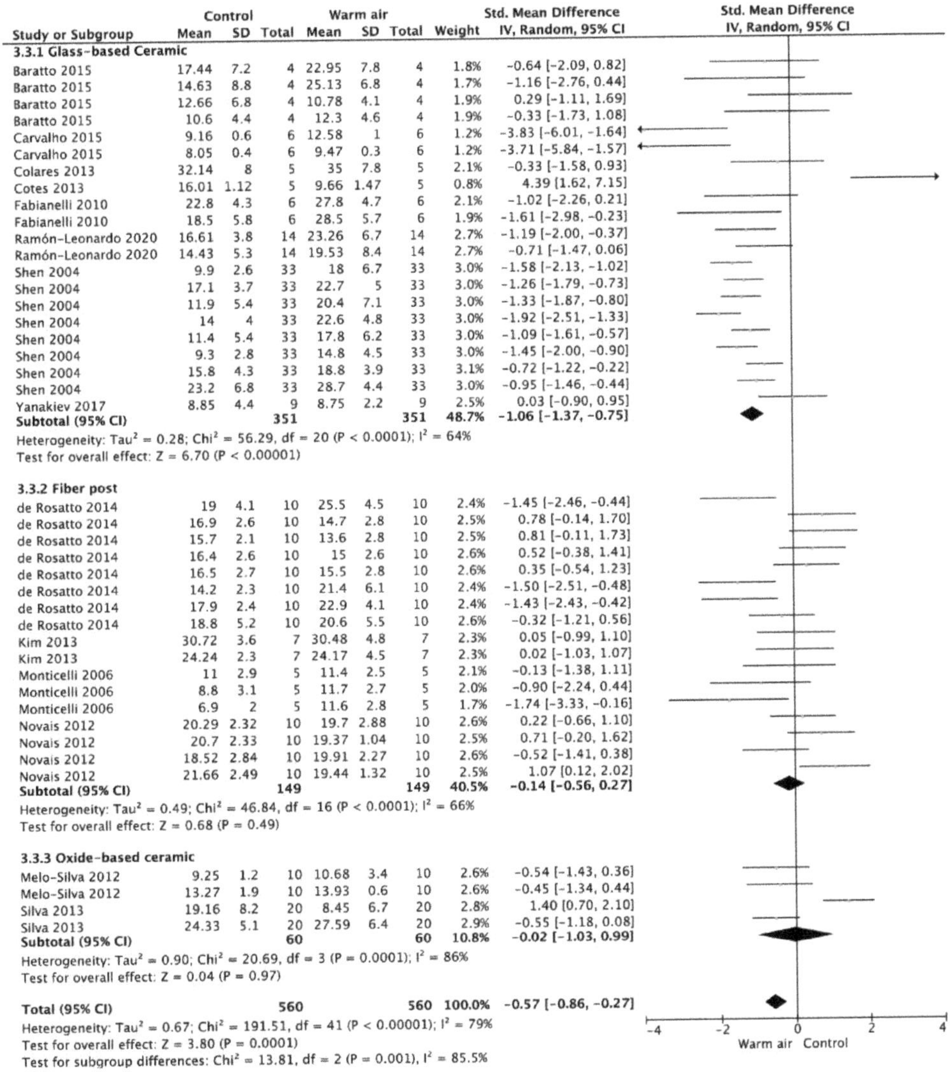

Figure 4. Forest plot showing the effect of warm air stream on the bond strength for silane-based materials.

4. Discussion

This article aimed to evaluate the effect of different warm-air temperatures used for solvent evaporation on the bond strength of resin-based materials to dental and nondental substrates. For ER or SE adhesives, and for silane as well, the use of a warm air stream for solvent evaporation enhanced the bond strength of direct and indirect restorations. This improvement is directly related to the adhesive system's base solvent and the tested substrate to which the resin-based materials adhere. The bond strength was enhanced when using a warm air stream for water/alcohol-based systems. On the other hand, the same result was not witnessed for the acetone-based adhesive. For glass-based ceramics, the bond strength was enhanced when a warm air stream was used. In contrast, this enhancement was not observed for fiber post and oxide-based ceramics. Based on the obtained results, the bond strength of the tested resin-based materials improved when using a warm air stream for solvent evaporation. Thus, the null hypothesis tested in this systematic review and meta-analysis was rejected.

Solvent evaporation is a very crucial step in any adhesive system application [13]. That said, solvents and water should be properly eliminated from the dental surface prior to light polymerization. To achieve this evaporation, researchers suggest a process known as air drying for dental adhesion. However, such a task is complicated since the density of the monomer rapidly rises as solvents and water are evaporating [57]. This limits the evaporation capacity and results in reduced water and solvent elimination [58]. However, the exact amount of unevaporated solvents is unspecified; yet, what is known is that this excess can lead to the deterioration of the hybrid layer, as well as a negative impact on the long-term bond performance [59,60]. Other complications include nanoleakage due to the pathways created, in addition to a reduced polymerization of resin monomers [61,62].

The ability of solvents to evaporate is reliant on vapor pressure (mmHg), which is defined as the value of the pressure needed for a liquid to transform into its gaseous state [63,64]. The vapor pressure of dissimilar solvents differs based on the value of the vapor pressure for each solvent, for example, the vapor pressure of acetone was higher (184 mm Hg at 20 °C) than that of water (17.5 mm Hg) and ethanol (43.9 mm Hg) [1,61,65]. It is worth noting that solvents can be bound to the collagen matrix, which reduces evaporation. This explains why ethanol-based bonding systems can retain a bigger number of organic/water mixtures than acetone-based bonding systems following evaporation [66]. Yet, assuming the temperature of air drying was the same for all the solvents, the boiling temperature can still affect the volatility of solvents; in this case, at a lower boiling temperature, ethanol-based adhesive systems will be easier to store than acetone-based adhesive systems [42,67].

Adhesive systems containing a mixture of water and alcohol showed an improvement in the bond strength after a stream of warm air was applied. Hydroxyl group interactions occurring between water and ethanol lead to higher boiling temperatures, which hinder the evaporation of the solvents [61]. A previous study deduced that the percentage of the unevaporated mixture is nearly 13% after a full minute of evaporation [61]. The excess solvents alter the bond strength and the degree of conversion of the adhesive, acting like a plasticizer [68]. Additionally, air drying at a higher temperature increases the kinetic energy of the molecules found inside the adhesive system, which promotes a stronger molecular vibration and alters the bonding process between the polar groups of the resin monomer and the solvent [69]. This in turn enhances solvent evaporation, facilitates polymerization, and strengthens the bond between the adhesive system and the dental substrate [70].

Aside from the air-drying temperature, prolonging the air-drying time also plays an important part in improving the evaporation of the solvents [60,71]. An insufficient air-drying time weakens the bond strength due to the presence of residual solvents such as water and ethanol, which inhibit monomer penetration and polymerization [72,73]. However, the concentration of the solvent does not have a direct impact on the degree of cure of the adhesive [74]. This shows that there should be a threshold level of residual solvents to encourage an improved conversion. Ethanol-based adhesives are a good example of the previous statement. Regarding the amount of solvent, an ideal percentage would be less than 10%. This percentage is not estimated as a remaining solvent after evaporation, but rather incorporated purposefully inside the adhesive bottle depending on the desired amount [47]. The assumption should be that the ideal amount required for the cure is less than that of the remaining solvents after evaporation [75]. All in all, warm air drying decreases the viscosity of the monomers and increases their temperature, which facilitates their diffusion into the dental substrate and leads to a stronger resin bond, justified by the findings of this meta-analysis.

Regarding acetone-based adhesives, warm air stream application did not enhance the bond strength. Acetone is known for its poor hydrogen bonding ability to water and to monomers inside the adhesive bottles [61,76,77]. This is due to its high volatility and vapor pressure [78]. Thus, acetone can be easily eliminated from the adhesives, especially after repeated usage and after storage, which alters its shelf-life [79,80]. Normally, adhesives containing acetone possess a lower monomer/solvent ratio, which dictates multiple layer applications for a better bond integrity [58]. Although, it should be noted that acetone evaporates 5 min after the application, but this is clinically unacceptable. Still, other components can also influence the evaporation time despite the solvent being a crucial factor [21]. Moreover, experimental acetone-based adhesives evaporated at the same rate after a room-temperature or 40 °C air stream was applied [69]. This might prove why the application of warm air did not enhance the bond strength of the acetone-based adhesives in a significant manner.

For the glass-based ceramics, the bond strength was improved when a warm air stream was used. On the contrary, this enhancement was not observed for the fiber post and oxide-based ceramics. The presence of a consistent bond is one of the main aspects of a successful all-ceramic restoration [81]. This requires the integration of all the different parts into one comprehensible and clear system [82]. Etching with hydrofluoric acid (HF) is the preferred process to condition the ceramic restoration surface [83], after which a silane coupling agent is applied to ensure a strong bond [81,84–86]. HF dissolves the glassy part of the ceramic network by generating surface pits [87]. The application of silane on the etched ceramic surface enhances the wettability and produces a covalent bond between the luting cement and the ceramic [88]. Both mechanisms can lead to micromechanical attachment as well as chemical bonding [44,89]. Silane can provide a mean for a chemical bond to occur at the silica-based ceramic surface by binding inorganic components to organic ones to achieve "coupling" [90]. Silanol is produced after the inorganic components of silane hydrolyze. This produces a metal hydroxide, also known as a siloxane bond, with the inorganic group. A covalent bond is produced when the organic component of silane causes a reaction with the organic material. Consequently, the organic and the inorganic materials will be bound to each other after heat treatment [91].

Heat treatment can evaporate alcohol and water as well as other products, while aiding with the silane–silica condensation reaction and the covalent bond formation [88,89,92]. This enhancement in the procedure of silane bonding can lead to a reliable and durable bond to the indirect substrate, making micromechanical retention unnecessary [89]. HF can be avoided due to reasons that include its high toxicity [93], its ability to create insoluble silica-fluoride salts which might be retained on the surface and weaken the bond strength [50,94], in addition to the fact that some ceramic structures do not require HF [95]. Despite that, the silane bond should be adequate to justify the absence of HF. This can be performed using heat treatment. After silane application, the restoration can be heat-treated for 2 min at 100 °C in an oven, allowing the removal of solvents along with other by-products on the ceramic surface, thus fulfilling the reaction of condensation between silane and silica. After that, the adhesion will be stronger and more effective due to the formation of many covalent bonds on the surface of the ceramic [46,96]. The same result can be obtained by the application of hot air at 50 ± 5 °C for 15 s [97]. A previous study attempted to determine which silane heat treatment process yields a better bond strength value. The researchers of that study found that even without etching, the heat treatment of silane containing a functional monomer called 10-Methacryloyloxydecyl dihydrogen phosphate (10-MDP) enhanced the bond strength between the ceramic and the luting agent, whether it was heated in an oven (100 °C for 2 min) or with hot air (50 ± 5 °C) [40]. Undergoing heat treatment at a temperature of 100 °C showed that when applying three layers of the silane on the ceramic surface, these multiple layers become consolidated into a monolayer, which increases the bond strength of the composite to the ceramic [88,92]. This heat treatment is also capable of evaporating the solvent and the volatile products formed after the condensation reaction of the silanol groups [88]. Another study carried out by Moniticelli et al. proved that warm air drying at a temperature of 38 °C increased the efficiency of the silane coupling agents when bonding ceramic to composite resin [50]. Additionally, a study concluded that, after treatment through surface roughening, the shear bond strength was significantly improved with heat treatment at a temperature of 100 °C for 60 s [88]. Likewise, Shen et al. discovered that silane drying with a stream of warm air (45 °C) improved the bond strength of glass ceramic [29]. Thus, the bond between the ceramic and the resin is significantly strengthened when a drying step is added to the application of silane at 100 °C [46]. Heat treatment at high temperatures (70–80 °C) for silane coupling agents might not be suitable for dental chair-side operations. Yet, a warm air stream (38 °C) can be tolerated to a better extent by the patient [29]. Warm air streams can aid in solvent evaporation while promoting the condensation reaction of silanol groups. As demonstrated in this meta-analysis, warm air drying may stimulate the formation of a chemical bond, not only in ceramic, but in the silane group as well [29]. Thus, the possession of a mini blow dryer may be viable.

When the tooth is endodontically treated and an essential part of the coronal tooth substance is lost, in addition to being prone to masticatory forces, a post is required inside the root canal to retain the end restoration [98]. Fiberglass has been used in recent years as a type of post since it can improve the performance of the restoration with a better stress distribution on the residual tooth when compared to ceramic or metallic posts [99]. However, a previous study concluded that the most common reason for the clinical failure of this restoration is the detachment of the fiberglass post from the root dentin [100].

A vital step in any clinical situation is to undergo silanization on the post surface after treating it chemically or mechanically [101]. The key factors influencing the efficiency include: the silane type, the pH, the solvent's content, the molecule of the silane, the molecule dimensions, and the mode of application such as drying settings, duration between the application of the silane and the adhesive resin, and the environmental humidity and temperature [102]. The rule is that the molecules known as organosilanes are bifunctional; one end is able to react with the inorganic part of the fiberglass and the other end can copolymerize with the resin which is organic [102].

The bond strength to fiberglass posts might be influenced by the composition of silane and the temperature used for air drying [88]. The temperature will act as a catalyst for the silane reaction, which accelerates the chemical interaction that occurs between the silane and the inorganic surface. Moreover, solvent evaporation plays a significant role in silane performance by encouraging silane wetting, but, an inadequate solvent evaporation may negatively affect the interaction with fiberglass [103]. Warm air thus aids with the aforementioned process [29]. Multiple opinions exist regarding the use of warm air. One study showed that the application of warm air on silane at a temperature of 38 °C improved the bond strength of the composite resin to the fiber post [50]. However, previously, it was argued that air drying at room temperature (23 °C) is more effective than at a high temperature (60 °C) when bonding resin cement to fiber-reinforced composite posts [48]. The main speculation is that despite the improvement that heating can have on solvent evaporation, the extent to which the evaporation is enhanced is possibly linked to the volatility of the specified solvent, such that the bond strength may be reduced after heating [51]. When the temperature is raised on the fiber post, the reaction may be finalized and the remaining solvent may be evaporated adequately, thus enhancing the bond strength of resin-based materials to indirect restoration, and which does not support the findings of this review [45].

For oxide-based ceramics, the use of warm air did not improve the bond strength of indirect restorations. Few articles have researched this topic [49,54]. The researchers involved in this field experimented on the effect of silane air-drying temperature (room temperature and 70 °C) on the adhesion of oxide-based ceramics (feldspar and yttrium-stabilized zirconia) and found that the increased air-drying temperature enhanced the bond resistance of zirconia ceramics, but not feldspar ceramics [49]. This is justified by the different structures of both materials tested. Different to ceramic, zirconia is comprised of small particles with the absence of a glassy stage which does not retain the resin bonding agents in a satisfactory way [104]. Alternative methods to enhance the adhesion of zirconia include aluminum oxide gushing and silica tribochemical coating [105]. The latest technique is warm air which did not match with the findings of this meta-analysis as no improvement in the bond strength was observed [49].

Finally, it should be highlighted that the use of a warm air stream for solvent evaporation can enhance the bond strength of adhesive systems and silane-based materials; thus, this review might identify that in a clinical situation, this variable should be explored, leading to long-term direct and indirect adhesive restorations. Scientists must be encouraged to design randomized clinical trials in order to explore if the use of warm air for restorative procedures is reliable.

5. Conclusions

Within the limitation of this systematic review of in vitro studies, it could be concluded that the use of a warm air stream for solvent evaporation enhanced the bonding performance of alcohol-/water-based adhesive systems when applied to dentin. According to the studies examined, the gold-standard temperature of a warm air stream should be around 50 and 60 °C when applied to dentin. In addition, this improvement seems to be similar when a silane coupling agent is submitted to heat treatment before the cementation of a glass-based ceramic. For indirect substrates, a temperature above 60 °C is recommended for the warm air stream.

Author Contributions: Conceptualization, L.H. and R.B.; methodology, L.H., R.B., Y.H. and C.E.C.-S.; software, L.H., R.B., Y.H. and C.E.C.-S.; validation, F.S., N.K. and D.M.; formal analysis, L.H., R.B., Y.H. and C.E.C.-S.; investigation, L.H., R.B., Y.H., N.K. and C.E.C.-S.; resources, L.H., R.B., Y.H., N.K., F.S. and C.E.C.-S.; data curation, L.H., R.B., Y.H., D.M., F.S. and C.E.C.-S.; writing—original draft preparation, L.H., R.B. and C.E.C.-S.; writing—review and editing, L.H., R.B., N.K., Y.H. and C.E.C.-S.; visualization, D.M.; supervision, L.H. and Y.H.; project administration, L.H. All authors have read and agreed to the published version of the manuscript.

Funding: This research received no external funding.

Data Availability Statement: The data presented in this study are available on reasonable request from the authors (R.B. and L.H.).

Acknowledgments: Authors Louis Hardan and Rim Bourgi would like to recognize the Saint-Joseph University of Beirut, Lebanon. Additionally, the authors would also recognize the University of Hidalgo State, Mexico and the University of Strasbourg for supporting this research.

Conflicts of Interest: The authors declare no conflict of interest.

References

1. Hardan, L.; Bourgi, R.; Cuevas-Suárez, C.E.; Zarow, M.; Kharouf, N.; Mancino, D.; Villares, C.F.; Skaba, D.; Lukomska-Szymanska, M. The Bond Strength and Anti-bacterial Activity of the Universal Dentin Bonding System: A Systematic Review and Meta-Analysis. *Microorganisms* **2021**, *9*, 1230. [CrossRef]
2. Sofan, E.; Sofan, A.; Palaia, G.; Tenore, G.; Romeo, U.; Migliau, G. Classification Review of Dental Adhesive Systems: From the IV Generation to the Universal Type. *Ann. Stomatol.* **2017**, *8*, 1–17.
3. Kameyama, A.; Haruyama, A.; Abo, H.; Kojima, M.; Nakazawa, Y.; Muramatsu, T. Influence of Solvent Evaporation on Ultimate Tensile Strength of Contemporary Dental Adhesives. *Appl. Adhes. Sci.* **2019**, *7*, 4. [CrossRef]

4. Ferreira, J.C.; Pires, P.T.; Azevedo, A.F.; Oliveira, S.A.; Melo, P.R.; Silva, M.J. Influence of Solvents and Composition of Etch-and-Rinse and Self-Etch Adhesive Systems on the Nanoleakage within the Hybrid Layer. *J. Contemp. Dent. Pract.* **2013**, *14*, 691. [CrossRef]
5. Hardan, L.; Bourgi, R.; Kharouf, N.; Mancino, D.; Zarow, M.; Jakubowicz, N.; Haikel, Y.; Cuevas-Suárez, C.E. Bond Strength of Universal Adhesives to Dentin: A Systematic Review and Meta-Analysis. *Polymers* **2021**, *13*, 814. [CrossRef] [PubMed]
6. El-Din, A. Effect of Changing Application Times on Adhesive Systems Bond Strengths. *Am. J. Dent.* **2002**, *15*, 321–324. [PubMed]
7. Hardan, L.; Bourgi, R.; Cuevas-Suárez, C.E.; Devoto, W.; Zarow, M.; Monteiro, P.; Jakubowicz, N.; Zoghbi, A.E.; Skaba, D.; Mancino, D.; et al. Effect of Different Application Modalities on the Bonding Performance of Adhesive Systems to Dentin: A Systematic Review and Meta-Analysis. *Cells* **2023**, *12*, 190. [CrossRef] [PubMed]
8. Reis, A.; Pellizzaro, A.; Dal-Bianco, K.; Gomes, O.; Patzlaff, R.; Loguercio, A.D. Impact of Adhesive Application to Wet and Dry Dentin on Long-Term Resin-Dentin Bond Strengths. *Oper. Dent.* **2007**, *32*, 380–387. [CrossRef]
9. Hardan, L.; Orsini, G.; Bourgi, R.; Cuevas-Suárez, C.E.; Nicastro, M.; Lazarescu, F.; Filtchev, D.; Cornejo-Ríos, E.; Zamarripa-Calderón, J.E.; Sokolowski, K. Effect of Active Bonding Application after Selective Dentin Etching on the Immediate and Long-Term Bond Strength of Two Universal Adhesives to Dentin. *Polymers* **2022**, *14*, 1129. [CrossRef]
10. Cadenaro, M.; Antoniolli, F.; Sauro, S.; Tay, F.R.; Di Lenarda, R.; Prati, C.; Biasotto, M.; Contardo, L.; Breschi, L. Degree of Conversion and Permeability of Dental Adhesives. *Eur. J. Oral Sci.* **2005**, *113*, 525–530. [CrossRef]
11. Carvalho, C.N.; Lanza, M.D.S.; Dourado, L.G.; Carvalho, E.M.; Bauer, J. Impact of Solvent Evaporation and Curing Protocol on Degree of Conversion of Etch-and-Rinse and Multimode Adhesives Systems. *Int. J. Dent.* **2019**, *2019*, 5496784. [CrossRef]
12. Reis, A.; Klein-Junior, C.A.; de Souza, F.C.; Stanislawczuk, R.; Loguercio, A.D. The Use of Warm Air Stream for Solvent Evaporation: Effects on the Durability of Resin-Dentin Bonds. *Oper. Dent.* **2010**, *35*, 29–36. [CrossRef]
13. Klein-Junior, C.A.; Zander-Grande, C.; Amaral, R.; Stanislawczuk, R.; Garcia, E.J.; Baumhardt-Neto, R.; Meier, M.M.; Loguercio, A.D.; Reis, A. Evaporating Solvents with a Warm Air-Stream: Effects on Adhesive Layer Properties and Resin–Dentin Bond Strengths. *J. Dent.* **2008**, *36*, 618–625. [CrossRef] [PubMed]
14. Taguchi, K.; Hosaka, K.; Ikeda, M.; Kishikawa, R.; Foxton, R.; Nakajima, M.; Tagami, J. The Effect of Warm Air-Blowing on the Microtensile Bond Strength of One-Step Self-Etch Adhesives to Root Canal Dentin. *J. Prosthodont. Res.* **2018**, *62*, 330–336. [CrossRef] [PubMed]
15. Trujillo, M.; Newman, S.M.; Stansbury, J.W. Use of Near-IR to Monitor the Influence of External Heating on Dental Composite Photopolymerization. *Dent. Mater.* **2004**, *20*, 766–777. [CrossRef]
16. Moura, S.K.; Murad, C.G.; Reis, A.; Klein-Júnior, C.A.; Grande, R.H.M.; Loguercio, A.D. The Influence of Air Temperature for Solvent Evaporation on Bonding of Self-Etch Adhesives to Dentin. *Eur. J. Dent.* **2014**, *8*, 205–210. [CrossRef]
17. Reis, A.; Klein-Júnior, C.A.; Accorinte, M.d.L.R.; Grande, R.H.M.; dos Santos, C.B.; Loguercio, A.D. Effects of Adhesive Temperature on the Early and 6-Month Dentin Bonding. *J. Dent.* **2009**, *37*, 791–798. [CrossRef]
18. Klein-Junior, C.A.; Sobieray, K.; Zimmer, R.; Portella, F.F.; Reston, E.G.; Marinowic, D.; Hosaka, K. Effect of Heat Treatment on Cytotoxicity and Polymerization of Universal Adhesives. *Dent. Mater. J.* **2020**, *39*, 970–975. [CrossRef]
19. Malekipour, M.R.; Shirani, F.; Ebrahimi, M. The Effect of Washing Water Temperature on Resin-Dentin Micro-Shear Bond Strength. *Dent. Res. J.* **2016**, *13*, 174.
20. Ogura, Y.; Shimizu, Y.; Shiratsuchi, K.; Tsujimoto, A.; Takamizawa, T.; Ando, S.; Miyazaki, M. Effect of Warm Air-Drying on Dentin Bond Strength of Single-Step Self-Etch Adhesives. *Dent. Mater. J.* **2012**, *31*, 507–513. [CrossRef]
21. Marsiglio, A.A.; Almeida, J.C.F.; Hilgert, L.A.; D'Alpino, P.H.P.; Garcia, F.C.P. Bonding to Dentin as a Function of Air-Stream Temperatures for Solvent Evaporation. *Braz. Oral Res.* **2012**, *26*, 280–287. [CrossRef] [PubMed]
22. Bourgi, R.; Hardan, L.; Rivera-Gonzaga, A.; Cuevas-Suárez, C.E. Effect of Warm-Air Stream for Solvent Evaporation on Bond Strength of Adhesive Systems: A Systematic Review and Meta-Analysis of in Vitro Studies. *Int. J. Adhes. Adhes.* **2021**, *105*, 102794. [CrossRef]
23. Yanakiev, S.; Yordanov, B.; Dikov, V. Influence of Silane Heat Treatment on the Tensile Bond Strength between EX-3 Synthetic Veneering Porcelain and Composite Resin Using Five Different Activation Temperatures. *J. IMAB—Annu. Proceeding Sci. Pap.* **2017**, *23*, 1456–1459. [CrossRef]
24. Ramón-Leonardo, P.-G.; Juan-Norberto, C.-R.; Wahjuningrum, D.A.; Tanzil, M.I.; Alberto-Carlos, C.-G. Effect of a Thermal Treatment of Two Silanes on the Bond Strength between a Lithium Disilicate and a Resin Cement. *J. Int. Dent. Med. Res.* **2020**, *13*, 868–872.
25. Monticelli, F.; Ferrari, M.; Toledano, M. Cement System and Surface Treatment Selection for Fiber Post Luting. *Med. Oral Patol. Oral Cir. Bucal* **2008**, *13*, 214.
26. Goracci, C.; Raffaelli, O.; Monticelli, F.; Balleri, B.; Bertelli, E.; Ferrari, M. The Adhesion between Prefabricated FRC Posts and Composite Resin Cores: Microtensile Bond Strength with and without Post-Silanization. *Dent. Mater.* **2005**, *21*, 437–444. [CrossRef]
27. Zakir, M.; Ashraf, U.; Tian, T.; Han, A.; Qiao, W.; Jin, X.; Zhang, M.; Tsoi, J.K.-H.; Matinlinna, J.P. The Role of Silane Coupling Agents and Universal Primers in Durable Adhesion to Dental Restorative Materials-a Review. *Curr. Oral Health Rep.* **2016**, *3*, 244–253. [CrossRef]
28. de Carvalho, R.F.; Martins, M.E.M.N.; de Queiroz, J.R.C.; Leite, F.P.P.; Oezcan, M. Influence of Silane Heat Treatment on Bond Strength of Resin Cement to a Feldspathic Ceramic. *Dent. Mater. J.* **2011**, *30*, 392–397. [CrossRef]

29. Shen, C.; Oh, W.; Williams, J.R. Effect of Post-Silanization Drying on the Bond Strength of Composite to Ceramic. *J. Prosthet. Dent.* **2004**, *91*, 453–458. [CrossRef]
30. Kaykhine, P.; Tichy, A.; Abdou, A.; Hosaka, K.; Foxton, R.M.; Sumi, Y.; Nakajima, M.; Tagami, J. Long-Term Evaluation of Warm-Air Treatment Effect on Adaptation of Silane-Containing Universal Adhesives to Lithium Disilicate Ceramic. *Dent. Mater. J.* **2021**, *40*, 379–384. [CrossRef]
31. Moreno, M.B.P.; Murillo-Gómez, F.; de Goes, M.F. Physicochemical and Morphological Characterization of a Glass Ceramic Treated with Different Ceramic Primers and Post-Silanization Protocols. *Dent. Mater.* **2019**, *35*, 1073–1081. [CrossRef] [PubMed]
32. Kay Khine, P.P.; Tichy, A.; Abdou, A.; Hosaka, K.; Sumi, Y.; Tagami, J.; Nakajima, M. Influence of Silane Pretreatment and Warm Air-Drying on Long-Term Composite Adaptation to Lithium Disilicate Ceramic. *Crystals* **2021**, *11*, 86. [CrossRef]
33. De Figueiredo, V.M.G.; Corazza, P.H.; Lepesqueur, L.S.S.; Miranda, G.M.; Pagani, C.; De Melo, R.M.; Valandro, L.F. Heat Treatment of Silanized Feldspathic Ceramic: Effect on the Bond Strength to Resin after Thermocycling. *Int. J. Adhes. Adhes.* **2015**, *63*, 96–101. [CrossRef]
34. Dal Piva, A.M.; Carvalho, R.L.; Lima, A.L.; Bottino, M.A.; Melo, R.M.; Valandro, L.F. Silica Coating Followed by Heat-treatment of MDP-primer for Resin Bond Stability to Yttria-stabilized Zirconia Polycrystals. *J. Biomed. Mater. Res. B Appl. Biomater.* **2019**, *107*, 104–111. [CrossRef]
35. Sutil, B.G.d.S.; Susin, A.H. Dentin Pretreatment and Adhesive Temperature as Affecting Factors on Bond Strength of a Universal Adhesive System. *J. Appl. Oral Sci.* **2017**, *25*, 533–540. [CrossRef]
36. Corazza, P.H.; Cavalcanti, S.C.; Queiroz, J.R.; Bottino, M.A.; Valandro, L.F. Effect of Post-Silanization Heat Treatments of Silanized Feldspathic Ceramic on Adhesion to Resin Cement. *J. Adhes. Dent.* **2013**, *15*, 473–479.
37. Papacchini, F. *A Study into the Materials and Techniques for Improving the Composite-Repair Bond*; University of Siena: Siena, Italy, 2006.
38. Al-Salamony, H.; Naguibb, E.A.; Hamzac, H.S.; Younisd, S.H. The Effect of Warm Air Solvent Evaporation on Microtensile Bond Strength of Two Different Self-Etch Adhesives to Dentin (In Vitro Study). *Sylwan* **2020**, *5*, 479–497.
39. Baratto, S.S.P.; Spina, D.R.F.; Gonzaga, C.C.; da Cunha, L.F.; Furuse, A.Y.; Baratto Filho, F.; Correr, G.M. Silanated Surface Treatment: Effects on the Bond Strength to Lithium Disilicate Glass-Ceramic. *Braz. Dent. J.* **2015**, *26*, 474–477. [CrossRef]
40. Carvalho, R.F.; Cotes, C.; Kimpara, E.T.; Leite, F.P.P. Heat Treatment of Pre-Hydrolyzed Silane Increases Adhesion of Phosphate Monomer-Based Resin Cement to Glass Ceramic. *Braz. Dent. J.* **2015**, *26*, 44–49. [CrossRef]
41. Carvalho, M.P.M.; Rocha, R.d.O.; Krejci, I.; Bortolotto, T.; Bisogno, F.E.; Susin, A.H. Influence of a Heating Device and Adhesive Temperature on Bond Strength of a Simplified Ethanol-Based Adhesive System. *Rev. Odontol. UNESP* **2016**, *45*, 97–102. [CrossRef]
42. Chen, Y.; Yan, X.; Li, K.; Zheng, S.; Sano, H.; Zhan, D.; Fu, J. Effect of Air-Blowing Temperature and Water Storage Time on the Bond Strength of Five Universal Adhesive Systems to Dentin. *Dent. Mater. J.* **2021**, *40*, 116–122. [CrossRef]
43. Colares, R.C.R.; Neri, J.R.; de Souza, A.M.B.; Pontes, K.M.d.F.; Mendonca, J.S.; Santiago, S.L. Effect of Surface Pretreatments on the Microtensile Bond Strength of Lithium-Disilicate Ceramic Repaired with Composite Resin. *Braz. Dent. J.* **2013**, *24*, 349–352. [CrossRef]
44. Cotes, C.; de Carvalho, R.F.; Kimpara, E.T.; Leite, F.P.; Ozcan, M. Can Heat Treatment Procedures of Pre-Hydrolyzed Silane Replace Hydrofluoric Acid in the Adhesion of Resin Cement to Feldspathic Ceramic. *J. Adhes. Dent.* **2013**, *15*, 569–574. [PubMed]
45. de Rosatto, C.M.P.; Roscoe, M.G.; Novais, V.R.; Menezes, M.d.S.; Soares, C.J. Effect of Silane Type and Air-Drying Temperature on Bonding Fiber Post to Composite Core and Resin Cement. *Braz. Dent. J.* **2014**, *25*, 217–224. [CrossRef]
46. Fabianelli, A.; Pollington, S.; Papacchini, F.; Goracci, C.; Cantoro, A.; Ferrari, M.; van Noort, R. The Effect of Different Surface Treatments on Bond Strength between Leucite Reinforced Feldspathic Ceramic and Composite Resin. *J. Dent.* **2010**, *38*, 39–43. [CrossRef] [PubMed]
47. Garcia, F.C.; Almeida, J.C.; Osorio, R.; Carvalho, R.M.; Toledano, M. Influence of Drying Time and Temperature on Bond Strength of Contemporary Adhesives to Dentine. *J. Dent.* **2009**, *37*, 315–320. [CrossRef]
48. Kim, H.-D.; Lee, J.-H.; Ahn, K.-M.; Kim, H.-S.; Cha, H.-S. Effect of Silane Activation on Shear Bond Strength of Fiber-Reinforced Composite Post to Resin Cement. *J. Adv. Prosthodont.* **2013**, *5*, 104–109. [CrossRef]
49. Melo-Silva, C.L.; de Carvalho, C.F.; Santos, C.; Lins, J.F.C. Evaluation of the Influence of the Silane Drying Temperature on the Feldspar and Zirconia-Based Ceramics Surfaces. *Trans Tech Publ.* **2012**, *2012*, 826–830. [CrossRef]
50. Monticelli, F.; Toledano, M.; Osorio, R.; Ferrari, M. Effect of Temperature on the Silane Coupling Agents When Bonding Core Resin to Quartz Fiber Posts. *Dent. Mater.* **2006**, *22*, 1024–1028. [CrossRef]
51. Novais, V.R.; Simamotos Júnior, P.C.; Rontani, R.M.P.; Correr-Sobrinho, L.; Soares, C.J. Bond Strength between Fiber Posts and Composite Resin Core: Influence of Temperature on Silane Coupling Agents. *Braz. Dent. J.* **2012**, *23*, 08–14. [CrossRef]
52. Riad, M.; Othman, H.I.; Mohamed, H.R. Influence of Diode Laser and Warm Air Drying on the Shear Bond Strength of Lithium Di-Silicate to Dentin. An in-Vitro Study. *Braz. Dent. Sci.* **2022**, *25*, e2782. [CrossRef]
53. Shiratsuchi, K.; Tsujimoto, A.; Takamizawa, T.; Furuichi, T.; Tsubota, K.; Kurokawa, H.; Miyazaki, M. Influence of Warm Air-drying on Enamel Bond Strength and Surface Free-energy of Self-etch Adhesives. *Eur. J. Oral Sci.* **2013**, *121*, 370–376. [CrossRef] [PubMed]
54. Silva, E.A.; Trindade, F.Z.; Reskalla, H.N.J.F.; de Queiroz, J.R.C. Heat Treatment Following Surface Silanization in Rebonded Tribochemical Silica-Coated Ceramic Brackets: Shear Bond Strength Analysis. *J. Appl. Oral Sci.* **2013**, *21*, 335–340. [CrossRef] [PubMed]

55. Yonekura, K.; Hosaka, K.; Tichy, A.; Taguchi, K.; Ikeda, M.; Thanatvarakorn, O.; Prasansuttiporn, T.; Nakajima, M.; Tagami, J. Air-Blowing Strategies for Improving the Microtensile Bond Strength of One-Step Self-Etch Adhesives to Root Canal Dentin. *Dent. Mater. J.* **2020**, *39*, 892–899. [CrossRef]
56. Zimmer, R.; Leite, M.; de Souza Costa, C.; Hebling, J.; Anovazzi, G.; Klein, C.; Hosaka, K.; Reston, E. Effect of Time and Temperature of Air Jet on the Mechanical and Biological Behavior of a Universal Adhesive System. *Oper. Dent.* **2022**, *47*, 87–96. [CrossRef]
57. Fu, J.; Pan, F.; Kakuda, S.; Sidhu, S.K.; Ikeda, T.; Nakaoki, Y.; Selimovic, D.; Sano, H. The Effect of Air-Blowing Duration on All-in-One Systems. *Dent. Mater. J.* **2012**, *31*, 1075–1081. [CrossRef] [PubMed]
58. Reis, A.F.; Oliveira, M.T.; Giannini, M.; De Goes, M.F.; Rueggeberg, F.A. The Effect of Organic Solvents on One-Bottle Adhesives' Bond Strength to Enamel and Dentin. *Oper. Dent.* **2003**, *28*, 700–706.
59. Hardan, L.; Lukomska-Szymanska, M.; Zarow, M.; Cuevas-Suárez, C.E.; Bourgi, R.; Jakubowicz, N.; Sokolowski, K.; D'Arcangelo, C. One-Year Clinical Aging of Low Stress Bulk-Fill Flowable Composite in Class II Restorations: A Case Report and Literature Review. *Coatings* **2021**, *11*, 504. [CrossRef]
60. Fu, J.; Saikaew, P.; Kawano, S.; Carvalho, R.M.; Hannig, M.; Sano, H.; Selimovic, D. Effect of Air-Blowing Duration on the Bond Strength of Current One-Step Adhesives to Dentin. *Dent. Mater.* **2017**, *33*, 895–903. [CrossRef]
61. Yiu, C.K.; Pashley, E.L.; Hiraishi, N.; King, N.M.; Goracci, C.; Ferrari, M.; Carvalho, R.M.; Pashley, D.H.; Tay, F.R. Solvent and Water Retention in Dental Adhesive Blends after Evaporation. *Biomaterials* **2005**, *26*, 6863–6872. [CrossRef]
62. Carvalho, R.M.; Manso, A.P.; Geraldeli, S.; Tay, F.R.; Pashley, D.H. Durability of Bonds and Clinical Success of Adhesive Restorations. *Dent. Mater.* **2012**, *28*, 72–86. [CrossRef]
63. Pashley, D.H.; Agee, K.; Nakajima, M.; Tay, F.; Carvalho, R.; Terada, R.; Harmon, F.; Lee, W.; Rueggeberg, F. Solvent-induced Dimensional Changes in EDTA-demineralized Dentin Matrix. *J. Biomed. Mater. Res. Off. J. Soc. Biomater. Jpn. Soc. Biomater. Aust. Soc. Biomater. Korean Soc. Biomater.* **2001**, *56*, 273–281. [CrossRef]
64. Pashley, D.H.; Carvalho, R.M.; Tay, F.R.; Agee, K.A.; Lee, K.W. Solvation of Dried Dentin Matrix by Water and Other Polar Solvents. *Am. J. Dent.* **2002**, *15*, 97–102.
65. Moszner, N.; Salz, U.; Zimmermann, J. Chemical Aspects of Self-Etching Enamel–Dentin Adhesives: A Systematic Review. *Dent. Mater.* **2005**, *21*, 895–910. [CrossRef]
66. Nihi, F.M.; Fabre, H.S.C.; Garcia, G.; Fernandes, K.B.P.; Ferreira, F.B.d.A.; Wang, L. In Vitro Assessment of Solvent Evaporation from Commercial Adhesive Systems Compared to Experimental Systems. *Braz. Dent. J.* **2009**, *20*, 396–402. [CrossRef]
67. Miyazaki, M.; Tsujimoto, A.; Tsubota, K.; Takamizawa, T.; Kurokawa, H.; Platt, J.A. Important Compositional Characteristics in the Clinical Use of Adhesive Systems. *J. Oral Sci.* **2014**, *56*, 1–9. [CrossRef] [PubMed]
68. Boeira, P.O.; Meereis, C.T.W.; Suárez, C.E.C.; de Almeida, S.M.; Piva, E.; da Silveira Lima, G. Coumarin-Based Iodonium Hexafluoroantimonate as an Alternative Photoinitiator for Experimental Dental Adhesives Resin. *Appl. Adhes. Sci.* **2017**, *5*, 2. [CrossRef]
69. Bail, M.; Malacarne-Zanon, J.; Silva, S.; Anauate-Netto, A.; Nascimento, F.; Amore, R.; Lewgoy, H.; Pashley, D.H.; Carrilho, M. Effect of Air-Drying on the Solvent Evaporation, Degree of Conversion and Water Sorption/Solubility of Dental Adhesive Models. *J. Mater. Sci. Mater. Med.* **2012**, *23*, 629–638. [CrossRef]
70. Daronch, M.; Rueggeberg, F.A.; De Goes, M.; Giudici, R. Polymerization Kinetics of Pre-Heated Composite. *J. Dent. Res.* **2006**, *85*, 38–43. [CrossRef] [PubMed]
71. Saikaew, P.; Fu, J.; Chowdhury, A.A.; Carvalho, R.M.; Sano, H. Effect of Air-Blowing Time and Long-Term Storage on Bond Strength of Universal Adhesives to Dentin. *Clin. Oral Investig.* **2019**, *23*, 2629–2635. [CrossRef] [PubMed]
72. Paul, S.; Leach, M.; Rueggeberg, F.; Pashley, D.H. Effect of Water Content on the Physical Properties of Model Dentine Primer and Bonding Resins. *J. Dent.* **1999**, *27*, 209–214. [CrossRef]
73. Miyazaki, M.; Platt, J.; Onose, H.; Moore, B. Influence of Dentin Primer Application Methods on Dentin Bond Strength. *Oper. Dent.* **1996**, *21*, 167–172. [PubMed]
74. Jacobsen, T.; Söderholm, K. Effect of Primer Solvent, Primer Agitation, and Dentin Dryness on Shear Bond Strength to Dentin. *Am. J. Dent.* **1998**, *11*, 225–228. [PubMed]
75. Cadenaro, M.; Breschi, L.; Antoniolli, F.; Navarra, C.O.; Mazzoni, A.; Tay, F.R.; Di Lenarda, R.; Pashley, D.H. Degree of Conversion of Resin Blends in Relation to Ethanol Content and Hydrophilicity. *Dent. Mater.* **2008**, *24*, 1194–1200. [CrossRef] [PubMed]
76. Amaral, R.; Stanislawczuk, R.; Zander-Grande, C.; Gagler, D.; Reis, A.; Loguercio, A.D. Bond Strength and Quality of the Hybrid Layer of One-Step Self-Etch Adhesives Applied with Agitation on Dentin. *Oper. Dent.* **2010**, *35*, 211–219. [CrossRef]
77. Irmak, Ö.; Baltacıoğlu, İ.H.; Ulusoy, N.; Bağış, Y.H. Solvent Type Influences Bond Strength to Air or Blot-Dried Dentin. *BMC Oral Health* **2016**, *16*, 77. [CrossRef]
78. Abate, P.; Rodriguez, V.; Macchi, R. Evaporation of Solvent in One-Bottle Adhesives. *J. Dent.* **2000**, *28*, 437–440. [CrossRef]
79. Iliev, G.; Hardan, L.; Kassis, C.; Bourgi, R.; Cuevas-Suárez, C.E.; Lukomska-Szymanska, M.; Mancino, D.; Haikel, Y.; Kharouf, N. Shelf Life and Storage Conditions of Universal Adhesives: A Literature Review. *Polymers* **2021**, *13*, 2708. [CrossRef]
80. Cuevas-Suárez, C.E.; Ramos, T.S.; Rodrigues, S.B.; Collares, F.M.; Zanchi, C.H.; Lund, R.G.; da Silva, A.F.; Piva, E. Impact of Shelf-Life Simulation on Bonding Performance of Universal Adhesive Systems. *Dent. Mater.* **2019**, *35*, e204–e219. [CrossRef]
81. Özcan, M.; Vallittu, P.K. Effect of Surface Conditioning Methods on the Bond Strength of Luting Cement to Ceramics. *Dent. Mater.* **2003**, *19*, 725–731. [CrossRef]

82. Davidson, C.L. Luting Cement, the Stronghold or the Weak Link in Ceramic Restorations? *Adv. Eng. Mater.* **2001**, *3*, 763–767. [CrossRef]
83. Canay, Ş.; Hersek, N.; Ertan, A. Effect of Different Acid Treatments on a Porcelain Surface 1. *J. Oral Rehabil.* **2001**, *28*, 95–101. [CrossRef] [PubMed]
84. Zarow, M.; Hardan, L.; Szczeklik, K.; Bourgi, R.; Cuevas-Suárez, C.E.; Jakubowicz, N.; Nicastro, M.; Devoto, W.; Dominiak, M.; Pytko-Polończyk, J.; et al. Porcelain Veneers in Vital vs. Non-Vital Teeth: A Retrospective Clinical Evaluation. *Bioengineering* **2023**, *10*, 168. [CrossRef] [PubMed]
85. Harouny, R.; Hardan, L.; Harouny, E.; Kassis, C.; Bourgi, R.; Lukomska-Szymanska, M.; Kharouf, N.; Ball, V.; Khairallah, C. Adhesion of Resin to Lithium Disilicate with Different Surface Treatments before and after Salivary Contamination—An In-Vitro Study. *Bioengineering* **2022**, *9*, 286. [CrossRef]
86. Hardan, L.; Devoto, W.; Bourgi, R.; Cuevas-Suárez, C.E.; Lukomska-Szymanska, M.; Fernández-Barrera, M.á.; CornejoRíos, E.; Monteiro, P.; Zarow, M.; Jakubowicz, N.; et al. Immediate Dentin Sealing for Adhesive Cementation of Indirect Restorations: A Systematic Review and Meta-Analysis. *Gels* **2022**, *8*, 175. [CrossRef] [PubMed]
87. Al Edris, A.; Al Jabr, A.; Cooley, R.L.; Barghi, N. SEM Evaluation of Etch Patterns by Three Etchants on Three Porcelains. *J. Prosthet. Dent.* **1990**, *64*, 734–739. [CrossRef]
88. Roulet, J.; Söderholm, K.; Longmate, J. Effects of Treatment and Storage Conditions on Ceramic/Composite Bond Strength. *J. Dent. Res.* **1995**, *74*, 381–387. [CrossRef]
89. Hooshmand, T.; van Noort, R.; Keshvad, A. Bond Durability of the Resin-Bonded and Silane Treated Ceramic Surface. *Dent. Mater.* **2002**, *18*, 179–188. [CrossRef]
90. Bertolotti, R.L. Adhesion to Porcelain and Metal. *Dent. Clin. N. Am.* **2007**, *51*, 433–451. [CrossRef]
91. Moon, J.H.; Shul, Y.G.; Hong, S.Y.; Choi, Y.S.; Kim, H.T. A Study on UV-Curable Adhesives for Optical Pick-Up: II. Silane Coupling Agent Effect. *Int. J. Adhes. Adhes.* **2005**, *25*, 534–542. [CrossRef]
92. Barghi, N. To Silanate or Not to Silanate: Making a Clinical Decision. *Compend. Contin. Educ. Dent.* **2000**, *21*, 659–662.
93. Bertolini, J.C. Hydrofluoric Acid: A Review of Toxicity. *J. Emerg. Med.* **1992**, *10*, 163–168. [CrossRef]
94. Shimada, Y.; Yamaguchi, S.; Tagami, J. Micro-Shear Bond Strength of Dual-Cured Resin Cement to Glass Ceramics. *Dent. Mater.* **2002**, *18*, 380–388. [CrossRef]
95. Valandro, L.F.; Della Bona, A.; Bottino, M.A.; Neisser, M.P. The Effect of Ceramic Surface Treatment on Bonding to Densely Sintered Alumina Ceramic. *J. Prosthet. Dent.* **2005**, *93*, 253–259. [CrossRef] [PubMed]
96. Moharamzadeh, K.; Hooshmand, T.; Keshvad, A.; Van Noort, R. Fracture Toughness of a Ceramic–Resin Interface. *Dent. Mater.* **2008**, *24*, 172–177. [CrossRef]
97. Abduljabbar, T.; AlQahtani, M.A.; Al Jeaidi, Z.; Vohra, F. Influence of silane and heated silane on the bond strength of lithium disilicate ceramics-An in vitro study. *Pak. J. Med. Sci.* **2016**, *32*, 550–554. [CrossRef] [PubMed]
98. Schwartz, R.S.; Robbins, J.W. Post Placement and Restoration of Endodontically Treated Teeth: A Literature Review. *J. Endod.* **2004**, *30*, 289–301. [CrossRef] [PubMed]
99. Grandini, S.; Goracci, C.; Tay, F.R.; Grandini, R.; Ferrari, M. Clinical Evaluation of the Use of Fiber Posts and Direct Resin Restorations for Endodontically Treated Teeth. *Int. J. Prosthodont.* **2005**, *18*, 399–404.
100. Cagidiaco, M.C.; Goracci, C.; Garcia-Godoy, F.; Ferrari, M. Clinical Studies of Fiber Posts: A Literature Review. *Int. J. Prosthodont.* **2008**, *21*, 328–336.
101. Vano, M.; Goracci, C.; Monticelli, F.; Tognini, F.; Gabriele, M.; Tay, F.; Ferrari, M. The Adhesion between Fibre Posts and Composite Resin Cores: The Evaluation of Microtensile Bond Strength Following Various Surface Chemical Treatments to Posts. *Int. Endod. J.* **2006**, *39*, 31–39. [CrossRef]
102. Abel, M.-L.; Allington, R.; Digby, R.; Porritt, N.; Shaw, S.; Watts, J. Understanding the Relationship between Silane Application Conditions, Bond Durability and Locus of Failure. *Int. J. Adhes. Adhes.* **2006**, *26*, 2–15. [CrossRef]
103. Madruga, E.L.; Fernández-García, M. A Kinetic Study of Free-radical Copolymerization of Butyl Acrylate with Methyl Methacrylate in Solution. *Macromol. Chem. Phys.* **1996**, *197*, 3743–3755. [CrossRef]
104. Luthardt, R.; Holzhüter, M.; Sandkuhl, O.; Herold, V.; Schnapp, J.; Kuhlisch, E.; Walter, M. Reliability and Properties of Ground Y-TZP-Zirconia Ceramics. *J. Dent. Res.* **2002**, *81*, 487–491. [CrossRef] [PubMed]
105. Cavalcanti, A.N.; Pilecki, P.; Foxton, R.M.; Watson, T.F.; Oliveira, M.T.; Gianinni, M.; Marchi, G.M. Evaluation of the Surface Roughness and Morphologic Features of Y-TZP Ceramics after Different Surface Treatments. *Photomed. Laser Surg.* **2009**, *27*, 473–479. [CrossRef] [PubMed]

Journal of
Functional
Biomaterials

MDPI

Article

Biomimetic Liquid Crystal-Modified Mesoporous Silica—Based Composite Hydrogel for Soft Tissue Repair

Xiaoling Li [1,†], Lei Wan [1,†], Taifu Zhu [1], Ruiqi Li [1], Mu Zhang [1] and Haibin Lu [1,2,*]

1 Stomatological Hospital, School of Stomatology, Southern Medical University, Guangzhou 510280, China
2 The Fifth Affiliated Hospital, Southern Medical University, Guangzhou 510900, China
* Correspondence: luhaibin007@smu.edu.cn
† These authors contributed equally to this work.

Abstract: The reconstruction of blood vessels plays a critical role in the tissue regeneration process. However, existing wound dressings in tissue engineering face challenges due to inadequate revascularization induction and a lack of vascular structure. In this study, we report the modification of mesoporous silica nanospheres (MSNs) with liquid crystal (LC) to enhance bioactivity and biocompatibility in vitro. This LC modification facilitated crucial cellular processes such as the proliferation, migration, spreading, and expression of angiogenesis—related genes and proteins in human umbilical vein endothelial cells (HUVECs). Furthermore, we incorporated LC—modified MSN within a hydrogel matrix to create a multifunctional dressing that combines the biological benefits of LC—MSN with the mechanical advantages of a hydrogel. Upon application to full—thickness wounds, these composite hydrogels exhibited accelerated healing, evidenced by enhanced granulation tissue formation, increased collagen deposition, and improved vascular development. Our findings suggest that the LC—MSN hydrogel formulation holds significant promise for the repair and regeneration of soft tissues.

Keywords: liquid crystal; mesoporous silica nanosphere; hydrogel; angiogenesis; tissue repair

Citation: Li, X.; Wan, L.; Zhu, T.; Li, R.; Zhang, M.; Lu, H. Biomimetic Liquid Crystal-Modified Mesoporous Silica—Based Composite Hydrogel for Soft Tissue Repair. *J. Funct. Biomater.* **2023**, *14*, 316. https://doi.org/10.3390/jfb14060316

Academic Editors: Evangelos Manias and Jinwoo Lee

Received: 14 April 2023
Revised: 22 May 2023
Accepted: 5 June 2023
Published: 8 June 2023

1. Introduction

The oral and maxillofacial region plays a pivotal role in both functional and aesthetic aspects, housing numerous vital organs and structures that serve as a "window" to society [1]. However, the maxillofacial skin, being highly exposed, is susceptible to damage from surgical procedures, accidents, and burns [2]. While natural repair processes are effective for superficial wounds, extensive, full—thickness wound healing poses significant clinical challenges, often leading to severe infections and endangering lives [3]. Therefore, there is a pressing need for a multifunctional wound dressing. Ideally, such a dressing should (1) absorb wound exudate, (2) provide protection against infection and mechanical damage, (3) exhibit excellent tissue compatibility, and, most importantly, (4) accelerate vascularization and tissue regeneration [4–6].

Liquid crystals (LCs) are intermediate phases of matter, also known as a mesogens, which possess microstructures between isotropic liquid and crystalline solid states [7,8]. LC phases are observed in various biological compounds, including polypeptides, lipids, nucleic acids, and polysaccharides within organisms [9]. The conformation of cell membranes closely resembles the LC state, suggesting its significance in normal physiological activities [10]. Conversely, deviations from the LC state contribute to diseases such as atherosclerosis and gallstones [11,12]. Leveraging their biomimetic properties, different LC elastomers have shown the ability to promote cell attachment, proliferation, and alignment [9,13]. In the context of wound dressing, impaired angiogenesis poses a critical limitation to soft tissue regeneration, as it hampers nutrient supply and waste removal [14,15]. While exogenous protein growth factors and therapeutic stem cells have been explored for promoting angiogenesis, challenges regarding cost and vascular endothelial growth factor (VEGF) stability in proteolytic environments persist [16,17]. Recent studies have suggested the potential of LC in promoting angiogenesis, with earlier reports demonstrating LC−induced ring formation in endothelial cells [16]. In our latest study, a multifunctional hydrogel containing LC was successfully devised which exhibited satisfactory neovascularization via providing an anisotropic viscoelastic microenvironment [18]. However, the key challenge lies in effectively delivering LC to tissues, as the direct use of polymer LCs is impractical and challenging to mold.

Mesoporous silica nanoparticles (MSNs) have garnered significant attention in the field of biomedicine due to their unique physicochemical characteristics, including a uniform pore size distribution, large specific surface area, high pore volume, tunable pore size, and favorable biocompatibility [19–22]. Over the past few decades, MSN has been extensively explored as a versatile drug delivery system for various therapeutic agents in the treatment of diseases such as cancer, diabetes, inflammation, and tissue disorders [23–25]. In particular, MSNs have shown promise as a carrier for osteogenesis/angiogenesis−stimulating drugs, cytokines, and diverse nanoparticles to facilitate bone tissue repair [26–29]. Thus, MSN emerges as a prospective bioactive material for tissue engineering.

Hydrogels are of great interest in the field of wound dressings due to their unique properties. With a typical three−dimensional cross−linked structure and high water content, they can create a moist environment that softens the wound surface and provides a physical barrier [30,31]. The incorporation of various functional additives, such as antibacterial agents, growth factors, nanoparticles, and magnetically responsive biomaterials, can produce binary or ternary composite multifunctional hydrogels that enhance the mechanical and biological functions of hydrogels. For instance, researchers have demonstrated that polyacrylamide−chitosan composite hydrogels can significantly enhance tensile strength and toughness and can effectively accelerate skin repair. Despite these promising results, the antibacterial performance and biocompatibility of hydrogel dressings remain suboptimal, limiting their application [32]. Further studies are needed to optimize the biocompatibility and stability of ternary composite hydrogels to ensure their safety and efficacy. Recently, some researchers have used molecular dynamics to optimize drug delivery, enabling hydrogels to achieve better physical properties and more precise drug release [33,34]. Among the naturally derived hydrogels, polysaccharides such as hyaluronic acid and chitosan (CS) are commonly employed. CS in particular is highly favored due to its good biocompatibility and degradability as well as its antibacterial and anti−inflammatory properties, which can accelerate the wound healing process. The combination of chitosan hydrogels with excellent physical properties and biomaterials with angiogenic properties holds significant promise for promoting tissue regeneration.

In this study, our primary objective was to propose a convenient and effective method for enhancing the angiogenic activity of MSN through the bioactive modification of LC. To achieve this, MSN and cholesteric LC were separately prepared and subsequently combined through ontology polymerization, resulting in a novel bionic composite material (Scheme 1). Systematic characterization was conducted to optimize the morphology and relevant physical properties of both the LC and the LC−MSN composite system. In addition, the cell affinity and angiogenic capacity of the LC−MSN material were investigated for human umbilical vein endothelial cells (HUVECs). Based on the great cytocompatibility and exciting angiogenesis results in vitro, LC−MSN was processed with CS into hydrogel form for wound healing. The present study aims to develop a novel dressing that induces angiogenesis and promotes tissue regeneration in full−thickness skin wound models.

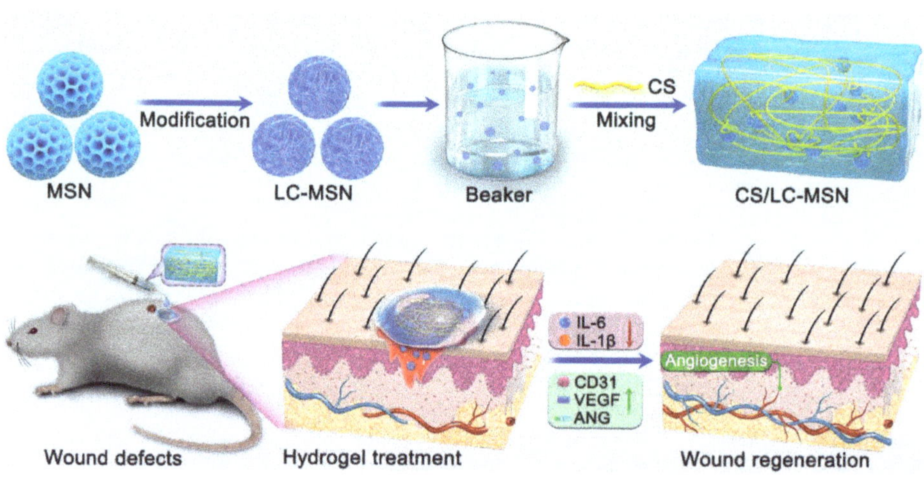

Scheme 1. The fabrication of CS/LC−MSN composite scaffolds by simply mixing of CS with LC−MSN. The CS/LC−MSN gel could be facilely administrated via injection to treat wound defects.

2. Materials and Methods

2.1. Chemicals and Materials

Cholesterol (degree of deacetylation ≈ 95%) was sourced from Aladdin Biochemical Technology Co., Ltd. (Shanghai, China). Undecylenic acid, chloroplatinic acid ($H_2PtCl_6{\cdot}6H_2O$), and polymethylhydrogensiloxane (PMHS) were obtained from Shanghai Macklin Biochemical Co. (Shanghai, China). Cetyltrimethylammonium bromide (CTAB), tetraethoxysilane (TEOS), azobisisobutyronitrile (AIBN), methyl acetoacetate (MAA), ethylene dimethacrylate (EDMA), and all other analytical reagents used in the experiments were procured from Tianjin Damao Chemical Reagent Factory.

2.2. Synthesis of LC

The detailed experimental method for LC synthesis has been previously described in the literature [8]. In brief, cholesterol and undecylenic acid underwent a simple esterification reaction to produce the undecylcholesteryl ester monomer LC. Subsequently, the side−chain polymer LC (P−UChol) was obtained through hydrosilylation of the aforementioned monomer with PMHS in toluene.

2.3. Synthesis of MSN and LC−Modified MSN (LC−MSN)

The classical Stoeber strategy [35] was employed for the synthesis of MSN. Ammonium hydroxide and CTAB were added to distilled water and stirred at 40 °C for 1 h. Next, TEOS dissolved in ethanol was added to the above aqueous solution and stirred at 40 °C for 5 h. In this study, the complexes based on liquid crystals and MSN were prepared using the ontology polymerization method. Initially, MSN, LC, MPDE, and AIBN were weighed in a synthesis vial, followed by the addition of acetonitrile and toluene, which were dissolved through ultrasound. Subsequently, MAA and EDMA were added and further sonicated for mixing. The mixture was then subjected to nitrogen gas to remove dissolved oxygen and reacted in a water bath at 53 °C for 4 h. After completion of the reaction, the synthesized materials were eluted with methanol/acetic acid (v/v, 9/1) using a Soxhlet extractor for 72 h until no template molecules or unreacted monomers were detected by UV spectrophotometry. The remaining unreacted substances were washed off with methanol, and the compound was dried under vacuum condition. Subsequently, a systematic examination of the characterization and physicochemical properties of LC/LC−MSN was conducted as follows.

The chemical structure of composites was analyzed by attenuated total reflectance–Fourier transformation infrared (ATR–FTIR, Bruker Tensor27, Karlsruhe, Germany). The samples were mixed with potassium bromide then pressed into discs and measured in the wavelength range 4000–400 cm^{-1}.

The intermolecular and intramolecular interactions were recorded with an NMR spectrometer (Bruker, AVANCEIII400MHz, Karlsruhe, Germany). The samples were fully dissolved in deuterated chloroform and transferred to the NMR tube for scanning.

The morphology and structure of MSN and LC–MSN were characterized by transmission electron microscopy (TEM, Philips CM–120, Amsterdam, The Netherlands) at an accelerating voltage of 3 kV on a copper platform under vacuum condition.

The surface area and pore size were evaluated by the automated surface area and pore size analyzer (Quantachrome Autosorb–iQ–2, Boynton Beach, FL, USA). The surface areas were obtained by the Brunauer–Emmet–Teller (BET) method, and the pore size distributions were estimated by the Barrett–Joyner–Halenda (BJH) method. The hydrodynamic size of the samples was determined using a dynamic light scattering analyzer (DLS, SZ–100–Z, Horiba Ltd., Kyoto, Japan). The surface charges of MSN and LC–MSN in distilled water were detected by Zeta–potential (Malvern Panalytical Zetasizer Nano ZS90, Worcester, UK). The sample suspension was ultrasonically dispersed well prior to analysis. The clean point and the liquid crystal phase transition temperature of LC–MSN were measured by differential scanning calorimetry (DSC, Netzsch STA449F5, Selb, Germany) at a temperature range of 0–200 °C with a heating rate of 10 °C min^{-1} under nitrogen protection.

The texture of optical anisotropic LC–MSN composites was viewed and recorded with polarizing optical microscopy (POM, Wetzlar Leica DMRX, Wetzlar, Germany) with a heating stage. A small amount of the sample was placed on a slide, pressed firmly with a cover glass, and observed on a hot platform at room temperature.

2.4. *Fabrication and Characterization of CS/LC–MSN Composite Hydrogel*

CS was dispersed in HCl for amino protonation, after which 1, 2–propylene glycol in an amount equal to HCl was added and stirred for 1 day to obtain a homogeneous 2 wt% CS solution. The resultant LC–MSN (0.25, 0.50, 0.75, or 1.0%) was dissolved with alcohol and then dispersed into the aforementioned CS solution by equal volume under rapid stirring. The solution was poured into a mold followed by dehydration at 50 °C to obtain CS/LC–MSN alcogels. These alcogels were then immersed in NaOH for 1 day at a concentration of 1 mol/L and then washed thoroughly with phosphate buffer saline (PBS) for purification. The XRD patterns of the lyophilized CS and CS/LC–MSN specimens were acquired by using a Bruker D8–Advance powder diffract meter (Karlsruhe, Germany) equipped with Cu radiation. X–ray photoelectron spectroscopy (XPS) spectra were obtained using a Thermo Scientific ESCALAB 250 XI. The microstructures of the composite hydrogel samples were observed via scanning electron microscopy (SEM, LEO1530 VP, Philips, Amsterdam, The Netherlands). The compressive test of hydrogel samples was conducted with a universal mechanical test machine (AGI–1, Shimadzu, Kyoto, Japan).

To investigate biodegradation, the CS/LC–MSN hydrogel samples were cut into discs of approximately equal mass (500 ± 10 mg) and incubated in simulated body fluid (SBF) at 37 °C. At 1, 3, 7, 10, and 14 days, the accumulated silica ion concentrations released from composites wre detected using the inductively coupled plasma optical emission spectrometer (ICP–AES, iCAP 6300Duo, Thermo Fisher Scientific, Waltham, MA, USA). Subsequently, the discs were removed, gently washed with deionized water, dried, and observed for morphological changes using TEM during the degradation process.

2.5. *Cellular Behavior on LC–MSN In Vitro*

2.5.1. Culture of the Human Umbilical Vein Endothelial Cells

Human umbilical vein endothelial cells (HUVECs; Cat. No. iCell–h110, iCell, Shanghai, China) were utilized to assess biocompatibility, cell migration, and angiogenesis capacity. The HUVECs were cultured in endothelial cell medium (ECM; Cat. No. 1001,

Sciencell, Carlsbad, CA, USA) without glutamine (ECM−NG), supplemented with 5 vol.% FBS and 1 vol.% endothelial cell growth supplements (ECGs). Cells at passages 3–5 were employed for the cell experiments, and they were maintained in a humidified incubator at 37 °C with 5% CO_2.

2.5.2. Cell Proliferation Assay

The biocompatibility of LC−MSN was evaluated using a Cell Counting Kit−8 (CCK−8, Dojindo, Japan) assay. HUVECs were seeded in 96−well plates (Jet Biofil, China) at a density of 5000 cells per well. After overnight incubation, the ECM culture medium was replaced with various dilutions of CS/LC−MSN (LC−MSN concentrations: 0, 4.7, 9.4, 18.8, 37.5, 75, 150 μg/mL). The cells were further cultured for 12, 24, and 48 h, respectively. At each time point, the residual medium was removed, and the cells were washed twice with PBS. A CCK−8 working solution, prepared by mixing the CCK−8 solution and fresh medium at a volume ratio of 1:10, was added to each well. The cells were then incubated for 2 h at 37 °C. Finally, the absorbance at 450 nm for each well was measured using an enzyme−linked immunoassay (Bio−rad iMark, Hercules, CA, USA). Each group consisted of five parallel samples.

2.5.3. Live/Dead Assay

To assess the cytotoxicity of LC−MSN on HUVECs, a live/dead assay was conducted using calcein AM/PI staining. HUVECs were seeded on 12−well plates at a density of 5×10^4 cells/well. After incubation for 24 and 48 h, the medium was removed, and the cells were washed twice with PBS. Subsequently, calcein AM/PI reagent (Meilunbio, Dalian, China) was added to the wells following the protocol and incubated at 37 °C for 30 min in the dark. The images were then observed using an inverted fluorescence microscope (Leica DMi8, Wetzlar, Germany). Finally, the living and dead cells were counted using ImageJ software (V1.8.0.112, Media Cybernetics, Rockville, MD, USA).

2.5.4. Scratch Migration Assay

The scratch assay was performed to evaluate the effect of LC−MSN on endothelial cell migration. HUVECs were seeded in 6−well plates at a density of 6×10^5 cells/well and cultured in an incubator. After 24 h of incubation, a confluent monolayer of cells was formed. Two parallel scratches were created vertically across the center of the well using a 1000 μL pipette tip. The wells were then rinsed gently three times with PBS to remove free cells. Next, 2000 μL of culture medium containing LC−MSN at concentrations of 0, 9.4, 18.8, and 37.5 μg/mL was added. After incubation for 0, 24, and 48 h, images were captured using a bright−field microscope at 10× magnification (Leica DMIL−LED, Wetzlar, Germany), and the unrecovered area was calculated using ImageJ software. The experiments were performed in triplicate. The simulated wound closure was quantified using the following equation:

$$\text{Wound area closure (\%)} = (S_0 - S_t)/S_0 \times 100\% \quad (1)$$

where S_0 represents the original area of the scratch, and S_t represents the area at the observation times.

2.5.5. Cell Morphology

The cellular morphology of HUVECs was observed using F−actin cytoskeleton stains. After 24 and 48 h of incubation, cells were rinsed with PBS and fixed in 4% paraformaldehyde (Meilunbio, Dalian, China) for 20 min at room temperature. Subsequently, the cells were permeabilized with PBS containing 0.1% Triton X−100 (Beyotime, Shanghai, China). Intracellular F−actin filaments were then stained for 1 h using phalloidin conjugated to TRITC (1:200 dilution in PBS, Solarbio, Beijing, China). Additionally, the nuclei were stained with 300 nM DAPI (Solarbio, Beijing, China) for 30 s. After rinsing, the cell morphology was visualized using a confocal laser scanning microscope (CLSM, Leica STELLARIS5, Wetzlar, Germany).

2.6. Angiogenesis Assay on LC−MSN In Vitro

2.6.1. Tube Formation Assay

To evaluate the impact of LC−MSN on the angiogenesis function of HUVECs, a tube formation assay was conducted. Firstly, 10 μL of BD Matrigel TM (Corning, NY, USA) was transferred to wells of the ibidi μ−plate and incubated at 37 °C for 1 h to form a gel substrate. Subsequently, HUVECs cultured in ECM or treated with LC−MSN for 24 or 48 h were digested and resuspended in ECM at a density of 1×10^5 cells/mL, respectively. Then, 50 μL of the cell suspension was added onto the Matrigel. After 6 h of culture, the cells on the Matrigel were observed using an optical microscope (Leica DMIL−LED, Wetzlar, Germany) under bright field illumination, and five random microscopic areas were selected for photography. The obtained images were processed using ImageJ software, and the meshes, nodes, and total length formed by the cells were counted.

2.6.2. qPCR

The expression of angiogenesis−related genes in HUVECs was assessed using qPCR. Cells were seeded in a 6−well plate at a density of 5×10^5 cells per well and cultured in ECM or the medium containing LC−MSN for 48 h. Total RNAs of the cells in each group were extracted using the TRIzol reagent (Invitrogen, Carlsbad, CA, USA) and reverse−transcribed into cDNA using an Evo M−MLV RT Premix kit (Accurate Biology, Changsha, China). RT−qPCR was performed using the SYBR Green QPCR Master Mix (APE × BIO, Houston, TX, USA) with a real−time qPCR system (Roche LightCycler−96, Switzerland) for the following genes: CD31, VEGF, ANG, and GAPDH. GAPDH was used as the endogenous control. The expression levels of the target genes were calculated and normalized using the $2^{-\Delta\Delta Ct}$ method. The primer sequences are listed in Table 1, and there were three parallel samples in each group.

Table 1. Primers used in the qRT−PCR of HUVECs.

Gene	Forward Primer Sequence (5′−3′)	Reverse Primer Sequence (3′−5′)
CD 31	AACAGTGTTGACATGAAGAGCC	TGTAAAACAGCACGTCATCCTT
VEGF	GAGCAGAAAGCCCATGAAGTG	ACTCCAGGGCTTCATCATTGC
ANG	CTGGGCGTTTTGTTGTTGGTC	CGCACAGTCTCCTTCACAGA
GAPDH	GGAGCGAGATCCCTCCAAAAT	GGCTGTTGTCATACTTCTCATGG

2.6.3. Western Blot

The expression of angiogenesis−related proteins in HUVECs was analyzed using western blot. The cell seeding and treatment procedures were the same as those for qPCR. After 48 h, the cells were lysed on ice for 30 min with RIPA lysis buffer containing a protease inhibitor. The lysate was then centrifuged at 12,000× *g* for 15 min at 4 °C, and the supernatant was collected. The protein concentration was determined using a BCA protein assay kit (GLPBIO, Montclair, CA, USA). Loading buffer (NCM, Suzhou, China) was added to the samples at a ratio of 5:1, followed by heating at 100 °C for 10 min. Subsequently, 20 μg of the samples was separated by polyacrylamide gel electrophoresis and transferred to a PVDF membrane (Merck Millipore, Darmstadt, Germany). The membrane was blocked with 5% BSA for 30 min and then incubated with primary antibodies against CD31 (Abcam, ab76533), VEGF (Abcam, ab46154), ANG (Abcam, ab276132), or GAPDH (MBL, M171−7) at 4 °C overnight. After washing the membrane with TBST three times, it was incubated with a horseradish peroxidase (HRP)−conjugated secondary antibody at room temperature for 2 h. The protein bands were visualized using chemiluminescence with the ECL detection kit (NCM, Suzhou, China), and the intensity of the bands was analyzed using ImageJ software. All experiments were performed with three replicates.

2.7. Animal Study

2.7.1. In Vivo Wound Healing

A total of fifteen healthy Sprague–Dawley rats (7–8 weeks old) were included in the approved in vivo study, authorized by the Animal Ethics Committee of Southern Medical University under approval/accreditation number SCXK(YUE)2021−0041. To assess the effect of the LC−MSN−incorporated hydrogel on wound healing, three subgroups were established: the control, CS hydrogel, and CS/LC−MSN hydrogel groups. The rats were anesthetized by intraperitoneal injection of pentobarbital sodium (3% *w*/*v*, 40 mg/mL). Following shaving, four full−thickness skin wounds (round, diameter = 6 mm) were created symmetrically on the dorsal skin using a biopsy punch. The hydrogels were then applied to the wounds, while the control wounds remained untreated but were covered with Tegaderm transparent dressing (3M Health Care, Germany) to prevent infection. The healing progress of each wound was observed and photographed at 0, 3, 7, and 14 days. The degree of wound closure was quantitatively determined by measuring the wound area using ImageJ software and calculated using the following equation:

$$\text{Wound healing rate (\%)} = (A_0 - A_t)/A_0 \times 100\% \qquad (2)$$

where A_0 represents the initial wound area and A_t is the wound area at the specific time points (0, 3, 7, and 14 days). At the end of the experiment, the rats were euthanized using an isoflurane overdose. The wound sites were excised for subsequent histological and immunohistochemistry analyses.

2.7.2. Histological Analysis

Wound tissue samples were collected for histological evaluation. The samples were fixed in 4% paraformaldehyde, dehydrated through an ethanol series, embedded in paraffin, and then sectioned into 5 μm thick slices using a microtome. The paraffin sections were mounted on slides and subjected to staining with hematoxylin−eosin (H&E) and Masson's trichrome stains according to the manufacturer's instructions. These stains allowed for the examination of regenerated tissue and collagen deposition, respectively. Digital images of all images of all sections were captured using a digital pathology slide scanner (Leica, Aperio VERSA, Wetzlar, Germany).

2.7.3. Immunohistochemistry Staining

Immunohistochemistry staining was performed to evaluate neovascularization and inflammation levels. Paraffin−embedded tissue sections were dewaxed, rehydrated, and then blocked with 5% bovine serum albumin for 1 h. Next, the sections were incubated overnight at 4 °C with primary antibodies: CD31 (ab76533, Abcam), interleukin−1beta (IL−1β) (ab5694, Abcam), and interleukin−6 (IL−6) (ab2105, Abcam). After three washes with PBS, the sections were incubated with a goat−anti−rabbit secondary antibody solution for 30 min. All slides were examined using a light microscope (Leica DMi8, Wetzlar, Germany).

2.8. Statistical Analysis

Each experiment was conducted with at least three parallel samples. Data analysis was performed using GraphPad Prism 8 (GraphPad Software, San Diego, CA, USA). The results are presented as the mean ± standard deviation (SD). Statistical differences between groups were evaluated using one−way analysis of variance and Student's *t*−test. A *p*−value less than 0.05 was considered statistically significant.

3. Results

3.1. Synthesis and Characterization of LC−MSN

The successful modification of LC on MSN was confirmed through ATR−FTIR analysis (Figure 1A). The spectrum of LC−MSN exhibited two new peaks at 1724 cm^{-1} and

1640 cm^{-1}, indicating the completion of the esterification reaction, a crucial step in LC preparation. Furthermore, the $^{1}H-NMR$ spectrum showed new resonances at approximately 6.4 and 6.1 ppm (a, b), attributed to the C=C bonds (Figure 1B). Additionally, a signal at around 4.60 ppm (c) indicated the characteristic hydrogen on the ring carbon bound to the newly formed C=O. Therefore, the consistent results from ATR−FTIR and $^{1}H-NMR$ confirmed that the synthesized LC−MSN conformed to the expected molecular design [8,36,37].

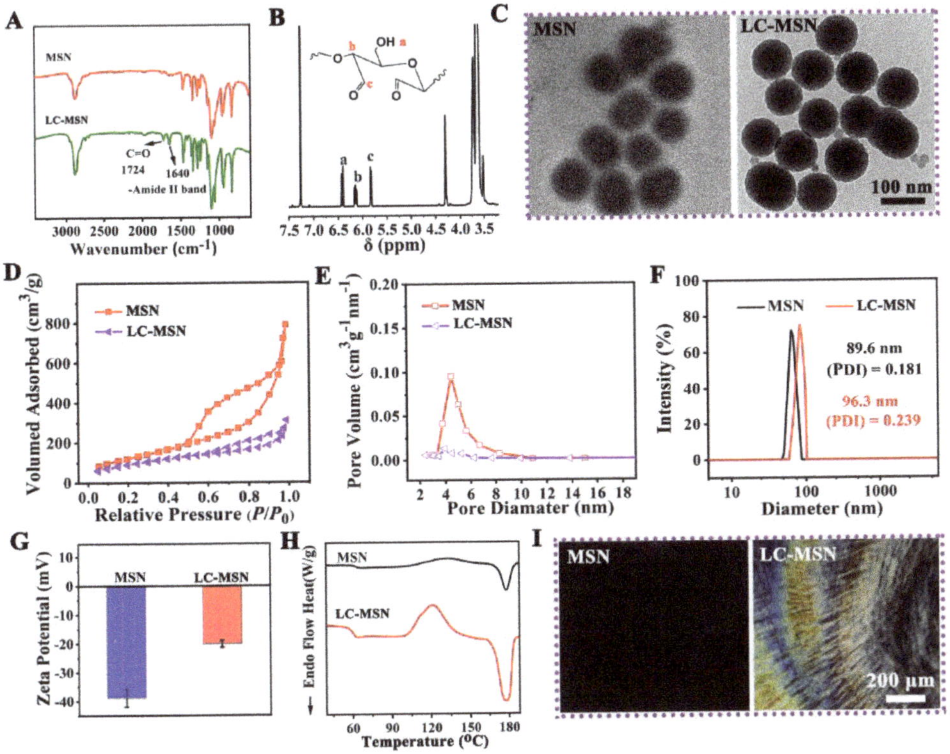

Figure 1. Characterization of LC−MSN and MSN; (**A**) FTIR spectra of MSN and LC−MSN highlighting the chemical structures (indicated by the black arrow); (**B**) ^{1}H NMR of resulting LC−MSN; (**C**) TEM images of MSN and LC−MSN; (**D**,**E**) N_2 adsorption–desorption isotherm and pore distribution of MSN and LC−MSN; (**F**,**G**) DLS characterization and surface Zeta potential of MSN and LC−MSN; (**H**) DSC thermal curves of MSN and LC−MSN; (**I**) the optical texture of MSN and LC−MSN.

The morphology of MSN and LC−MSN was observed using TEM, revealing that LC−MSN maintained a spherical shape with uniform particle size and good mono−dispersibility, similar to MSN (Figure 1C). These findings suggested that LC modification did not significantly affect the average particle size and dispersion of MSN.

The N_2 adsorption/desorption isotherms of MSN exhibited a typical type IV isotherm, indicating its mesoporous nature (Figure 1D). The average BET surface area and pore diameter of MSN were measured as 609.6 m^2/g and 5.29 nm, respectively. After loading LC, these values decreased to 349.4 m^2/g and 4.21 nm, respectively (Figure 1E).

To further characterize the nanoscale structure changes of MSN, DLS and Zeta potential measurements were conducted. DLS analysis revealed that the average size of MSN was 89.6 ± 5.7 nm, with a slight increase of 6.3 nm upon the addition of LC (Figure 1F). Moreover, the polymer dispersity index (PDI) in the aqueous phase was less than 0.3, indicating good dispersion stability of the nanocomposite. The Zeta potential of MSN was -38.9 ± -3.1 mV, which changed to -20.1 ± 2.1 mV upon LC incorporation (Figure 1G), suggesting the presence of a positively charged carbonyl group.

DSC thermal analysis revealed an additional endothermic peak at approximately 120 °C, corresponding to the glass transition temperature (Tg) of LC−MSN, indicating the crystalline properties of LC (Figure 1H). According to the POM images, the LC−MSN showed the typical cholesteric texture and color of anisotropic materials under a concentration of 2.5% in contrast to the MSN, and the optical field of LC−MSN under POM gradually became brighter with more distinct optical textures (Figure 1I). These results confirmed that the modification of LC into LC−MSN did not affect the properties of the LC phase.

3.2. Morphology and Biodegradation of CS/LC−MSN Composite Hydrogel

Considering the excellent biodegradability, biocompatibility, antibacterial, and anti−inflammatory properties of natural CS [5], the CS was used as a matrix polymer to prepare composite hydrogel scaffolds. In this work, the LC−MSN was incorporated into the CS polymer (the final concentration was 1.0%), and the CS/LC−MSN composite hydrogel system containing 0.25, 0.50, 0.75, or 1.0% of LC−MSN was fabricated by the electrostatic and intermolecular hydrogen bonding. We evaluated the mechanical properties of the different hydrogel samples through a compression test. Figure S4 displays the compressive strength of the composite hydrogel enhancement with the increased concentration of LC−MSN added. Moreover, the composite scaffolds showed a decline in ductility with the introduction of 1.0% LC−MSN, despite high compressive stress. As a result, CS/LC−MSN containing 0.75% of LC−MSN was deemed to be the ideal component for the studies that followed.

The microstructure of the cross−section of CS or CS/LC−MSN hydrogels was observed by SEM. As revealed in Figure 2A,B, both the CS and CS/LC−MSN scaffolds showed a typical 3D porous morphology, while the CS/LC−MSN sample displayed interconnected porous structures with a denser, cross−linked network compared to that of the neat CS hydrogel due to the incorporation of LC−MSN, leading to the enhancement of the CS polymer network. The FTIR, XRD, and XPS analyses were further exploited to determine the chemical structure and interactions of CS/LC−MSN. Figure 2C shows the FTIR spectra of the CS and CS/LC−MSN hydrogels. The characteristic absorption peaks of −OH and N−H for LC−MSN were observed at 3380 cm^{-1}, while the two absorption peaks appearing at positions 1627 and 1525 cm^{-1} corresponded to the stretching vibrations of the amide I and amide II bands in the CS chain. As for the inconspicuous characteristic peak at 2610 cm^{-1} and the broad peak at 1000–1100 cm^{-1}, they belong to the Si−O−Si multiple absorption peaks of LC−MSN [38]. Further XPS survey spectra of the CS/LC−MSN composite hydrogels also confirmed the above findings (Figure S1). These outcomes demonstrated the successful formation of the CS/LC−MSN composite hydrogel.

From the XRD patterns of the CS and CS/LC−MSN specimens (Figure S2), the diffraction peaks of LC−MSN were observed around 16.7 and 19.3, which corresponded to the (200)/(110) and (203) planes of the liquid crystal [39], respectively. The biomineralization ability of the CS/LC−MSN hydrogels was evaluated in a 1.5 SBF solution. Figure S3 showed that the number and size of minerals on the CS/LC−MSN scaffolds increasingly grew as the mineralization time prolonged, indicating the favorable mineralization induction performance in 1.5 SBF.

In addition, to determine the biodegradation of hydrogels, hydrogels with LC−MSN were incubated in SBF. The degradation analysis (i.e., accumulated degraded Si) demonstrated that the degradation profile of the CS/LC−MSN hydrogel was in an acceptable range of 45 to 89% during the 14 days (Figure 2E), revealing the favorable degradation ability of the composite scaffolds in vitro. It is noted that CS/LC−MSN degrades more slowly than pristine CS due to its higher cross−linking density than the CS hydrogel. The degradation rate of hydrogels depends on the density, mass, and hydrophilicity behavior of the polymers, as well as the number of linkage bonds.

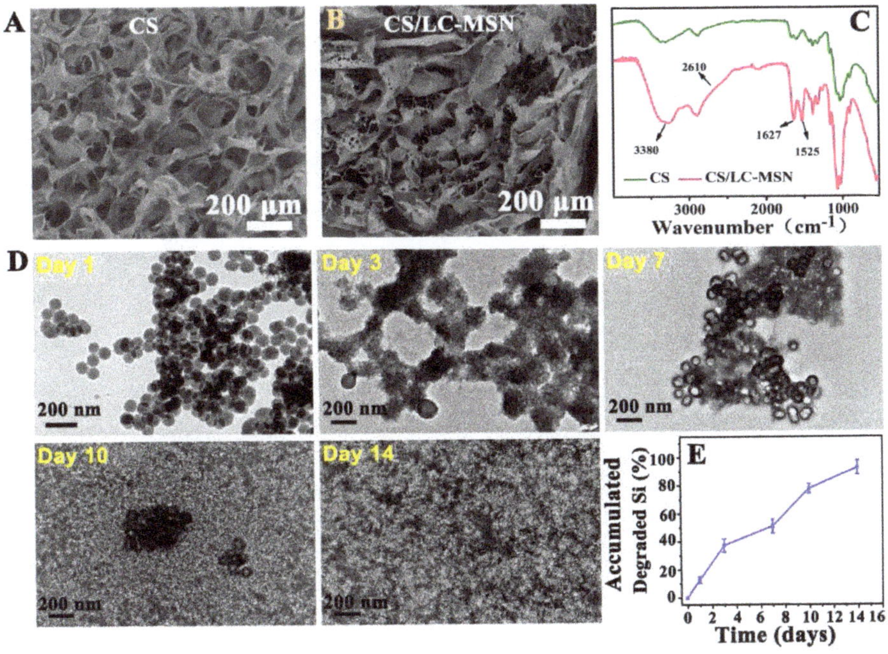

Figure 2. (**A**,**B**) SEM images of lyophilized CS or CS/LC−MSN specimens. (**C**) FTIR spectra of CS or CS/LC−MSN scaffolds. (**D**,**E**) In vitro evaluation of biodegradation and degradation behavior of CS/LC−MSN composite hydrogel incubated with SBF solution for 1, 3, 7, 10, or 14 days.

To get more insights into the release of LC−MSN from composite hydrogels, the morphological changes were observed by TEM during the degradation process. The morphology of LC−MSN was observed at higher magnification, confirming the retention of its spherical shape and uniform morphology. Upon incubation, the assemblies underwent significant swelling after 1 day, leading to a loosening of the compact network structure and increased swelling over the next 14 days. Throughout this process, the amount of unreleased LC−MSN within the hydrogel gradually decreased, while maintaining good morphology and dispersion.

3.3. *Cytocompatibility of CS/LC−MSN In Vitro*

To investigate the hypothesis that the LC−MSN composite creates a favorable microenvironment for skin regeneration and cell proliferation, various assays were performed, including live/dead staining, wound scratch testing, and a cell morphology assay.

To evaluate cell proliferation, HUVECs treated with different groups were examined for up to 48 h using a CCK−8 assay. The results, shown in Figure 3C, demonstrate that LC−MSN exhibited similar cytocompatibility to the normal complete culture medium (control) in terms of HUVEC proliferation, except at concentrations higher than 75 μg/mL, which inhibited HUVEC growth on day 2. Considering that biocompatibility is essential for biomaterials intended for tissue regeneration, concentrations higher than 75 μg/mL were not used in subsequent experiments. The cytotoxicity of HUVECs was further assessed using a calcein−AM/PI kit, which stained live cells green and dead cells red. The results of live/dead staining (Figure 3A,B) aligned with the cell proliferation assay, indicating that low concentrations of LC−MSN did not affect HUVEC viability. The fluorescent intensity of the LC−MSN groups remained relatively unchanged after 24 and 48 h of incubation compared to the control group.

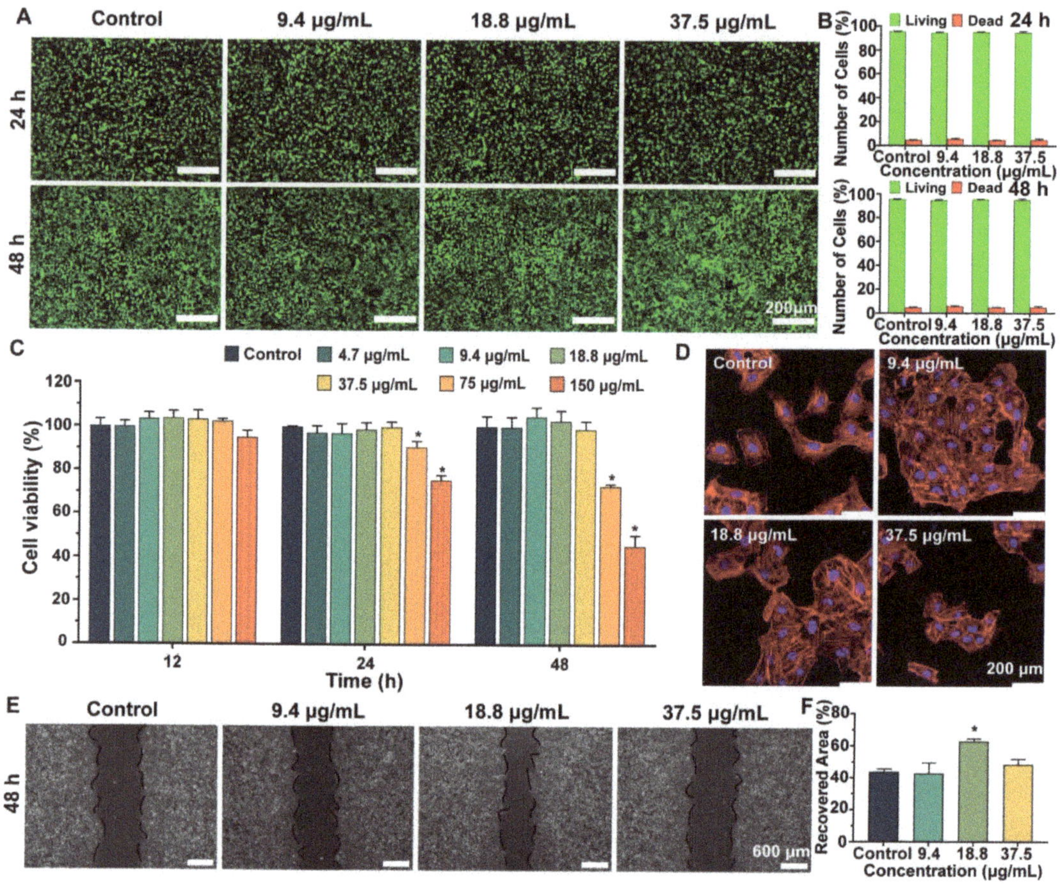

Figure 3. Effect of LC−MSN on the cell behavior of HUVECs. (**A**) Live/dead assays of HUVECs co−incubated with complete culture medium or LC−MSN and (**B**) its quantitative analysis of live/dead cells. (**C**) CCK−8 assays of the proliferation of HUVECs after culturing for 12, 24, and 48 h (* $p < 0.05$). (**D**) CLSM images of stained HUVECs showing morphology cultured in concentrations of LC−MSN for 48 h. (**E**,**F**) Migration assays of HUVECs cultured in the four groups (* $p < 0.05$).

Next, we performed a scratch assay to examine the migration of endothelial cells from the edge to the center area, which is a critical step in the angiogenesis process. The migration behavior of each group was captured in images (Figure 3E,F). After 48 h of culturing, distinct cell migration patterns were observed in the experimental groups. The wound closure in the normal culture medium and LC−MSN at 18.8 μg/mL reached 75% and 80%, respectively. Furthermore, LC−MSN at 9.4 μg/mL and 37.5 μg/mL did not exhibit any migration inhibition.

Finally, since cell migration ability is integrally linked to the formed stress fiber and microfilament containing actin, the cellular morphology of HUVECs was observed. In comparison to the control group, the LC−MSN group at 18.8 μg/mL displayed more prominent stress fibers, actin−containing microfilaments, and fully expanded cytoskeletons (Figure 3D). Importantly, the cells exhibited well−elongated and distributed morphology, with filopodia extension and cellular spreading fronts. These observations provide a clear explanation for the beneficial effects of LC−MSN on the interconnection and migration of HUVECs. In this regard, it is speculated that the flow and order of LC helped cells to guide the extension of the cellular pseudopods. Based on the obtained results, LC−MSN at low concentrations showed no inhibition of cytocompatibility, while 18.8 μg/mL exhibited the most effective promotion of HUVEC migration.

3.4. Angiogenesis Assay of HUVECs In Vitro

In order to assess the angiogenic potential of LC−MSN, a tube formation assay was conducted as it is a crucial step in angiogenesis [40]. Representative images revealed that HUVECs cultured with LC−MSN at concentrations of 9.4 μg/mL and 18.8 μg/mL exhibited a higher tendency to self−assemble into capillary−like structures on Matrigel compared to those cultured with ECM alone (Figure 4A). Quantitative analysis was performed to evaluate the node number, mesh number, and total length (Figure 4B). The average numbers of nodes and meshes were significantly higher in the LC−MSN groups compared to the control group ($p < 0.001$), particularly in the 18.8 μg/mL treatment group, which exhibited nearly twice the number of nodes as the control group ($p < 0.0001$). Additionally, HUVECs stimulated with LC−MSN at 18.8 μg/mL showed a slight increase in total length ($p < 0.001$). Given that angiogenesis plays a crucial role in tissue regeneration, LC−MSN at 18.8 μg/mL demonstrated a potential advantage in wound healing.

To gain further insight into the angiogenic properties of LC−MSN, qRT−PCR and Western blot analysis were conducted. Following coculturing with LC−MSN for 48 h, the expression of several angiogenesis−related genes, including CD31, VEGF, and ANG, in HUVECs (Figure 4C) showed a significant upregulation compared to that of the control groups ($p < 0.01$). The Western blot analysis results (Figure 4D) were consistent with the qRT−PCR findings ($p < 0.01$). Collectively, these results indicate that LC−MSN at an appropriate concentration can enhance the proliferation, migration, and angiogenic ability of HUVECs, thus contributing to skin regeneration.

3.5. In Vivo Skin Regeneration of Full−Thickness Wounds

To evaluate the therapeutic efficacy of the LC−MSN hydrogels on soft tissue repair, full−thickness skin defects were developed on rats. The LC−MSN hydrogel−treated group demonstrated a significantly higher percentage of wound closure compared to both the blank hydrogel−treated and non−treated groups in the in vivo study of full−thickness wounds. By day 14, the LC−MSN hydrogel−treated wounds exhibited superior healing compared to the non−treated group, with almost complete wound repair in the composite gel−treated group, while approximately 3% wound closure remained in the control groups (Figure 5A,B). The histological analyses by H&E staining and Masson's trichrome staining were further conducted after 7 or 14 days of treatment. The histological analysis using H&E staining revealed that wounds treated with the LC−MSN hydrogel displayed enhanced granulation tissue formation compared to the control groups (Figure 5C). Specifically, at 14 days post−operation, the composite gel−treated group exhibited a well−healed epithelial layer with a tightly regenerated dermis underneath, containing ample appendages. Masson's trichrome staining further demonstrated increased collagen deposition and more organized fiber alignment, resembling the density of normal skin in the LC−MSN hydrogel−treated group (Figure 5D).

Notably, the control group exhibited a presence of injured epithelialized tissue after 14 days of treatment, along with evident dermal congestion and significant inflammation. In contrast, the CS/LC−MSN hydrogel−treated group displayed the formation of complete epithelial and dermal structures, abundant skin appendages in the healed tissue, and noticeably thicker granulation compared to the CS and control groups (Figure 5E). These findings indicate that the CS/LC−MSN composites, when used as a novel wound dressing, can accelerate epidermal remodeling and promote tissue repair.

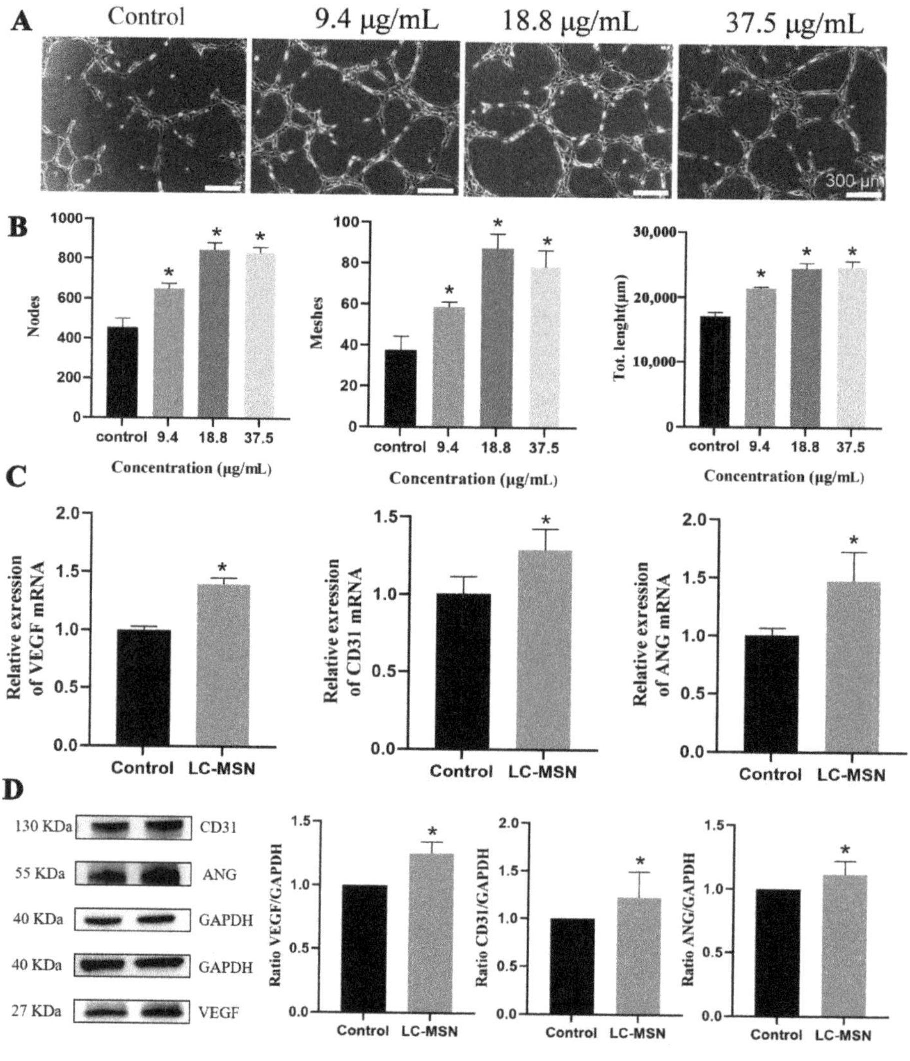

Figure 4. Angiogenesis assay of HUVECs in vitro. (**A**) Tube formation (CTF) assays of different concentrations of LC−MSN and (**B**) the corresponding quantitative analysis (* $p < 0.05$). (**C**) The expression levels of VEGF, CD31, and ANG are based on the qPCR technique (* $p < 0.05$). (**D**) Western blot analysis and the corresponding quantification of VEGF, CD31, or ANG expression in HUVEC cells incubated with various formulations (* $p < 0.05$).

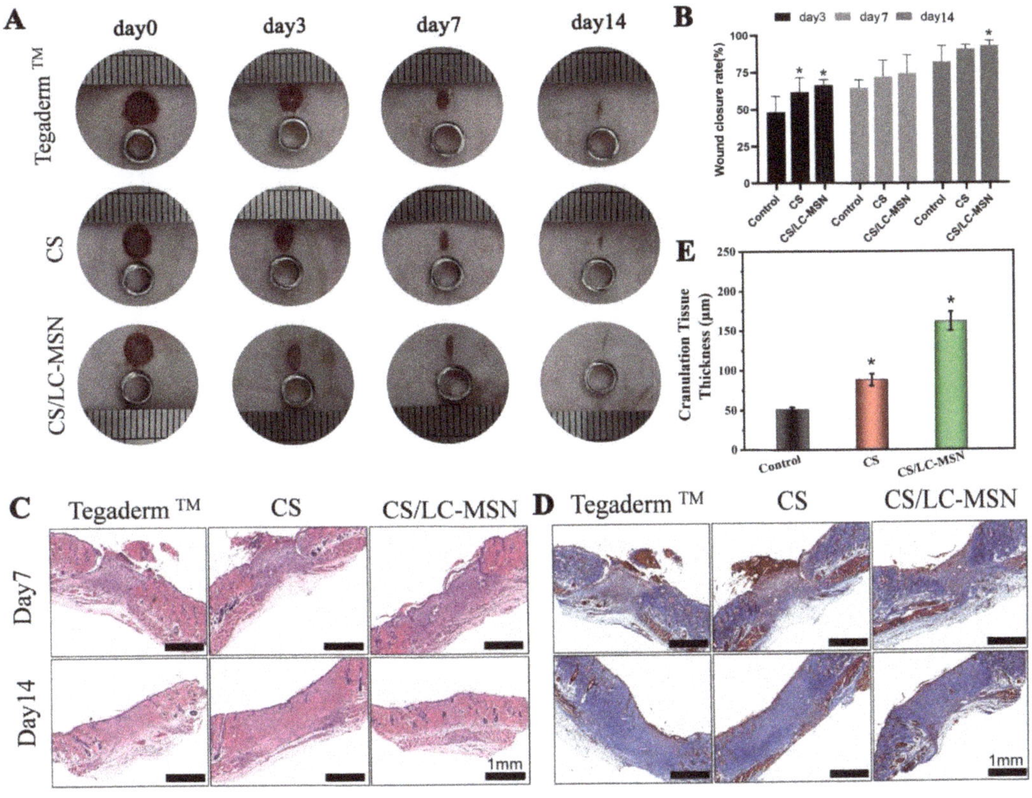

Figure 5. Morphologies and histological assessments of wounds. (**A**) Digital photos of skin wounds that were treated with various treatments within 14 days. (**B**) Wound closure rates during the 14−day post−surgery period (* $p < 0.05$). (**C**,**D**) Histological analysis of wound defects treated with Tegaderm™, CS, and CS/LC−MSN hydrogels using H&E staining and Masson staining at 7 and 14 days. (**E**) Corresponding quantitative analysis of the granulation tissue thicknesses at 14 days (* $p < 0.05$).

Immunohistochemical staining of IL−1β (early inflammatory cytokine markers) and IL−6 (intermediate inflammatory cytokine markers) was performed to evaluate the inflammatory response during the cutaneous wound healing process. The goal was to further investigate the mechanisms by which CS/LC−MSN promotes wound healing. As anticipated, the CS− and CS/LC−MSN−treated groups exhibited a lower expression of IL−1β or IL−6 (brown staining) in the wound tissue compared to the control groups (Figure 6). This This demonstrates that CS significantly inhibits the inflammatory reaction at the wound site.

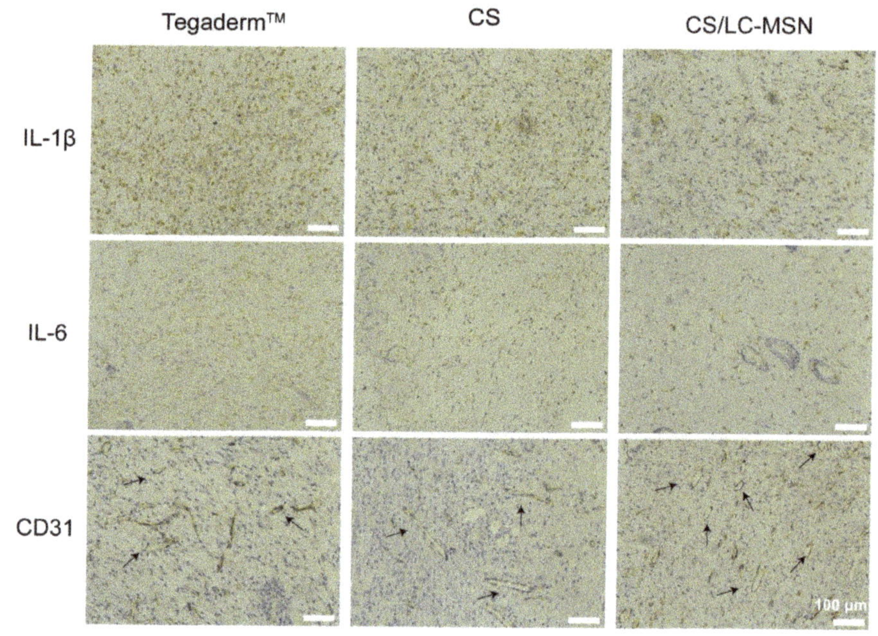

Figure 6. Images of immunohistochemical staining for IL−1β or IL−6 and CD31 of the full−thickness wounds after 7−day implantation of Tegaderm™, CS, and CS/LC−MSN hydrogels.

Furthermore, at day 7 post−treatment, neovascularization around the wound was assessed through the immunohistochemical staining of CD31. The LC−MSN hydrogel−treated group displayed a significantly higher density of blood vessels with larger diameters and thicker walls compared to the other groups (Figure 6), indicating the formation of new blood vessels. Although the blank hydrogel group showed a slight increase in the number of blood vessels compared to the control group, the difference was not significant. Hydrogels loaded with LC−MSN demonstrated favorable properties in promoting vascularization during full−thickness skin repair. Consequently, the resulting CS/LC−MSN composite scaffolds exert a dual function of inhibiting the inflammatory microenvironment and promoting angiogenesis, thereby exhibiting synergistic effects in triggering skin regeneration.

4. Discussion

Injuries to the oral and maxillofacial tissues resulting from burns, surgery, and trauma can compromise their protective barrier function and impede proper healing. The promotion of neovascularization by endothelial cells at the wound site, facilitating nutrient supply and waste removal, is crucial for collagen deposition, epithelial regeneration, and wound closure. Consequently, enhancing early vascularization and creating a favorable microenvironment for tissue repair have become key areas of research interest. Various hydrogel−based strategies have been evaluated for tissue engineering. While growth factors with pro−angiogenic properties have potential clinical applications, their high cost and instability pose significant limitations. In addition, novel hydrogels modified with nanomaterials, especially magnetic nanoparticles (MNPs), offer many advantages, such as improved mechanical and biological properties [41,42]. However, future research should focus on developing more efficient preparation methods and addressing the challenges associated with optimizing the magnetic responsiveness and biocompatibility [43]. LC materials possess desirable properties such as accessibility, stability, and low toxicity, making them promising biomaterials. LCs have shown potential in improving angiogenesis by facilitating the formation of endothelial cell rings. However, the direct utilization of liquid crystals poses challenges. To address this, a novel biomimetic material was developed by combining the mechanical properties and mesoporous structure of MSN with LCs.

In this study, the LC−MSN composite system was successfully prepared. It was reported that mesoporous materials have a pore size distribution in the range of 2–50 nm [44]. The N_2 adsorption/desorption isotherms of the MSN synthesized in this experiment showed a typical type IV isotherm, indicating its mesoporosity. The calculations demonstrated that the MSN possessed a large surface area and porosity, enabling high drug−carrying capacity. Following LC modification, the surface area and porosity slightly decreased, suggesting that some LC molecules occupied interstitial spaces without replacing the Si sites in the loose nanoscale network. The stronger electrostatic effect of the LC resulted in a more compact network, leading to a smaller surface area. Chemical structure analysis, optical structure examination using polarized optical microscopy (POM), and crystalline characterization consistently confirmed the retention of liquid crystal properties in the LC−MSN composite. Moreover, the particle size of the synthesized LC−MSN remained stable in the solution, indicating good dispersion and stability without significant aggregation. The hydrogel formed by LC−MSN exhibited favorable biodegradability, meeting the requirements of biomaterials. Collectively, the characterization results confirmed that the synthesized LC−MSN composite aligned with the intended design.

In our study, HUVECs were utilized as the cell models due to their crucial role in angiogenesis, the process of forming tubular structures resembling blood vessels. HUVECs are widely used for studying their biological behavior and response to various stimuli, and they have been extensively employed to investigate the mechanisms involved in angiogenesis [45]. Since we prepared LC−MSN as a new biomaterial, our first priority was to verify its biocompatibility. Cell proliferation and viability studies were conducted to evaluate the cytocompatibility of LC−MSN at various concentrations for up to 48 h. The results demonstrated favorable cytocompatibility of LC−MSN at low concentrations. To ensure biocompatibility, concentrations over 37.5 μg/mL were excluded from subsequent experiments. Notably, when HUVECs were cultured with LC−MSN at a concentration of 18.8 μg/mL, we observed significant promotion of cell extension and migration. These findings provide further evidence of the bioactive environment created by LC−MSN. The ability of endothelial cells to form tubular structures is crucial for angiogenesis, so we also assessed the potential of LC−MSN to promote vascularization using an in vitro tube formation assay. The results revealed that HUVECs cultured with LC−MSN formed more branches, nodes, and complete loops compared to the control group. The enhanced migration ability and actin microfilament performance of HUVECs explain why LC−MSN facilitated the formation of interconnected tubular networks in a shorter time. To further investigate the role of LC−MSN in promoting angiogenesis, qRT−PCR and Western blot analysis were performed. Encouragingly, the expression of several angiogenesis−related genes and proteins, including CD31, VEGF, and ANG, was upregulated in HUVECs cultured with LC−MSN. CD31, a key angiogenic factor, is primarily located at the junctions between endothelial cells and is likely involved in integrin activation for angiogenesis during wound healing. Moreover, angiogenic activity is regulated through complex signaling pathways, with the VEGF−VEGFR and Ang−Tie 2 axes playing significant roles [46]. ANG, acting in conjunction with VEGF, may have a bi−directional impact on angiogenesis. In our study, the increased expression of VEGF suggests that ANG and VEGF may synergistically promote angiogenesis. However, further research is needed to explore the underlying mechanisms.

To take full advantage of its bioactivity, LC−MSN was cross−linked with CS to develop a multifunctional composite hydrogel to promote tissue repair. In vitro experiments demonstrated that the application of the CS/LC−MSN hydrogel significantly reduced the size of wounds compared to the control group. Wound healing is a dynamic process involving numerous cells and bioactive signaling molecules [47]. It comprises sequential and overlapping stages, including hemostasis, inflammation, proliferation, and remodeling.

During the proliferation stage, resident cells such as fibroblasts and endothelial cells migrate to the wound site under the influence of host−mediated inflammatory responses [48]. However, the proliferation process is susceptible to interference from excessive inflammation [49]. The in vivo study revealed that CS/LC−MSN modulated the inflammatory response during the inflammation stage by downregulating the expression of IL−1β and IL−6, thereby promoting the transition of wound repair towards the proliferation stage. Key indicators of wound healing during this stage involve epithelial cell regeneration and collagen deposition. The evaluation of healed tissues through H&E−staining and Masson's staining demonstrated that the CS/LC−MSN hydrogel not only facilitated epithelial regeneration and collagen synthesis but also promoted the formation of dermal appendages, thus approximating the morphological characteristics of normal skin. In line with the results of qPCR and Western blot analysis in vitro, areas with more new vessels were identified in the CS/LC−MSN groups in vivo via a CD31 staining assessment. In summary, these promising findings can be attributed to the favorable moist environment provided by the hydrogel for wound healing, along with the active release of LC−MSN that promotes angiogenesis, collectively creating a conducive microenvironment.

Further studies are needed to investigate the potential mechanisms underlying the observations made in this study. It is also essential to optimize the hydrogel formulation to improve its clinical translation, taking into account long−term safety and potential toxicity concerns. Furthermore, exploring other potential applications of the LC−MSN system, such as its bio−mineralization and osteogenic properties, could further expand its scope of use in tissue engineering.

5. Conclusions

In conclusion, we successfully synthesized LC−MSN, which exhibited excellent cytocompatibility and angiogenesis potential. The LC−MSN was further incorporated into a hydrogel to create a multifunctional biomaterial. Through FTIR, 1H NMR, DLS, BET, and POM analyses, we confirmed the successful modification of LC onto MSN. In vitro cell behavior experiments demonstrated the biocompatibility of LC−MSN. Moreover, the biomaterial showed remarkable wound healing potential in terms of in vitro angiogenesis assays. The LC−MSN−incorporated hydrogel not only possessed the inherent advantages of hydrogels but also provided a biomimetic microenvironment due to the presence of LC. In vivo, the biodegradable hybrid hydrogel promoted the formation of granulation tissue, collagen deposition, and neovascularization, thereby enhancing early−stage full−thickness wound healing. Overall, we have great confidence that our study on the LC−MSN−incorporated hydrogel will contribute valuable insights into the development of angiogenic strategies for soft tissue diseases where angiogenesis plays a crucial role.

Supplementary Materials: The following supporting information can be downloaded at: https://www.mdpi.com/article/10.3390/jfb14060316/s1, Figure S1. XPS survey spectrum of CS/LC-MSN composite hydrogels; Figure S2. XRD pattern of CS and CS/LC-MSN specimens. The diffraction peaks of LC-MSN around 16.7 and 19.3 were observed, which corresponded to the o (200)/ (110) and (203) planes of crystalline, respectively; Figure S3. Morphologies of CS/LC-MSN composite hydrogel scaffolds after 2, 4, and 6 days of biomimetic mineralization; Figure S4. The compressive test of CS/LC-MSN composite hydrogels.

Author Contributions: X.L.: conceptualization, methodology, formal analysis, writing—original draft; L.W.: investigation, formal analysis, writing—original draft; T.Z.: resources, visualization; R.L.: validation, data curation; M.Z.: resources, software; H.L.: supervision, conceptualization, project administration, funding acquisition. All authors have read and agreed to the published version of the manuscript.

Funding: Science Foundation of Stomatological Hospital, School of Stomatology, Southern Medical University [PY2022012], Guangdong Medical Research Fund [A2023387], President Foundation of The Fifth Affiliated Hospital, Southern Medical University [YZ2022MS01], Scientific Research Program of Traditional Chinese Medicine of Guangdong Province, China [20222143], Science and Technology Projects in Guangzhou [202201011107].

Data Availability Statement: Data are available from the corresponding author upon request.

Conflicts of Interest: The authors declare no conflict of interest.

References

1. Cleveland, P.W.; Smith, J.E. Complications of Facial Trauma of the Fronto-orbital Region. *Facial Plast. Surg.* **2017**, *33*, 581–590. [CrossRef] [PubMed]
2. Huang, S.; Liu, H.; Liao, K.; Hu, Q.; Guo, R.; Deng, K. Functionalized GO Nanovehicles with Nitric Oxide Release and Photothermal Activity-Based Hydrogels for Bacteria-Infected Wound Healing. *ACS Appl. Mater. Interfaces* **2020**, *12*, 28952–28964. [CrossRef] [PubMed]
3. Robert, A.W.; Azevedo Gomes, F.; Rode, M.P.; Marques da Silva, M.; Veleirinho, M.; Maraschin, M.; Hayashi, L.; Wosgrau Calloni, G.; Stimamiglio, M.A. The skin regeneration potential of a pro-angiogenic secretome from human skin-derived multipotent stromal cells. *J. Tissue Eng.* **2019**, *10*, 2041731419833391. [CrossRef] [PubMed]
4. Ahn, S.; Ardoña, H.A.M.; Campbell, P.H.; Gonzalez, G.M.; Parker, K.K. Alfalfa Nanofibers for Dermal Wound Healing. *ACS Appl. Mater. Interfaces* **2019**, *11*, 33535–33547. [CrossRef] [PubMed]
5. Huang, L.; Zhu, Z.; Wu, D.; Gan, W.; Zhu, S.; Li, W.; Tian, J.; Li, L.; Zhou, C.; Lu, L. Antibacterial poly (ethylene glycol) diacrylate/chitosan hydrogels enhance mechanical adhesiveness and promote skin regeneration. *Carbohydr. Polym.* **2019**, *225*, 115110. [CrossRef]
6. Huang, L.; Li, W.; Guo, M.; Huang, Z.; Chen, Y.; Dong, X.; Li, Y.; Zhu, L. Silver doped-silica nanoparticles reinforced poly (ethylene glycol) diacrylate/hyaluronic acid hydrogel dressings for synergistically accelerating bacterial-infected wound healing. *Carbohydr. Polym.* **2023**, *304*, 120450. [CrossRef]
7. Sung, B.; Kim, M.H. Liquid-crystalline nanoarchitectures for tissue engineering. *Beilstein J. Nanotechnol.* **2018**, *9*, 205–215. [CrossRef]
8. Zhan, Y.; Deng, B.; Wu, H.; Xu, C.; Wang, R.; Li, W.; Pan, Z. Biomineralized Composite Liquid Crystal Fiber Scaffold Promotes Bone Regeneration by Enhancement of Osteogenesis and Angiogenesis. *Front. Pharmacol.* **2021**, *12*, 736301. [CrossRef]
9. Zheng, Z.; Wang, R.; Lin, J.; Tian, J.; Zhou, C.; Li, N.; Li, L. Liquid Crystal Modified Polylactic Acid Improves Cytocompatibility and M2 Polarization of Macrophages to Promote Osteogenesis. *Front. Bioeng. Biotechnol.* **2022**, *10*, 887970. [CrossRef]
10. Wu, S.; Yang, X.; Li, W.; Du, L.; Zeng, R.; Tu, M. Enhancing osteogenic differentiation of MC3T3-E1 cells by immobilizing RGD onto liquid crystal substrate. *Mater. Sci. Eng. C Mater. Biol. Appl.* **2017**, *71*, 973–981. [CrossRef]
11. Baumer, Y.; Dey, A.K.; Gutierrez-Huerta, C.A.; Khalil, N.O.; Sekine, Y.; Sanda, G.E.; Zhuang, J.; Saxena, A.; Stempinski, E.; Elnabawi, Y.A.; et al. Hyperlipidaemia and IFNgamma/TNFalpha Synergism are associated with cholesterol crystal formation in Endothelial cells partly through modulation of Lysosomal pH and Cholesterol homeostasis. *EBioMedicine* **2020**, *59*, 102876. [CrossRef]
12. Wang, B.; Huang, Y.; Huang, Z.; Wang, H.; Chen, J.; Pan, X.; Wu, C. Self-assembling in situ gel based on lyotropic liquid crystals containing VEGF for tissue regeneration. *Acta Biomater.* **2019**, *99*, 84–99. [CrossRef]
13. Price, J.C.; Roach, P.; El Haj, A.J. Liquid Crystalline Ordered Collagen Substrates for Applications in Tissue Engineering. *ACS Biomater. Sci. Eng.* **2016**, *2*, 625–633. [CrossRef]
14. Wang, W.; Liu, Y.; Yang, C.; Qi, X.; Li, S.; Liu, C.; Li, X. Mesoporous bioactive glass combined with graphene oxide scaffolds for bone repair. *Int. J. Biol. Sci.* **2019**, *15*, 2156–2169. [CrossRef]
15. Ma, J.; Qin, C.; Wu, J.; Zhang, H.; Zhuang, H.; Zhang, M.; Zhang, Z.; Ma, L.; Wang, X.; Ma, B.; et al. 3D Printing of Strontium Silicate Microcylinder-Containing Multicellular Biomaterial Inks for Vascularized Skin Regeneration. *Adv. Healthc. Mater.* **2021**, *10*, e2100523. [CrossRef]
16. Shan, L.; Fan, W.; Wang, W.; Tang, W.; Yang, Z.; Wang, Z.; Liu, Y.; Shen, Z.; Dai, Y.; Cheng, S.; et al. Organosilica-Based Hollow Mesoporous Bilirubin Nanoparticles for Antioxidation-Activated Self-Protection and Tumor-Specific Deoxygenation-Driven Synergistic Therapy. *ACS Nano* **2019**, *13*, 8903–8916. [CrossRef]
17. Li, Q.; Gong, S.; Yao, W.; Yang, Z.; Wang, R.; Yu, Z.; Wei, M. Exosome loaded genipin crosslinked hydrogel facilitates full thickness cutaneous wound healing in rat animal model. *Drug Deliv.* **2021**, *28*, 884–893. [CrossRef]
18. Lu, H.; Li, X.; Zhang, M.; Xu, C.; Li, W.; Wan, L. Antibacterial Cellulose Nanocrystal-Incorporated Hydrogels With Satisfactory Vascularization for Enhancing Skin Regeneration. *Front. Bioeng. Biotechnol.* **2022**, *10*, 876936. [CrossRef]
19. Huang, L.; Feng, J.; Zhu, J.; Yang, J.; Xiong, W.; Lu, X.; Chen, S.; Yang, S.; Li, Y.; Xu, Y.; et al. A Strategy of Fenton Reaction Cycloacceleration for High-Performance Ferroptosis Therapy Initiated by Tumor Microenvironment Remodeling. *Adv. Healthc. Mater.* **2023**, e2203362. [CrossRef]

20. Zhang, J.; Cai, L.; Tang, L.; Zhang, X.; Yang, L.; Zheng, K.; He, A.; Boccaccini, A.R.; Wei, J.; Zhao, J. Highly dispersed lithium doped mesoporous silica nanospheres regulating adhesion, proliferation, morphology, ALP activity and osteogenesis related gene expressions of BMSCs. *Colloids Surf. B Biointerfaces* **2018**, *170*, 563–571. [CrossRef]
21. Beck, G.R., Jr.; Ha, S.W.; Camalier, C.E.; Yamaguchi, M.; Li, Y.; Lee, J.K.; Weitzmann, M.N. Bioactive silica-based nanoparticles stimulate bone-forming osteoblasts, suppress bone-resorbing osteoclasts, and enhance bone mineral density in vivo. *Nanomedicine* **2012**, *8*, 793–803. [CrossRef] [PubMed]
22. Li, Z.; Barnes, J.C.; Bosoy, A.; Stoddart, J.F.; Zink, J.I. Mesoporous silica nanoparticles in biomedical applications. *Chem. Soc. Rev.* **2012**, *41*, 2590–2605. [CrossRef] [PubMed]
23. Huang, Y.; Chen, J.; Lin, J.; Lin, J.; Chen, X. Preparation of Vanillic Acid-Loaded Core-Shell Gold Nanospheres/Mesoporous Silica Nanoparticles for the Treatment of Orthopedic Infection. *ACS Omega* **2021**, *6*, 2899–2905. [CrossRef] [PubMed]
24. He, Q.; Shi, J. MSN anti-cancer nanomedicines: Chemotherapy enhancement, overcoming of drug resistance, and metastasis inhibition. *Adv. Mater.* **2014**, *26*, 391–411. [CrossRef]
25. Li, H.; Gu, J.; Shah, L.A.; Siddiq, M.; Hu, J.; Cai, X.; Yang, D. Bone cement based on vancomycin loaded mesoporous silica nanoparticle and calcium sulfate composites. *Mater. Sci. Eng. C Mater. Biol. Appl.* **2015**, *49*, 210–216. [CrossRef]
26. Shi, M.; Xia, L.; Chen, Z.; Lv, F.; Zhu, H.; Wei, F.; Han, S.; Chang, J.; Xiao, Y.; Wu, C. Europium-doped mesoporous silica nanosphere as an immune-modulating osteogenesis/angiogenesis agent. *Biomaterials* **2017**, *144*, 176–187. [CrossRef]
27. Gan, Q.; Zhu, J.; Yuan, Y.; Liu, H.; Qian, J.; Li, Y.; Liu, C. A dual-delivery system of pH-responsive chitosan-functionalized mesoporous silica nanoparticles bearing BMP-2 and dexamethasone for enhanced bone regeneration. *J. Mater. Chem. B* **2015**, *3*, 2056–2066. [CrossRef]
28. Slowing, I.I.; Vivero-Escoto, J.L.; Wu, C.W.; Lin, V.S. Mesoporous silica nanoparticles as controlled release drug delivery and gene transfection carriers. *Adv. Drug Deliv. Rev.* **2008**, *60*, 1278–1288. [CrossRef]
29. Huang, L.; Feng, J.; Fan, W.; Tang, W.; Rong, X.; Liao, W.; Wei, Z.; Xu, Y.; Wu, A.; Chen, X.; et al. Intelligent Pore Switch of Hollow Mesoporous Organosilica Nanoparticles for High Contrast Magnetic Resonance Imaging and Tumor-Specific Chemotherapy. *Nano Lett.* **2021**, *21*, 9551–9559. [CrossRef]
30. Liang, Y.; Zhao, X.; Hu, T.; Chen, B.; Yin, Z.; Ma, P.X.; Guo, B. Adhesive Hemostatic Conducting Injectable Composite Hydrogels with Sustained Drug Release and Photothermal Antibacterial Activity to Promote Full-Thickness Skin Regeneration During Wound Healing. *Small* **2019**, *15*, 1900046. [CrossRef]
31. Yue, X.; Zhang, X.; Wang, C.; Huang, Y.; Hu, P.; Wang, G.; Cui, Y.; Xia, X.; Zhou, Z.; Pan, X.; et al. A bacteria-resistant and self-healing spray dressing based on lyotropic liquid crystals to treat infected post-operative wounds. *J. Mater. Chem. B* **2021**, *9*, 8121–8137. [CrossRef]
32. Yang, Y.; Wang, X.; Yang, F.; Shen, H.; Wu, D. A Universal Soaking Strategy to Convert Composite Hydrogels into Extremely Tough and Rapidly Recoverable Double-Network Hydrogels. *Adv. Mater.* **2016**, *28*, 7178–7184. [CrossRef]
33. Koochaki, A.; Shahgholi, M.; Sajadi, S.M.; Babadi, E.; Inc, M. Investigation of the mechanical stability of polyethylene glycol hydrogel reinforced with cellulose nanofibrils for wound healing: Molecular dynamics simulation. *Eng. Anal. Bound. Elem.* **2023**, *151*, 1–7. [CrossRef]
34. Kocaaga, B.; Guner, F.S.; Kurkcuoglu, O. Molecular dynamics simulations can predict the optimum drug loading amount in pectin hydrogels for controlled release. *Mater. Today Commun.* **2022**, *31*, 103268. [CrossRef]
35. Wu, S.H.; Mou, C.Y.; Lin, H.P. Synthesis of mesoporous silica nanoparticles. *Chem. Soc. Rev.* **2013**, *42*, 3862–3875. [CrossRef]
36. Larkin, P. *Infrared and Raman Spectroscopy: Principles and Spectral Interpretation*; Elsevier: Amsterdam, The Netherlands, 2017.
37. Rosan, A.M. *Organic Structures from Spectra*; Field, L.D., Sternhell, S., Kalman, J.R., Eds.; John Wiley & Sons: Hoboken, NJ, USA, 2002.
38. Bera, T.; Freeman, E.J.; McDonough, J.A.; Clements, R.J.; Aladlaan, A.; Miller, D.W.; Malcuit, C.; Hegmann, T.; Hegmann, E. Liquid Crystal Elastomer Microspheres as Three-Dimensional Cell Scaffolds Supporting the Attachment and Proliferation of Myoblasts. *ACS Appl. Mater. Interfaces* **2015**, *7*, 14528–14535. [CrossRef]
39. Xie, W.; Ouyang, R.; Wang, H.; Li, N.; Zhou, C. Synthesis and cytotoxicity of novel elastomers based on cholesteric liquid crystals. *Liq. Cryst.* **2019**, *47*, 449–464. [CrossRef]
40. Mukherjee, S.; Sriram, P.; Barui, A.K.; Nethi, S.K.; Veeriah, V.; Chatterjee, S.; Suresh, K.I.; Patra, C.R. Graphene Oxides Show Angiogenic Properties. *Adv. Healthc. Mater.* **2015**, *4*, 1722–1732. [CrossRef]
41. Yan, T.; Rao, D.; Chen, Y.; Wang, Y.; Zhang, Q.; Wu, S. Magnetic nanocomposite hydrogel with tunable stiffness for probing cellular responses to matrix stiffening. *Acta Biomater.* **2022**, *138*, 112–123. [CrossRef]
42. Gila-Vilchez, C.; Mañas-Torres, M.C.; Contreras-Montoya, R.; Alaminos, M.; Duran, J.D.G.; de Cienfuegos, L.; Lopez-Lopez, M.T. Anisotropic magnetic hydrogels: Design, structure and mechanical properties. *Philos. Trans. A Math. Phys. Eng. Sci.* **2019**, *377*, 20180217. [CrossRef]
43. Pardo, A.; Gomez-Florit, M.; Barbosa, S.; Taboada, P.; Domingues, R.M.A.; Gomes, M.E. Magnetic Nanocomposite Hydrogels for Tissue Engineering: Design Concepts and Remote Actuation Strategies to Control Cell Fate. *ACS Nano* **2021**, *15*, 175–209. [CrossRef] [PubMed]
44. Wang, Y.; Zhao, Q.; Han, N.; Bai, L.; Li, J.; Liu, J.; Che, E.; Hu, L.; Zhang, Q.; Jiang, T.; et al. Mesoporous silica nanoparticles in drug delivery and biomedical applications. *Nanomedicine* **2015**, *11*, 313–327. [CrossRef] [PubMed]

45. Hauser, S.; Jung, F.; Pietzsch, J. Human Endothelial Cell Models in Biomaterial Research. *Trends Biotechnol.* **2017**, *35*, 265–277. [CrossRef] [PubMed]
46. Nicolò, M.; Ferro Desideri, L.; Vagge, A.; Traverso, C.E. Faricimab: An investigational agent targeting the Tie-2/angiopoietin pathway and VEGF-A for the treatment of retinal diseases. *Expert Opin. Investig. Drugs* **2021**, *30*, 193–200. [CrossRef]
47. Martin, P.; Nunan, R. Cellular and molecular mechanisms of repair in acute and chronic wound healing. *Br. J. Dermatol.* **2015**, *173*, 370–378. [CrossRef]
48. Krawczyk, W.S. A pattern of epidermal cell migration during wound healing. *J. Cell Biol.* **1971**, *49*, 247–263. [CrossRef]
49. Gurtner, G.C.; Werner, S.; Barrandon, Y.; Longaker, M.T. Wound repair and regeneration. *Nature* **2008**, *453*, 314–321. [CrossRef]

preparation using a double mix impression technique with vinyl siloxane ether material (Identium light and heavy, Kettenbach GmbH & Co. KG, Eschenburg, Germany) under relative isolation. After gingival retraction, a low-viscosity material was injected around the prepared teeth while the custom tray was filled with a high-viscosity material. The tray was then removed and disinfected, and type IV super hard plaster (Fujirock EP Classic, GC Corporation, Tokyo, Japan) was poured into it by a dental technician, in accordance with the manufacturer's processing and setting times. The models were examined, trimmed, and digitally scanned with an optical lab scanner (D2000, 3Shape, Copenhagen, Denmark, Software 3Shape Version 2015), and the data were saved as STL files. The measurements of the models were performed by S.P., who has over 10 years of experience in intraoral scanning and conventional impression techniques.

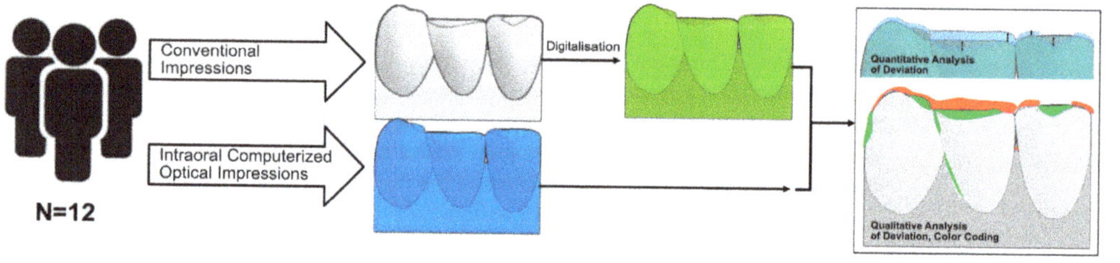

Figure 1. Schematic representation of the study design.

2.3. *Intraoral Computerized Optical Impression*

COIM was performed using a commercially available intraoral scanner (True Definition Scanner, 3M, St. Paul, MN, USA, Software version 5.0.2-production-eu) via the application of a titanium dioxide (High-Resolution Scanning Spray, 3M, St. Paul, MN, USA) scanning powder applied using a battery-powered powder applicator. The True Definition Scanner uses an active wavefront sampling technique and captures data in real-time at a capture speed of 20 3D images per second. COIM was always performed directly after the conventional impression to obtain a second impression. Impressions of the partially edentulous patients' upper and lower jaws were obtained under relative isolation. The abutment teeth were anesthetized and prepared using the double cord technique before they were dusted with scanning powder. A standardized scan path was adhered to (Figure 2), and the abutment teeth were digitally isolated from the scan file. For the upper jaw, scanning proceeded from the back right to the front left, in a zigzag pattern, with missing areas added subsequently (Figure 2A). For the lower jaw, scanning proceeded from the back right to the front left in a zigzag pattern along the jaw ridge, and the vestibule and sublingual areas were then scanned (Figure 2B). Missing areas were added subsequently. The intraoral scans were performed by S.P., who is extensively trained and has over 10 years of experience in intraoral scanning procedures.

2.4. *Indirect Extraoral Digitization and Processing of Models*

Master models were scanned using an optical laboratory scanner (D2000, 3Shape). This lab scanner operates on the principle of structured light projection via a multiline blue LED and utilizes four 5.0-megapixel cameras. The manufacturer claims that it has an accuracy of 5 μm (according to ISO 12836) [16]. The scanning time for a complete dental arch was approximately 16 seconds. The abutment teeth scans were fully captured and auto-aligned. The scan files were uploaded to a proprietary portal and downloaded as STL files.

Several parameters were uniformly preset in the 3D evaluation software (Geomagic Control 2014, software version 2014.0.1.1671, Geomagic, Morrisville, NY, USA). The master and TrueDefinition datasets (obtained via COIM) were trimmed to eliminate plaster parts and artifacts, respectively. Four different comparison situations were assessed for each patient: Whole Jaw (WJ), Mucosa with Residual Teeth (M_RT), Isolated Mucosa (IM), and

Isolated Abutment Teeth (AT; Figure 3). Each situation involved different trimming and selection processes. In all cases, the cutting tool was carefully used to maintain model integrity. After processing, the individual models of each patient were combined and saved as WRP files, a proprietary 3D modeling format used by Geomagic. Using the 3D evaluation software, the scans of the master models were virtually superimposed (best-fit algorithm) and compared (3D comparisons) with the intraoral scans in the next step.

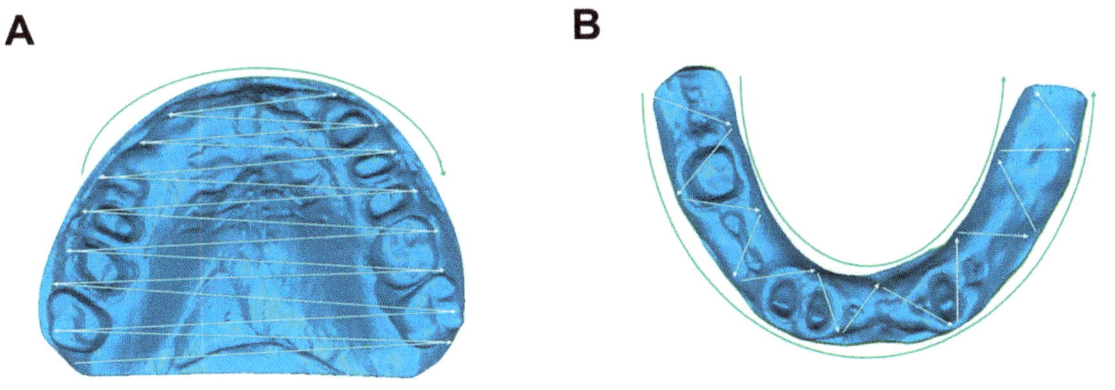

Figure 2. Schematic representation of the scan path in the (**A**) upper and (**B**) lower jaw.

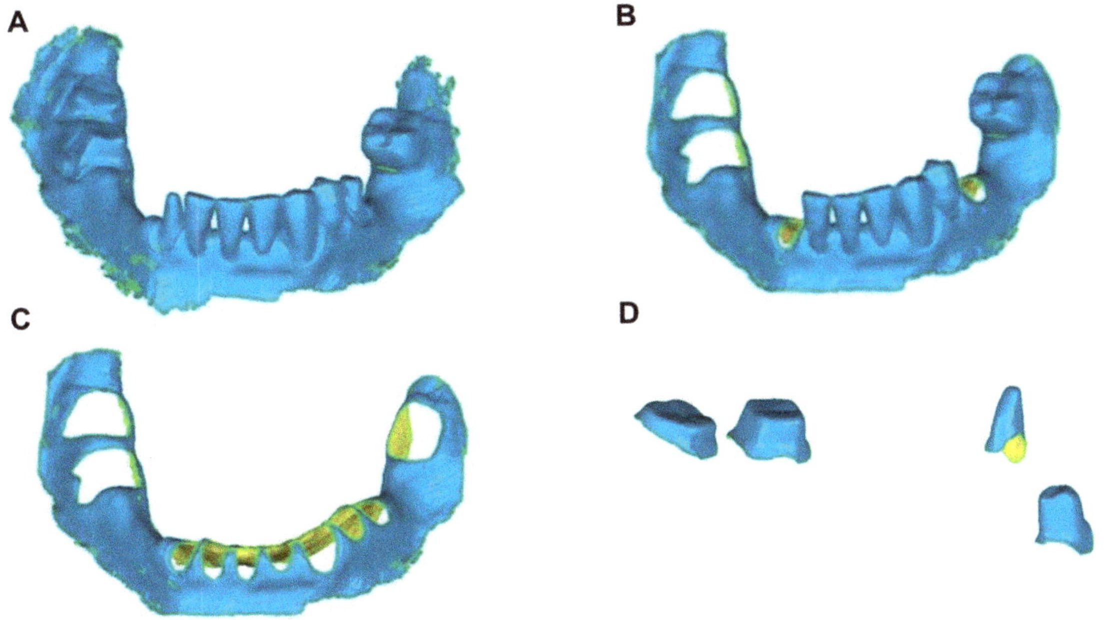

Figure 3. Illustration of the different cuts (TrueDefinition model). (**A**) Whole Jaw (WJ); (**B**) Mucosa with Residual Teeth (M_RT); (**C**): Isolated Mucosa (IM); and (**D**): Isolated Abutment Teeth (AT).

2.5. Superimposition and 3D Comparison

Following model preparation, corresponding segments from the COIs and CIs were superimposed using Geomagic Control 2014's Best-Fit-Alignment algorithm. This process, known as registration, utilizes the software's Best-Fit method to obtain the maximum congruence of the datasets to be compared. Parameters such as sample size and tolerance value were preset in the software, and registration was executed automatically. A 3D comparison was then performed, providing both quantitative and qualitative analyses. The quantitative assessment included maximum and minimum deviations, average positive and negative deviations, root mean squares, and standard deviations between the digitized and digital models. The qualitative analysis involved a color-coded image illustrating the deviation distribution between the reference and digital models. The areas marked in green showed no deviations. The areas that changed into the positive range were colored from yellow to orange to red, and the areas that moved into the negative range were colored from light blue to dark blue. The settings for Whole Jaw (WJ), Mucosa with Residual Teeth (M_RT), Isolated Mucosa (IM), and Isolated Abutment Teeth (AT) were slightly modified for each patient due to greater variability. The tolerance value was increased, the sample size was broadened for a more accurate fit, and the spectrum for color coding was expanded.

2.6. Statistical Analyses

The collected data were tabulated using spreadsheet software (Microsoft Excel for Microsoft 365, Version 2111, Microsoft, Redmond, Washington, DC, USA). The maximum and minimum average deviations, the mean values, and the standard deviations (SDs) (in µm) were included in the table. The statistical evaluation was carried out with a statistics program (STATA, Version 16.1, StataCorp LLC, College Drive, TX, USA). Descriptive statistics were obtained by calculating the medians and mean values ± standard deviations. A linear mixed regression model was used to evaluate both the dependence of the mean deviation on the used sections and to distinguish the influence of the number of abutment teeth on the mean deviation within the subgroups. To solve the problem of multiple testing in several pairwise comparisons, Scheffé's method was used to adjust the p-values. Box plots were created for the graphical representation of the data. A p-value < 0.05 was considered significant.

3. Results

3.1. Enrollment

During the recruitment period (October 2015—July 2016), 24 individuals expressed their interest in participating in the study. However, only 12 patients met the inclusion criteria and were finally enrolled. Out of the initially considered 24 participants, 12 withdrew their participation after being informed about the study's content, or because they did not need prosthetic treatment, as per the inclusion criteria. Among the final 12 participants, 8 received prosthetic treatment in the lower jaw, 3 received prosthetic treatment in the upper jaw, and 1 received prosthetic treatment in both jaws. Consequently, 13 conventional impressions and 13 computer-aided optical impressions were taken. A total of 48 3D comparisons were made (Table 1). Due to the insufficient number of identical abutment teeth, individual abutment teeth were not compared statistically.

Table 1. Number of comparisons stratified by section, jaw, and abutment teeth.

Jaw	**WJ (n = 13)**	**M_RT (n = 9)**	**IM (n = 13)**	**AT (n = 13)**
Maxilla	4	2	4	4
Mandible	9	7	9	9
Abutment teeth	**WJ (n = 13)**	**M_RT (n = 9)**	**IM (n = 13)**	**AT (n = 13)**
2	4	4	4	4
3	3	2	3	3
4	4	2	4	4
5	2	1	2	2

Note: WJ = Whole Jaw, M_RT = Mucosa with Residual Teeth, IM = Isolated Mucosa, and AT = Isolated Abutment Teeth.

3.2. Comparisons across Different Sections

The descriptive statistical evaluation revealed that the smallest mean deviations were in the AT sections. The largest mean deviations were found in the IM sections, with the WJ and M_RT sections lying in between (Figure 4). The 13 WJ sections resulted in a median of 461.5 µm and a mean of 542.3 ± 189.5 µm. For the nine M_RT sections, a median of 599.5 µm and a mean of 633.4 ± 230.2 µm were found. For the 13 IM sections, a median of 697.5 µm and a mean of 736.8 ± 247.7 µm were calculated. For the 13 AT sections, a median of 53.5 µm and a mean of 67.3 ± 28.8 µm were found (Table 2).

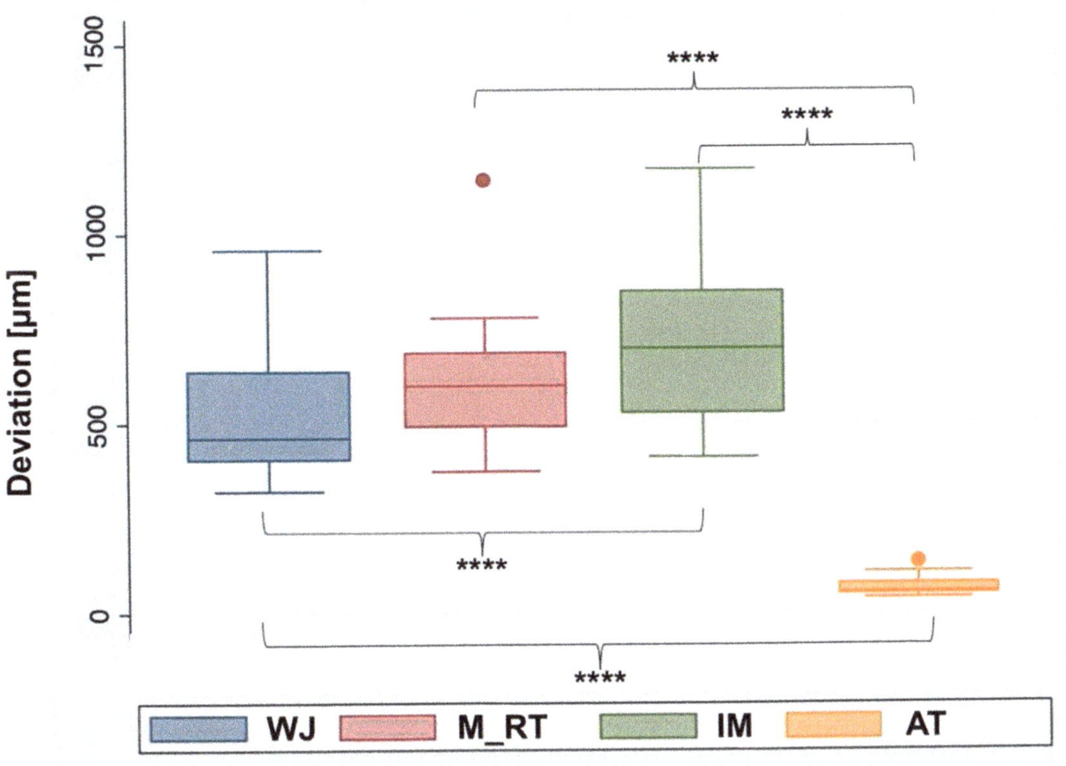

Figure 4. Boxplot comparison of mean deviations of different sections. WJ = Whole Jaw, M_RT = Mucosa with Residual Teeth, IM = Isolated Mucosa, and AT = Isolated Abutment Teeth. Circles indicate outliers. **** $p < 0.0001$.

Table 2. Descriptive statistics for each section.

Section	Median (µm)	Mean (µm)	SD (µm)
WJ	461.5	542.3	189.5
M_RT	599.5	633.4	230.2
IM	697.5	736.8	247.7
AT	53.5	67.3	28.8

Note: WJ = Whole Jaw, M_RT = Mucosa with Residual Teeth, IM = Isolated Mucosa, and AT = Isolated Abutment Teeth.

3.3. Comparisons across Different Sections Stratified by Jaw

When they were divided according to maxilla and mandible, a similar pattern emerged for the individual sections as that presented in Section 3.2. The values of the measured mean deviations in the maxilla were greater than those in the mandible, with the exception of the AT sections (Figure 5). However, none of the comparisons revealed any statistical significance. In the maxilla, the four WJ sections yielded a median of 655 μm and a mean of 662.1 (SD ± 268.7) μm. For the two M_RT sections, a median of 833 μm and a mean of 833 (SD ± 437) μm were calculated. For the four IM sections, a median of 899.8 μm and a mean of 914.9 (SD ± 256.7) μm were found. For the four AT sections, the median was 47.3 μm and the mean was 46 (SD ± 5.8) μm. In the mandible, the nine WJ sections yielded a median of 452 μm and a mean of 489 (SD ± 128.2) μm. For the seven M_RT sections, a median of 599.5 μm and a mean of 576.4 (SD ± 147.6) μm were calculated. For the nine IM sections, the median was 662 μm and the mean was 657.6 (SD ± 210.8) μm. For the nine AT sections, a median of 69.5 μm and a mean of 76.8 (SD ± 30.1) μm were found (Table 3). Furthermore, a comparison was made between the maxilla and the mandible across all the sections. The difference between the maxilla and the mandible was 151.6 ± 77.9 μm ($p = 0.052$).

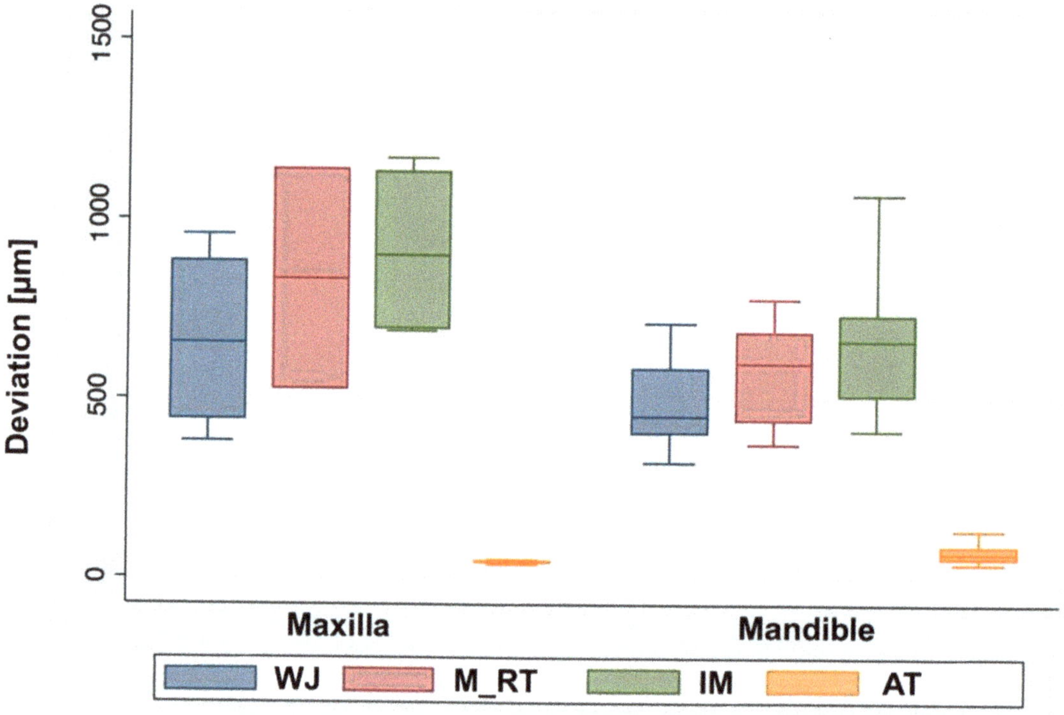

Figure 5. Boxplot comparison of mean deviations of different sections stratified by jaw. WJ = Whole Jaw, M_RT = Mucosa with Residual Teeth, IM = Isolated Mucosa, and AT = Isolated Abutment Teeth.

3.4. Comparisons across Different Sections Stratified by Abutment Teeth

The deviations for the M_RT and IM sections were higher than those for the WJ sections, although these deviations were not statistically significant. The smallest mean deviations were determined for the AT sections. Finally, the influence of the number of abutment teeth on the mean deviations of the COIs was assessed. There was no evidence of an association for the WJ sections, the M_RT sections, the IM sections, or the AT sections.

Table 3. Descriptive statistics for each section stratified by jaw.

Jaw	Section	Median (μm)	Mean (μm)	SD (μm)
Maxilla	WJ	655	662.1	268.7
Maxilla	M_RT	833	833	437
Maxilla	IM	899.8	914.9	256.7
Maxilla	AT	47.3	46	5.8
Mandible	WJ	452	489	128.2
Mandible	M_RT	599.5	576.4	147.6
Mandible	IM	662	657.6	210.8
Mandible	AT	69.5	76.8	30.1

Note: WJ = Whole Jaw, M_RT = Mucosa with Residual Teeth, IM = Isolated Mucosa, and AT = Isolated Abutment Teeth.

3.5. *Qualitative Analyses*

A visual analysis of the color-coded images revealed discrepancies in both the horizontal and vertical planes. The highest deviations are represented in dark blue and dark red in the color-coded images, while no or minor deviations are indicated by green, light yellow, and light blue, respectively. The mucosal areas showed the largest deviations. No remarkable differences were revealed between the different sections. High deviations (blue and red) and missing data (gray) were particularly noticeable in the sublingual areas, in the vestibule, and in the larger mucosal grooves (extraction sockets, palatal folds). The interdental papillae between the remaining teeth often showed large deviations. For the WJ sections, most of the deviations were in the light yellow to light blue range, i.e., between +130 μm and −130 μm. For the M_RT and IM sections, there were often larger scatterings. It was not possible to specify an exact μm range within which most of the deviations occurred (Figures 6 and 7).

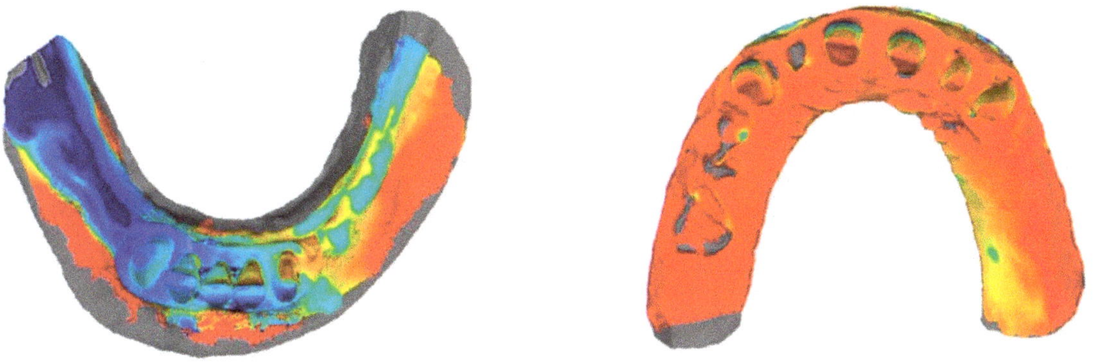

Figure 6. Exemplary color-coded representation of a 3D comparison of the mandible (**left**), which illustrates high deviations and missing data in the sublingual and vestibule areas, and the upper jaw (**right**), for which there is missing data in the larger mucosal groove areas.

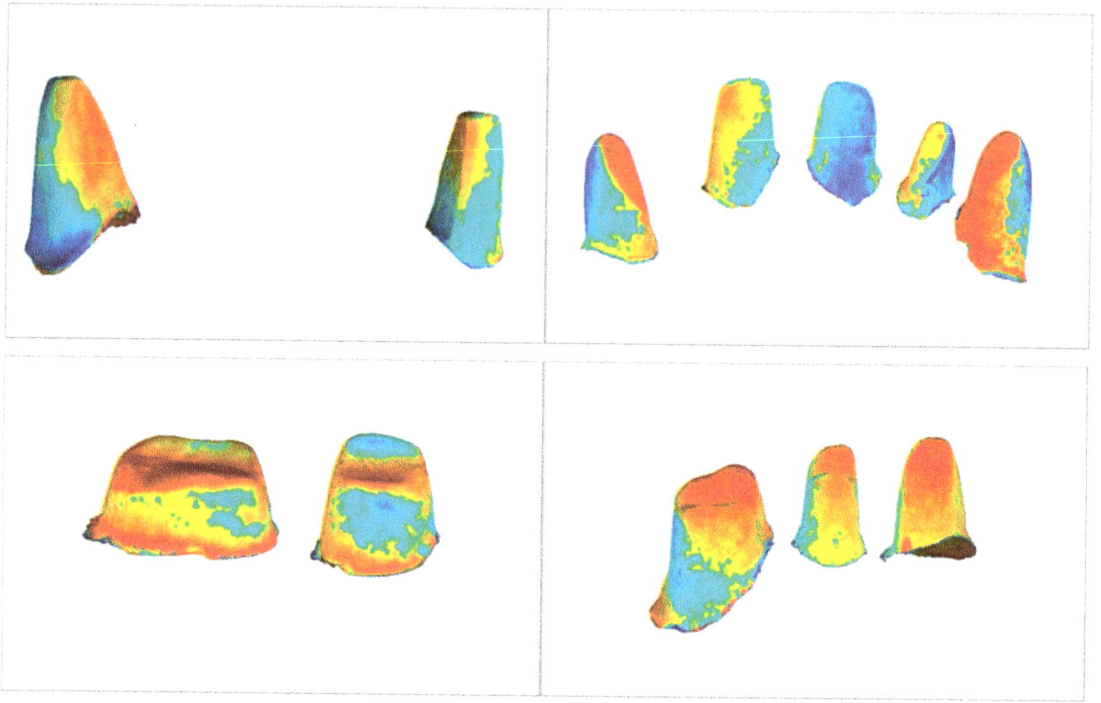

Figure 7. Exemplary color-coded representation of 3D comparisons of four sections of Isolated Abutment Teeth (AT).

4. Discussion

This in vivo study examined the accuracy of computer-assisted optical impressions of partially edentulous jaws compared with that of conventional impressions. Several studies have reported that COIM offers excellent accuracy when applied to fully edentulous jaws, with deviations ranging from 30 to 108 μm [17–19]. However, the present study found significantly higher deviations for the WJ, M_RT, and IM sections than for the AT section. This may be because intraoral scanners are optimized for capturing tooth structure and therefore produce higher deviations in areas with mucosal tissues. The present study also found that the number of abutment teeth did not significantly influence the accuracy of COIM. The visual analysis revealed that the mucosal areas, particularly the sublingual, vestibulum, and interdental regions, exhibited the largest deviations. These findings suggest that mucosal areas pose challenges for intraoral scanners, making it difficult to achieve accurate digital impressions. Our initial hypothesis posited that COIM would offer accuracy levels comparable to those of CIs in clinical scenarios involving partially edentulous jaws. Based on our results, this hypothesis was not fully supported; while COIM exhibited a similar degree of accuracy to that of the CI technique when capturing abutment teeth, significantly higher deviations were noted in mucosal areas.

When comparing our results with those of previous studies, it is important to note the discrepancies in the methodologies and equipment used. The existing literature includes numerous in vitro studies on impressions and scanning accuracy. However, most of these studies focused on fully dentate models [10,17,20] or patients with intact teeth [10–14], which do not adequately represent the typical cases that require treatment. Only a limited number of studies have specifically addressed the accuracy of impressions of partially edentulous jaws [21–24]. Furthermore, these studies primarily examined defect-free teeth and did not compare their results to those obtained using the current gold standard for complete-arch impressions, the conventional impression [21–24]. To the best of our knowledge, there are currently no in vivo studies on the accuracy of COIs of partially edentulous jaws, making direct comparisons difficult. In vitro studies have shown varying results. For instance, in the present study, the mean deviations for the IM sections were higher than the values reported by Hack et al., who applied the 3M system (True Definition Scanner and Lava C.O.S) to edentulous jaws [25]. These differences could be attributed to variations in operator experience, different reference scanners, or the use of different impression materials. It should be noted that the findings of the present study may not be directly comparable to those of in vitro studies due to the inability of in vitro studies to fully simulate clinical conditions. Factors such as patient movement, limited mouth opening, and the presence of blood and saliva can impact the accuracy of intraoral scanners. Future studies should therefore validate the present findings in vivo. A general overview of the advantages and disadvantages of COIM and the CI technique is provided in Table 4. Although our focus is on computer-assisted optical impressions, it is worth noting the broader advancements in technology that are applicable to the maxillofacial and dental fields, such as the growth in medical 3D printing, which is used for craniomaxillofacial surgery applications [26]. These innovations indicate a trend towards more precise and customizable treatment options and shed light on the need for ongoing research to optimize existing technologies, such as intraoral scanners. One critical advantage of COIM is its potential role in minimizing the spread of contagious diseases. Unlike conventional impression methods, which require direct physical contact and multiple exchanges of dental materials between clinicians and laboratories, digital impressions can be transferred electronically, thereby reducing the risk of cross-contamination. This is especially vital in times of global pandemics, during which infection control measures are a top priority. The implementation of COIM also offers benefits in terms of minimally invasive dentistry. The precise impressions made through COIM are particularly crucial for cases requiring minimally prepared dental bridges, such as inlay-retained bridges. A study by Tatarciuc et al. (2021) emphasizes that the proper design of such bridges, including the areas of maximum pressure on the supporting teeth, is crucial for their success [27]. Accurate digital impressions enable precise designs and contribute to the longevity of minimally invasive dental restorations.

Table 4. Advantages and disadvantages of computerized optical impression making (COIM) and conventional impressions (CIs).

Criteria	COIM	CIs
Accuracy	High precision, but this may vary with operator skill	Dependent on material and technique
Time-Efficiency	Generally quicker	May be time-consuming
Patient Comfort	Non-invasive, more comfortable	May be uncomfortable due to materials
Cost	Higher initial investment	Generally lower cost
Skill Level	Requires technical proficiency	Requires traditional expertise
Portability	Digital information is transferable	Transport of material required
Error Recovery	Easier to correct mistakes digitally	Errors often require reworking

For the COIM, an intraoral scanner operating on the active wavefront sampling principle was utilized. This scanner, which utilizes video technology and captures 20 frames per second, is suitable for recording smooth surfaces, such as mucosal areas in partially edentulous jaws [25]. The application of titanium dioxide powder generates small surface points that enable the assembly of 3D video recordings. Given its essential role in the active wavefront sampling capture technique, any negative impact on the accuracy is considered minimal. However, it remains unclear whether powder application negatively affects accuracy [25]. Saliva and tongue movements make it difficult to maintain the powder layer, necessitating occasional re-powdering during scanning. In this context, powder-free systems may be easier to use. The 3M True Definition Scanner was selected for this study due to its proven in vitro accuracy and repeatability [28–31]. Notably, since there is no recommended manufacturer-specific scanning path for partially edentulous jaws, the scanning paths were based on those described in the literature [25,32]. A zigzag-shaped scanning path was utilized, and the abutment teeth were scanned using diffuse movements in all directions until all of the areas were captured, without any gaps. Following this standardized scanning path in all cases ensured consistency and minimized variations due to different scanning paths [33].

Nevertheless, the present study has certain limitations. To achieve a sufficiently high sample size, the study incorporated various patients with differing numbers of remaining teeth. This variability could have impacted the results and contributed to the scatter within the data. Given that this was a feasibility study, and one of the first of its kind, no a priori power analysis was performed to determine the sample size. Consequently, the study may be underpowered for detecting smaller effect sizes. Thus, the results should be interpreted as preliminary and exploratory rather than confirmatory. Nevertheless, despite the lack of formal sample size calculation, we consider our sample size to be in line with or exceeding those found in similar feasibility studies [17,21,34]. Teeth and jaws have individual anatomical variations, and patients differ in salivary flow and mouth opening. These factors could have led to increased scatter in the deviation data. Enhanced salivary flow makes both COIs and conventional impressions more difficult to obtain. However, hybrid impression materials exhibit excellent hydrophilicity, suggesting that enhanced salivary flow has a minimal impact on the accuracy of conventional impressions. Conversely, studies have shown that the presence of saliva affects the accuracy of intraoral scanners [35,36]. Saliva can wash away the applied scan powder, simplifying the detection of wet areas but necessitating additional powder during scanning [36]. In this study, an experienced dentist performed the COIM under relatively dry conditions (i.e., minimal salivary contamination). However, complete prevention was not guaranteed. Importantly, the titanium dioxide powder used in this study had a specific role: it created small dots on the surface that were instrumental for the merging of the 3D video images. Given that our scanning technology operated on the principle of active wavefront sampling, this approach was deemed advantageous, particularly for the acquisition of smooth surfaces, such as mucosal areas in partially edentulous jaws [25]. Furthermore, the plaster models were digitized using an optical laboratory scanner. Although optical laboratory scanners achieve acceptable accuracy, discrepancies between the plaster models and the digitized models can occur [37,38]. The manufacturer of the laboratory scanner used in this study specifies an accuracy of 5 μm (ISO 12836) [16].

Notably, the cropping of the datasets to the analysis area was performed by one individual, who followed the same principle for each patient in order to ensure consistent cropping. Nevertheless, discrepancies in the manual cropping may have occurred due to variations in the selection of the areas. It is also important to note that the accuracy analysis was conducted using only one intraoral scanner, and that the results therefore cannot be generalized to all optical systems, especially to current systems. Additionally, the visual analyses of the upper jaws were challenging due to the absence of a hard palate in the reference models. This limited the accuracy of the visual assessments of the complete hard palates.

From a clinical standpoint, the findings of our study have significant implications for prosthetic treatment decisions concerning partially edentulous patients. The higher deviations observed in the mucosal areas via COIM underscore the need for caution when choosing an impression method for prosthetics involving these tissues. Practitioners should be aware that while COIM may provide excellent results for teeth and hard tissues, its accuracy in mucosal areas may be compromised, potentially leading to ill-fitting prosthetics and subsequent clinical complications. This is particularly pertinent in the design and manufacture of partial dentures, which require precise impressions for a clinically acceptable fit. Our results also call attention to the potential for higher deviations when scanning mucosal tissues, which could affect the longevity and clinical success of the prosthetic appliance.

While conventional impressions may remain the preferred method for capturing these areas, our study illuminates the need for technological advancements in intraoral scanners to improve their performance in capturing soft tissue details.

5. Conclusions

This study offers the first in vivo evaluation of the accuracy of COIs of partially edentulous jaws compared with CIs. Our findings reveal that while COIM shows high accuracy when capturing hard tissues such as abutment teeth, it exhibits greater deviations in mobile mucosal areas, underscoring its limitations when capturing soft tissues. From a clinical perspective, this is a crucial consideration for dental practitioners when selecting impression techniques for prosthetic applications involving soft tissues. While COIM may offer certain advantages, its shortcomings when capturing mucosal areas cannot be ignored and may result in ill-fitting prosthetics and potential clinical complications. Thus, our data suggest that conventional impressions may currently be more reliable for capturing complex soft tissue details in partially edentulous jaws. It is essential to acknowledge the limitations of our study. We used a single type of intraoral scanner, and the sample patients had varying numbers of remaining teeth, which might have introduced variability. Additionally, our study did not account for factors such as patient movement and saliva flow, which could potentially have influenced the accuracy of both the COIs and the CIs. Future research should focus not only on evaluating contemporary intraoral scanners, but also on studying how variables such as patient movement and saliva flow could impact accuracy. Our work serves as a foundational study in this research avenue, setting the stage for further in vivo investigations to refine and validate these findings.

Author Contributions: Conceptualization, S.B.M.P.; formal analysis, K.V. and S.B.M.P.; investigation, J.I.; methodology, S.B.M.P.; project administration, S.B.M.P.; supervision, R.J.K.; validation, B.S., J.I. and S.B.M.P.; visualization, B.S. and K.V.; writing—original draft preparation, B.S., J.I. and S.B.M.P.; writing—review and editing, B.S., J.I., K.V., R.J.K. and S.B.M.P. All authors have read and agreed to the published version of the manuscript.

Funding: This research received no external funding.

Data Availability Statement: The data presented in this study are available on request from the corresponding author.

Acknowledgments: The article processing charge was paid by the Baden-Wuerttemberg Ministry of Science, Research and Art and the University of Freiburg via the funding program Open Access Publishing.

Conflicts of Interest: The authors declare no conflict of interest.

References

1. Nagarkar, S.R.; Perdigão, J.; Seong, W.-J.; Theis-Mahon, N. Digital versus Conventional Impressions for Full-Coverage Restorations: A Systematic Review and Meta-Analysis. *J. Am. Dent. Assoc.* **2018**, *149*, 139–147.e1. [CrossRef] [PubMed]
2. Ting-Shu, S.; Jian, S. Intraoral Digital Impression Technique: A Review. *J. Prosthodont.* **2015**, *24*, 313–321. [CrossRef] [PubMed]
3. Saravi, B.; Vollmer, A.; Hartmann, M.; Lang, G.; Kohal, R.-J.; Boeker, M.; Patzelt, S.B.M. Clinical Performance of CAD/CAM All-Ceramic Tooth-Supported Fixed Dental Prostheses: A Systematic Review and Meta-Analysis. *Materials* **2021**, *14*, 2672. [CrossRef] [PubMed]
4. Cicciù, M.; Fiorillo, L.; D'Amico, C.; Gambino, D.; Amantia, E.M.; Laino, L.; Crimi, S.; Campagna, P.; Bianchi, A.; Herford, A.S.; et al. 3D Digital Impression Systems Compared with Traditional Techniques in Dentistry: A Recent Data Systematic Review. *Materials* **2020**, *13*, 1982. [CrossRef]
5. Waldecker, M.; Rues, S.; Trebing, C.; Behnisch, R.; Rammelsberg, P.; Bömicke, W. Effects of Training on the Execution of Complete-Arch Scans. Part 2: Scanning Accuracy. *Int. J. Prosthodont.* **2021**, *34*, 27–36. [CrossRef] [PubMed]
6. Joda, T.; Lenherr, P.; Dedem, P.; Kovaltschuk, I.; Bragger, U.; Zitzmann, N.U. Time Efficiency, Difficulty, and Operator's Preference Comparing Digital and Conventional Implant Impressions: A Randomized Controlled Trial. *Clin. Oral Implants Res.* **2017**, *28*, 1318–1323. [CrossRef]
7. de Paris Matos, T.; Wambier, L.M.; Favoreto, M.W.; Rezende, C.E.E.; Reis, A.; Loguercio, A.D.; Gonzaga, C.C. Patient-Related Outcomes of Conventional Impression Making versus Intraoral Scanning for Prosthetic Rehabilitation: A Systematic Review and Meta-Analysis. *J. Prosthet. Dent.* **2021**, *130*, 19–27. [CrossRef]

8. Morsy, N.; El Kateb, M.; Azer, A.; Fathalla, S. Fit of Zirconia Fixed Partial Dentures Fabricated from Conventional Impressions and Digital Scans: A Systematic Review and Meta-Analysis. *J. Prosthet. Dent.* **2021**, *130*, 28–34. [CrossRef]
9. Memari, Y.; Mohajerfar, M.; Armin, A.; Kamalian, F.; Rezayani, V.; Beyabanaki, E. Marginal Adaptation of CAD/CAM All-Ceramic Crowns Made by Different Impression Methods: A Literature Review. *J. Prosthodont.* **2019**, *28*, e536–e544. [CrossRef]
10. Keul, C.; Güth, J.-F. Accuracy of Full-Arch Digital Impressions: An In Vitro and In Vivo Comparison. *Clin. Oral Investig.* **2020**, *24*, 735–745. [CrossRef]
11. Schmidt, A.; Klussmann, L.; Wöstmann, B.; Schlenz, M.A. Accuracy of Digital and Conventional Full-Arch Impressions in Patients: An Update. *J. Clin. Med.* **2020**, *9*, 688. [CrossRef] [PubMed]
12. Winkler, J.; Gkantidis, N. Trueness and Precision of Intraoral Scanners in the Maxillary Dental Arch: An in Vivo Analysis. *Sci. Rep.* **2020**, *10*, 1172. [CrossRef] [PubMed]
13. Iturrate, M.; Amezua, X.; Garikano, X.; Solaberrieta, E. Use of Measuring Gauges for in Vivo Accuracy Analysis of Intraoral Scanners: A Pilot Study. *J. Adv. Prosthodont.* **2021**, *13*, 191–204. [CrossRef]
14. Jorquera, G.J.; Sampaio, C.S.; Bozzalla, A.; Hirata, R.; Sánchez, J.P. Evaluation of Trueness and Precision of Two Intraoral Scanners and a Conventional Impression: An In Vivo Clinical Study. *Quintessence Int.* **2021**, *52*, 904–910. [CrossRef]
15. Waldecker, M.; Rues, S.; Awounvo Awounvo, J.S.; Rammelsberg, P.; Bömicke, W. In Vitro Accuracy of Digital and Conventional Impressions in the Partially Edentulous Maxilla. *Clin. Oral Investig.* **2022**, *26*, 6491–6502. [CrossRef] [PubMed]
16. *DIN EN ISO 12836:2015-11*; Zahnheilkunde—Digitalisierungsgeräte Für CAD/CAM-Systeme Für Indirekte Dentale Restaurationen—Prüfverfahren Zur Beurteilung Der Genauigkeit (Deutsche Fassung EN_ISO_12836:2015). Beuth Verlag GmbH: Berlin, Germany, 2015.
17. Ender, A.; Zimmermann, M.; Mehl, A. Accuracy of Complete- and Partial-Arch Impressions of Actual Intraoral Scanning Systems in Vitro. *Int. J. Comput. Dent.* **2019**, *22*, 11–19.
18. Reich, S.; Yatmaz, B.; Raith, S. Do "Cut out-Rescan" Procedures Have an Impact on the Accuracy of Intraoral Digital Scans? *J. Prosthet. Dent.* **2021**, *125*, 89–94. [CrossRef]
19. Kwon, M.; Cho, Y.; Kim, D.-W.; Kim, M.; Kim, Y.-J.; Chang, M. Full-Arch Accuracy of Five Intraoral Scanners: In Vivo Analysis of Trueness and Precision. *Korean J. Orthod.* **2021**, *51*, 95–104. [CrossRef]
20. Waldecker, M.; Rues, S.; Rammelsberg, P.; Bömicke, W. Accuracy of Complete-Arch Intraoral Scans Based on Confocal Microscopy versus Optical Triangulation: A Comparative in Vitro Study. *J. Prosthet. Dent.* **2021**, *126*, 414–420. [CrossRef]
21. Schimmel, M.; Akino, N.; Srinivasan, M.; Wittneben, J.-G.; Yilmaz, B.; Abou-Ayash, S. Accuracy of Intraoral Scanning in Completely and Partially Edentulous Maxillary and Mandibular Jaws: An in Vitro Analysis. *Clin. Oral Investig.* **2021**, *25*, 1839–1847. [CrossRef]
22. Lee, J.-H.; Yun, J.-H.; Han, J.-S.; Yeo, I.-S.L.; Yoon, H.-I. Repeatability of Intraoral Scanners for Complete Arch Scan of Partially Edentulous Dentitions: An In Vitro Study. *J. Clin. Med.* **2019**, *8*, 1187. [CrossRef] [PubMed]
23. Ren, S.; Morton, D.; Lin, W.-S. Accuracy of Virtual Interocclusal Records for Partially Edentulous Patients. *J. Prosthet. Dent.* **2020**, *123*, 860–865. [CrossRef] [PubMed]
24. Hayama, H.; Fueki, K.; Wadachi, J.; Wakabayashi, N. Trueness and Precision of Digital Impressions Obtained Using an Intraoral Scanner with Different Head Size in the Partially Edentulous Mandible. *J. Prosthodont. Res.* **2018**, *62*, 347–352. [CrossRef] [PubMed]
25. Hack, G.; Liberman, L.; Vach, K.; Tchorz, J.P.; Kohal, R.J.; Patzelt, S.B.M. Computerized Optical Impression Making of Edentulous Jaws—An In Vivo Feasibility Study. *J. Prosthodont. Res.* **2020**, *64*, 444–453. [CrossRef] [PubMed]
26. Modgill, V.; Balas, B.; Chi, M.; Honigmann, P.; Thieringer, F.M.; Sharma, N. Knowledge Domain and Innovation Trends Concerning Medical 3D Printing for Craniomaxillofacial Surgery Applications: A 30-Year Bibliometric and Visualized Analysis. *Craniomaxillofac. Res. Innov.* **2023**, *8*, 275284642311709. [CrossRef]
27. Tatarciuc, M.; Maftei, G.A.; Vitalariu, A.; Luchian, I.; Martu, I.; Diaconu-Popa, D. Inlay-Retained Dental Bridges—A Finite Element Analysis. *Appl. Sci.* **2021**, *11*, 3770. [CrossRef]
28. Stanley, M.; Paz, A.G.; Miguel, I.; Coachman, C. Fully Digital Workflow, Integrating Dental Scan, Smile Design and CAD-CAM: Case Report. *BMC Oral Health* **2018**, *18*, 134. [CrossRef]
29. Osnes, C.A.; Wu, J.H.; Venezia, P.; Ferrari, M.; Keeling, A.J. Full Arch Precision of Six Intraoral Scanners In Vitro. *J. Prosthodont. Res.* **2020**, *64*, 6–11. [CrossRef]
30. Elbashti, M.E.; Hattori, M.; Patzelt, S.B.; Schulze, D.; Sumita, Y.I.; Taniguchi, H. Feasibility and Accuracy of Digitizing Edentulous Maxillectomy Defects: A Comparative Study. *Int. J. Prosthodont.* **2017**, *30*, 147–149. [CrossRef]
31. Patzelt, S.B.M.; Bishti, S.; Stampf, S.; Att, W. Accuracy of Computer-Aided Design/Computer-Aided Manufacturing-Generated Dental Casts Based on Intraoral Scanner Data. *J. Am. Dent. Assoc.* **2014**, *145*, 1133–1140. [CrossRef]
32. Patzelt, S.B.M.; Vonau, S.; Stampf, S.; Att, W. Assessing the Feasibility and Accuracy of Digitizing Edentulous Jaws. *J. Am. Dent. Assoc.* **2013**, *144*, 914–920. [CrossRef] [PubMed]
33. Ender, A.; Mehl, A. Influence of Scanning Strategies on the Accuracy of Digital Intraoral Scanning Systems. *Int. J. Comput. Dent.* **2013**, *16*, 11–21. [PubMed]
34. Müller, P.; Ender, A.; Joda, T.; Katsoulis, J. Impact of Digital Intraoral Scan Strategies on the Impression Accuracy Using the TRIOS Pod Scanner. *Quintessence Int.* **2016**, *47*, 343–349. [CrossRef] [PubMed]

35. Camcı, H.; Salmanpour, F. Effect of Saliva Isolation and Intraoral Light Levels on Performance of Intraoral Scanners. *Am. J. Orthod. Dentofac. Orthop.* **2020**, *158*, 759–766. [CrossRef] [PubMed]
36. Kurz, M.; Attin, T.; Mehl, A. Influence of Material Surface on the Scanning Error of a Powder-Free 3D Measuring System. *Clin. Oral Investig.* **2015**, *19*, 2035–2043. [CrossRef]
37. Joós-Kovács, G.; Vecsei, B.; Körmendi, S.; Gyarmathy, V.A.; Borbély, J.; Hermann, P. Trueness of CAD/CAM Digitization with a Desktop Scanner—An In Vitro Study. *BMC Oral Health* **2019**, *19*, 280. [CrossRef]
38. Vandeweghe, S.; Vervack, V.; Vanhove, C.; Dierens, M.; Jimbo, R.; De Bruyn, H. Accuracy of Optical Dental Digitizers: An in Vitro Study. *Int. J. Periodontics Restor. Dent.* **2015**, *35*, 115–121. [CrossRef]

MDPI AG
Grosspeteranlage 5
4052 Basel
Switzerland
Tel.: +41 61 683 77 34

Journal of Functional Biomaterials Editorial Office
E-mail: jfb@mdpi.com
www.mdpi.com/journal/jfb

www.ingramcontent.com/pod-product-compliance
Lightning Source LLC
Chambersburg PA
CBHW061456210926
51850CB00008B/1735
9783725817658